INTERNATIONAL SERIES ON THE
STRENGTH AND FRACTURE OF MATERIALS AND STRUCTURES

Series Editor: D. M. R. TAPLIN

MECHANISMS OF DEFORMATION AND FRACTURE

ALSO PUBLISHED IN THIS SERIES

TAPLIN: Advances in Research on the Strength and Fracture of Materials

        Volume 1: An Overview
        Volume 2a: Physical Metallurgy of Fracture
        Volume 2b: Fatigue
        Volume 3a: Analysis and Mechanics
        Volume 3b: Applications and Non-Metals
        Volume 4: Fracture and Society

Related Journals published by Pergamon Press

Engineering Fracture Mechanics

Fatigue of Engineering Materials and Structures

MECHANISMS OF DEFORMATION AND FRACTURE

Proceedings of the Interdisciplinary Conference
held at the University of Luleå,
Luleå, Sweden, September 20-22, 1978

Editor

K. E. EASTERLING

University of Luleå, Sweden

Conference committee

R. Pusch, Department of Soil Mechanics,
University of Luleå, Sweden

K. E. Easterling, Department of Technology,
University of Luleå, Sweden

B. Lundberg, Department of Solid Mechanics,
University of Luleå, Sweden

O. Stephansson, Department of Rock Mechanics,
University of Luleå, Sweden

PERGAMON PRESS

OXFORD · NEW YORK · TORONTO · SYDNEY · PARIS · FRANKFURT

U.K.	Pergamon Press Ltd., Headington Hill Hall, Oxford OX3 0BW, England
U.S.A.	Pergamon Press Inc., Maxwell House, Fairview Park, Elmsford, New York 10523, U.S.A.
CANADA	Pergamon of Canada, Suite 104, 150 Consumers Road, Willowdale, Ontario M2J 1P9, Canada
AUSTRALIA	Pergamon Press (Aust.) Pty. Ltd., P.O. Box 544, Potts Point, N.S.W. 2011, Australia
FRANCE	Pergamon Press SARL, 24 rue des Ecoles, 75240 Paris, Cedex 05, France
FEDERAL REPUBLIC OF GERMANY	Pergamon Press GmbH, 6242 Kronberg-Taunus, Pferdstrasse 1, Federal Republic of Germany

Copyright © 1979 Pergamon Press Ltd.

All Rights Reserved. No part of this publication may be reproduced, stored in a retrieval system or transmitted in any form or by any means: electronic, electrostatic, magnetic tape, mechanical, photocopying, recording or otherwise, without permission in writing from the publishers.

First edition 1979

British Library Cataloguing in Publication Data

Mechanisms of Deformation and Fracture
(Conference), *University of Lulea, 1978*
Mechanisms of deformation and fracture
(International series on the strength and
fracture of materials and structures)
1. Deformations (Mechanics) - Congresses
2. Fracture mechanics - Congresses
I. Easterling, K E II. Series
620.1'123   TA417.6   79-40236
ISBN 0-08-024258-8

In order to make this volume available as economically and as rapidly as possible the authors' typescripts have been reproduced in their original forms. This method unfortunately has its typographical limitations but it is hoped that they in no way distract the reader.

Printed and bound at William Clowes & Sons Limited
Beccles and London

CONTENTS

SESSION 1

Introductory keynote lectures

B Broberg: The inter-relationship: Solid mechanics - earth sciences 3

P Feltham: The inter-relationship: Earth sciences - material sciences 29

J F Knott: The inter-relationship: Material sciences - solid mechanics 43

DISCUSSION 71

SESSION 2

Experimental studies

D Shockey, K C Dao & R L Jones: Effect of grain size on the static and dynamic fracture behavior of -titanium 77

D A Porter, K E Easterling & G D W Smith: The mechanism of deformation of pearlite 87

L Jilkén & J Bäcklund: Electromagnetic detection of low-cycle fatigue 97

L Magnusson: Low cycle behaviour of case hardened steel 105

L Jilkén, J Bäcklund & H Knutsson: Automatic fatigue threshold value testing 111

I J Smalley, D H Krinsley, C F Moon & S P Bentley: Processes of quartz fracture in nature and the formation of clastic sediments 119

P Smart & J W Dickson: Deformation and shear of normally consolidated flocculated kaolin 129

R Pusch: Cohesion in fine-grained soils 137

M S Paterson & I van der Molen: The deformation of granular masses with particular references to partially-melted granite 145

G Simmons, M Batzle, R Siegfried, M Feves & H Cooper: Characterization of microcracks 151

K Röshoff: Deformation structures in the Tännäs augen gneiss 159

O Alm & O Stephansson: Experimental deformation of augen-gneisses 173

G Swan : Rock fracture under superimposed static and dynamic loads 187

A W B Siddans: Deformation of muddy rocks at low metamorhpic grades 195

v

	Page
M Lähdeniemi, E Minni, L Ranta & E Suoninen: Study of fresh fracture surfaces in metals by electron spectroscopy	203
DISCUSSION	209

SESSION 3

Theoretical studies

J Hult: CDM - capabilities, limitations and promises	233
G Rydholm & B Fredriksson: Shakedown analysis in rolling contact problems	249
J Kratochvíl: Analytical modelling in inelasticity	257
H Andersson: Initiation of crack growth at full plasticity	269
T Johannesson & P Sjöblom: Stress and strain distribution in two-phase systems	277
K Berglund: Micromodels for granular composites	285
A Bodare: A calculation of static and dynamic elasticity constants in an elastic inhomogenous material	293
J Kubát, M Rigdahl & R Seldén: A cooperative model of stress relaxation kinetics in solids	297
G L Anderson: Damage in a pressure transducer due to a high speed load	307
I Rosenqvist: A general theory for the sensitivity of clays	315
S Hansbo: Mechanical behaviour of clay explained in microstructural terms	321
R Larsson: Shear strength of soils	329
L Börgesson: Stress-strain relationships for silty soils	343
R Pusch: Creep mechanisms in clay	351
G Ranalli: Plastic flow mechanisms and rheological properties of the Earth's mantle	361
C C Ferguson: The simple fluid with fading memory as a rheological model for steady-state flow of rocks	371
G W Weidman & W Döll: Crazes and the fracture of glassy thermoplastics	385
D W Durney: A theory of mass-transfer-buckling deformation in finite amplitude sinusoidal multilayers	393
L Keinonen: A thermodynamic model of consolidation in cohesive soils	407
DISCUSSION	411

PREFACE

A successful interdisciplinary conference is a comparatively rare event. It is well known that even conferences within single disciplines often tend to break down into a number of minor groups of specialized interests. In planning this conference the Organization Committee decided from the outset that any form of subject segregation was to be avoided, at least in the discussions. The conference is thus divided up into three general sessions: an introductory session in which three invited lecturers emphasize possible areas of mutual interest between the three sciences, and two main sessions in which contributed papers from all disciplines are divided up only according to their experimental or theoretical bias. The important features of the two main sessions are that delegates presenting papers were not obliged to emphasize the interdisciplinary aspects of their work, i.e. papers were to be read as if to experts in the field, but that the general discussions at the end of each session are essentially interdisciplinary in character (at least as far as the Chairmen were able to keep control of events!). In order to encourage these all-important discussion sessions the number of delegates was restricted to about 60 persons, many of whom were invited because of their interdisciplinary interests.

Looking back on the conference now and having heard a number of (mainly) positive comments from participants, it appears that the above formula proved fairly successful. Of course, reading through the discussion at the end of these proceedings does not tell the whole story, the behind-the-scene contacts being as usual just as valuable as the sessions themselves. Even the conference locality, a small university situated at the edge of the unspoilt wilds of Lapland, contributed somehow to the spirit of the event.

Finally I should like to thank, on behalf of the Organization Committee, the Statens Råd för Byggnadsforskning (BFR), The Ministry of Education and the University of Luleå for financial support, the Town of Luleå for the banquet, and the valuable assistance of Roland Hortlund, Inger Wallin and Agneta Engfors in many of the organizational and editorial details.

K.E. Easterling.

SESSION 1

Introductory Keynote Lectures

THE INTER-RELATIONSHIP:
SOLID MECHANICS-EARTH SCIENCES

K. B. Broberg

Department of Solid Mechanics
Lund Institute of technology
S-2207 Lund, Sweden

ABSTRACT

The uses of solid mechanics in earth sciences and in engineering are briefly compared. The principle difference seems to be one of scale. Specifically, tectonic earthquakes are compared with dynamic fracture propagation. Similarities and dissimilarities are pointed out.

1. INTRODUCTION

In principle there is little difference between the ways in which we use the different basic disciplines of physics (mechanics, solid mechanics, electrodynamics, nuclear physics, etc.) in earth sciences and in engineering. The essential difference seems to be one of scale. One would think, perhaps, that our confinement to the surface of the earth would require methods of (say) solid mechanics which would be unique for the earth sciences. However, many problems subject to this constraint are astonishingly similar to those in engineering applications. Thus, for instance, geophysical prospecting methods (searching for geological anomalies, ore bodies, petroleum deposits, salt-domes, etc.) or seismic studies at an earthquake event have their counterparts in non-destructive evaluation and testing (searching for porosities, cracks, slag particles, etc.) . Knowledge and experience in one field may promote advances in the other. Moreover, there is no distinct gap between the fields; to some extent they overlap. At an intermediate scale (for instance in the paricular sector of geophysical prospecting) the concern may be both a geological and an engineering one.

Very typical for earth sciences is the broad arsenal of disciplines used to attack the problems. Continental drift, for example, has been studied by using observations of land contours, plant and animal evolution, (paleo)climates, (paleo)magnetism, chemistry, thermodynamics, creep, satellite based optics, etc. The interiour of the earth is investigated by a combination of different observations, by considering the consequences of different hypothetical models of equatorial bulge, precession, Chandler wobble, tidal dissipation, geomagnetism, thermal fluxes, etc. A specific discipline, like solid mechanics, is thus usually used in combination with other disciplines in investigations of the earth.

The difference in scale between the earth and engineering structures can be looked upon from different points of view. Some examples can be considered. While static pressures exceeding a few hundred MPa hardly can be obtained at laboratory experiments, the internal pressure at the centre of the earth amounts to about 36 000 MPa (which is more than one fourth of the bulk modulus).

Appreciable heat conduction from the inner core to the surface of the earth would require longer time than the age of the earth. (Since the equation of heat conduction is parabolic, times scale as distances squared.) Creep processes in the upper mantle go so slowly that Fenno-Scandia still exhibits a postglacial rebound, amounting to as much as (or as little as, depending on comparisons made) 1 cm/year at the site of this symposium. However, even though this figure implies an uplift of about a tenth of a millimeter during the course of the symposium, the lunar tides will lift and sink us much more.
In this connection it is interesting to note that the scale effects vary drastically with the mechanisms of deformation. The lunar tide causes an elastic deformation of the earth as a whole, a mechanism that is much faster than creep. Since the period, about 13 hours, is much larger than the longest period of free oscillations of the earth (about 98 minutes) the deformation is essentially a static one. The damping of stress waves in the mantle is comparatively small so that the energy dissipation in the mantle due to the lunar tide is equivalent to the output of a few nuclear reactors, only. (The tidal dissipation in the sea is certainly much greater.)

Solid mechanics plays a significant role in studies of earthquakes and of creep in the earth mantle. It is my intention to illustrate a certain use of solid mechanics related to earthquake source physics and to show the close connection with the engineering field of fracture mechanics. Since I assume that many of the participants at this symposium are not familiar with earthquake physics, I will begin with a very brief description of the basic phenomena leading to tectonic earthquake events.

2. CONTINENTAL DRIFT AND TECTONIC EARTHQUAKES

The Wegener hypothesis from the 1920s of continental drift has become again reestablished under the general rubric of "plate tectonics". The earth's crust (or, rather the lithosphere) is split up into a number of plates that are moving in relation to each other. The thickness of the plates is believed to be 70 - 100 km, whereas the thickness of the earth's crust varies from about 40 km (continents) to about 5 km (oceanic areas). Fig. 1 shows schematically one such plate, consisting of two movable parts. By convection in the astenosphere*)(below the lithosphere) the two parts are driven in opposite directions, but since new material is supplied from below at their junction they do not separate. The ends of the plate dip under the "meeting" lithosphere. Along the sides, then, slip occurs between this moving plate and adjacent plates.

The picture sketched here is, naturally, simplified. In reality the plates are not rectangular, but still the main principle of movement remains the same as described.

In Figs. 2 - 6 examples are shown of some main features of the continental drift.

The relative motion between two adjacent plates amounts to a few centimeters per year. This is an annual average. Multiplication by 100 million years gives a few thousand kilometers of plate movement.

The motions are usually observed only as relative motions, but there seems to be a possibility to relate the plate movements to a reference frame, roughly speaking to the interiour of the earth. Rather deep-seated hot spots exist, giving rise to "plumes" from the mantle to the earth's crust. These hot spots are believed to be rather

---

*) This is the present hypothesis.

stationary, at least on a time scale of several tens of million years. They leave traces at the surface of the earth. Thus, for instance, the Hawaiian island chain has resulted as volcanic activity from a hot spot as the pacific plate has been passing over it. The lava material can be dated, showing the time scale of the movement (Fig. 7). On the contrary, on the African continent volcanoes created by hot spots are found, where lava of different ages are superposed at the same site, indicating that the African continent has been fairly stationary.

Tectonic earthquakes (as opposed to volcanic ones) originate at the interfaces between the different plates. One could describe the mechanism as a stick-slip effect. The driving convection causes large strains to develop at the interfaces. Eventually, then, slip occurs, thereby releasing the strains so that arrest occurs after a certain amount of slip. Such events may occur frequently, but then with small amounts of total slip at each event. In major earthquakes, which appear at intervals of around hundred years at each site, the amount of slip may be several meters and may propagate several hundred kilometers along the fault (the interface). Examples are shallow earthquakes (down to 10 - 20 km depth) along the San Andreas fault in California, separating the northwards moving pacific plate from the southwards moving continental plate. The present study will be restricted to this kind of earthquakes, i.e. horizontal slip along a vertical, straight fault. (This is of course not typical for all tectonic earthquakes, since many fault planes are not vertical and many earthquakes may develop at great depths, as great as 700 km.)

The annually released energy from earthquakes - dominated by the few major ones - is of the same order of magnitude as the planned output of nuclear energy in Sweden.

3. SLIP AND FRACTURE ALONG A FAULT

There are many interesting similarities and dissimilarities between the propagation of slip at an earthquake event and the propagation of fracture in an engineering catastrophy. One obvious dissimilarity is that an earthquake of the type to be studied here is a mode II event (the two mating surfaces move in opposite directions but parallel to the fault direction) while engineering fractures usually are mode I events (the mating surfaces move in opposite directions but perpendicularly to their planes, the "opening" mode). In engineering structures mode II is usually not directionally stable. I assume that the stabilizing factor is the high pressure in the earth's crust. The argument is that a change towards mode I is impeded by a high pressure, since this counteracts attempts to separate the mating surfaces. Essentially the same argument may be used to explain why a mode I fracture loses directional stability when a sufficiently large tensile stress is acting in a direction parallel with the original crack.

A difficult question concerns the mechanism of slip formation and propagation at an earthquake event. Is it a sliding motion of the same nature as friction sliding between two bodies at contact or is it a continuous material deterioration? The pressure at a depth of some few kilometers is so high that the shear stresses needed to cause a frictional sliding, calculated from a dry friction (Coulomb friction) model, exceed the estimated shear strength. Therefore it is not quite obvious that a usual friction model would be appropriate when describing the sliding motion at an earthquake fault.

The shear strength in a pre-existing fault may be substantially lower than that in their surrounding rock. San Andreas fault (or, rather, the main fault in the San Andreas fault system) is more than one kilometer wide. The rock material, as far as can be judged from an inspection at the surface, is very soft so that a decimeter-sized piece can be easily broken by hand. Sliding does not involve the whole width, but only an about one centimeter thick "slice", as can be clearly seen at inspection of the "recent trace" of a major earthquake.

I would assume that what actually happens in the region of sliding is that the boundaries between macroscopic particles (grains?) in the "slice" are debonded so that the particles can move with respect to each other, they can even rotate. (Professor Clarence Allen at the California Institute of Technology has told me that traces of rolling particles can be seen in the "slice".) It is logical, then, to assume that the debonding process requires much larger shear stress than the one needed for the subsequent movement between the bodies at either side of the "slice". This point of view, however, leads to a model very similar to the classical one of dry friction.

It remains to explain how debonding can occur at the leading edge. Since debonding certainly requires much higher shear stress than the over-all stress, one must assume that the stresses are raised locally to a high level, exceeding the shear strength. Such a stress concentration is, however, very naturally associated with the process of slip propagation. In classical fracture mechanics it is described as the square root singularity. Widening of the "slice" would reduce the stress concentration. On the other hand, due to scaling effects, the energy loss in the "slice" (assuming rotation of particles, the size of which are given by intrinsic dimensions of the material) may probably increase very much when the thickness of the "slice" is decreased so much that the number of rotating particles in the thickness direction becomes very small. Perhaps a "slice" width of about one centimeter, as observed, provides an optimal compromise. Theoretical models could easily be constructed to show that a certain thickness of the "slice" implies the highest rate of propagation of the leading edge, but the many uncertainties about the mechanism and the relevant material parameters would reduce such a performance to essentially an intellectual exercise.

The speculations about debonding in a "slice" of a certain thickness are related to experiences from mode I fractures. Decohesion takes place in a region, the thickness of which is given by intrinsic dimensions of the material, but widens as the crack tip velocity is increased [1].

The mechanism suggested so far describes slip propagation, i.e. it does not consider the manner in which this propagation once started. The connection with fracture mechanics is obvious here, too. Fracture often begins at a preexisting crack, providing the required high stresses (at the tip of the crack). A major earthquake event may start at a position in the fault where previous slip has been blocked. An example is the well known bend near Palmdale of the San Andreas fault, where presently slip seems to have been blocked. North of Palmdale, seismic activity, including slowly propagating slip (due to other mechanisms than those here discussed [2]), has produced a displacement between the mating sufaces, but south of Palmdale there has been very little seismic activity for a long period. This has given rise to the famous "uplift" at Palmdale (about 25 centimeters in a few decades), a natural consequence of a mode II deformation, since the southwards moving material is piled up. Locally very high stresses may (will) eventually give rise to a catastrophic event.

When dealing with tectonic earthquakes of the type considered, one difficulty arises from the variation with depth of pressures, shear stresses and material parameters. Where is the lower surface of the plate? What are the boundary conditions at this surface? It seems logical to assume that the lower surface (if a distinct surface can be recognized) cannot slide very fast över the underlying material. A model where the lower edge of the fault is pinned during an earthquake event would therefore be natural, assuming that the horizontal movements at the lower regions of the plate are not synchronous with the event but proceed much more slowly (for instance by creep processes) than those of the upper and intermediate regions. However, the mathematical model to be described neglects such considerations. Instead advantage is taken of the simplicity implied by the assumption of plane stress (or strain).

4. MATHEMATICAL MODEL OF A SLIDING MOTION

Two models of transient sliding motion were presented in a previous paper [3]. One considered uni-directional slip propagation, i.e. a slip between one leading and one trailing edge, both travelling at the same velocity.

The other regarded extensional slip propagation, involving two leading edges, travelling at the same velocity but in opposite directions. Only the uni-directional model will be described here.

A dry friction model is used, but, as previously explained, the physical mechanism is probably rather different from usual friction. The "static friction" may be roughly substituted by the "debonding stress". Formally (mathematically) speaking, the laws of dry friction should be appropriate. There exists no satisfactory mathematical theory of dry friction, however, which involves the initiation of the process. This does not matter very much, because one can certainly assume that the region where sliding is initiated at the leading edge - the tearing region - is much smaller than the length of the sliding region. Then mathematical approximations, greatly simplifying the treatment, are justified. Similarly, at the trailing edge, where sliding ceases, a healing region exists, the linear size of which is assumed to be much smaller than the length of the sliding region. These assumptions are, of course, analogous to those used in fracture mechanics, where the linear size of the process region is assumed to be much smaller than the crack length. Moreover, it is here assumed that non-linear deformations at the sides of the fault are concentrated to small regions near the leading and trailing edges. This means, in the terminology of fracture mechanics, that one is dealing with a case of small scale yielding. In fact, the model will be linearized so that all non-linear effects are concentrated to the fault plane, and the fault itself is assumed to be infinitesimally thin. Barenblatt [4] used this procedure of linearizing the problem in the theory of static equilibrium cracks. Together with

the assumption of smallness of the non-linear regions,
it leads to a simple integrated description of the non-
-linear processes in terms of single moduli, one for
the tearing region (the modulus of tearing) and one for
the healing region (the modulus of healing).

A strict analytical treatment can hardly be carried out
without assuming infinite length of the fault and homogeneous remote boundary conditions. The model consists,
therefore, of two semi-infinite plates in contact along
the edges. This assumption limits the possibilities of
drawing conclusions regarding, for instance, the strain
relaxation in the region near an earthquake fault during
an event. Studies of such matters can be done through
quasi-dynamic treatments of the kind suggested by
Bergkvist [5] for fast running cracks in finite bodies.

5. THE MATHEMATICAL TREATMENT

Mathematically, the problem is stated as follows.

A linearly elastic, homogeneous and isotropic semi-infi-
nite solid, $y > 0$, is subjected to a remote stress $\tau_{xy} = \tau_\infty > 0$ and to the stress $\tau_{yx} = \tau_o(x-Vt)$ on $y = 0$, $-b < x-Vt < b$, where V is a constant velocity and t the time.
The displacement in the positive x direction, u, is zero
on $y = 0$, $x-Vt > b$ and equals a constant on $y = 0$, $x-Vt < -b$.
The stress σ_y is constant and can be taken to be zero.
The stresses τ_{yx} and σ_x on $y = 0$ are sought as well as
the displacements u and v (the displacement in the positive y direction) in the solid. See Fig. 8.

Details of the treatment are given in [3]. The problem
is divided into two parts so that first a simpler problem
is solved. In this problem the semi-infinite solid is
subjected to the stresses $\tau_{yx} = P\delta(x-Vt)$, $\sigma_y = 0$ on $y = 0$,
where $\sigma(x)$ is the Dirac delta-function. The displacements
u and v are sought. See Fig. 9.

The first part of the problem is obviously a simple boundary value problem, whereas the full problem is a mixed boundary value problem. A straight-forward analysis yields the solution to the first part of the problem:

$$\frac{\partial u}{\partial x} = \frac{2P}{\pi G} \cdot \frac{a_2}{R(\beta)} \cdot u_o(x) \qquad (1)$$

$$\frac{\partial v}{\partial x} = \frac{2P}{\pi G} \cdot \frac{a_2}{R(\beta)} \cdot v_o(x) \qquad (2)$$

where x, for simplicity, as in the following, stands for x-Vt, G is the modulus of rigidity,

$$u_o(x) = \frac{x}{a_1^2 y^2 + x^2} - \frac{1+a_2^2}{2} \cdot \frac{x}{a_2^2 y^2 + x^2}$$

$$v_o(x) = \frac{a_1^2 y}{a_1^2 y^2 + x^2} - \frac{1+a_2^2}{2} \cdot \frac{y}{a_2^2 y^2 + x^2}$$

$$R(\beta) = 4a_1 a_2 - (1+a_2^2)^2$$

$$\beta = V/c$$

$$a_1^2 = 1-\beta^2, \quad a_1 > 0$$

$$a_2^2 = 1-\beta^2/k^2, \quad a_2 > 0$$

and k is the ratio between the propagation velocity kc of equivoluminal waves and the propagation velocity c of irrotational waves. It should be noted that the choice of elastic constants (k and G) is such that the results are valid for both plane stress and plane strain. For plane stress $k^2=(1-\nu)/2$, where ν is Poisson's ratio.

The original problem is now constructed by a superposition technique. Since the first problem gives the solution for given stresses $\tau_{yx}=P\delta(x)$, $\sigma_y=0$ on y=0, then for stresses $\tau_{yx}=\tau(x)$, $\sigma_y=0$ on y=0 one obtains

$$\frac{\partial u}{\partial x} = \frac{2}{\pi G} \cdot \frac{a_2}{R(\beta)} \int_{-\infty}^{+\infty} \tau(a) u_o(x-a) \, da \tag{3}$$

As $y \to 0$

$$\frac{\partial u}{\partial x} \to \frac{f(\beta)}{2\pi(1-k^2)G} \int_{-\infty}^{+\infty} \frac{\tau(a)}{x-a} \, da \tag{4}$$

where $f(\beta) = 2(1-k^2) a_2 (1-a_2^2)/R(\beta) \to 1$ as $\beta \to 0$.

The right member of (4) expresses $\partial u/\partial x$ on $y=0$ for a stress $\tau_{yx} = \tau(x)$ on $y=0$. $\tau(x)$ shall now be chosen so that the following boundary conditions are satisfied:

$$\begin{cases} \tau_{xy} = \tau_\infty & \text{as } x^2+y^2 \to \infty \tag{5} \\[6pt] \tau_{xy} = \tau_o(x) & \text{for } |x|<b, \, y=0 \tag{6} \\[6pt] \sigma_y = 0 & \text{for } y=0 \tag{7} \\[6pt] \partial u/\partial x = 0 & \text{for } |x|>b, \, y=0 \tag{8} \end{cases}$$

Introduce now the sectionally holomorphic function

$$F(z) = \begin{cases} F_*(z) & \text{for } y>0 \\ -\overline{F_*(\bar{z})} & \text{for } y<0 \end{cases}$$

where $z=x+iy$ and

$$F_*(z) = \frac{1}{2\pi i} \int_{-\infty}^{+\infty} \frac{\tau(a)}{a-z} \, da, \quad y>0$$

Then, after making the substitution

$$2i(1-k^2) Gu/f(\beta) = U$$

and using the notations

$$F_+(x) = \lim_{y \to +0} F(x+iy)$$

$$F_-(x) = \lim_{y \to -0} F(x+iy)$$

one obtains (by the Plemelj formulae):

$$F_+(x) - F_-(x) = \tau(x) \tag{9}$$

$$F_+(x) + F_-(x) = \partial U/\partial x \tag{10}$$

Since $\tau(x)$ is given for $|x|<b$ and $\partial U/\partial x$ for $|x|>b$ these equations can be combined to the equation

$$F_+(x) = G(x) F_-(x) + g(x) \tag{11}$$

where

$$g(x) = \begin{cases} \tau_o(x) & \text{for } |x|<b \\ 0 & \text{for } |x|>b \end{cases}$$

$$G(x) = \begin{cases} 1 & \text{for } |x|<b \\ -1 & \text{for } |x|>b \end{cases}$$

By (11) a Hilbert problem is formulated. The solution can be found by standard methods [6]. From (9) and (10), then, the stress τ_{yx} for $|x|>b$ and the displacement gradient $\partial u/\partial x$ for $|x|<b$ are found. Then stress continuity at $|x|=b$ is not automatically satisfied, but subject to a condition

$$\frac{1}{\pi} \int_{-b}^{b} \frac{\tau_o(s) \, ds}{(b^2-s^2)^{1/2}} = \tau_\infty \tag{12}$$

Now $\tau_o(x)$ is chosen so as to agree with a dry friction model, i.e.

$$\tau_o(x) = \tau_D$$

where τ_D is a constant (the dynamic friction). This relation should hold along the region of sliding except near the ends, i.e. in the regions of tearing ($b-d_T<x<b$) and healing ($-b<x<-b+d_H$). Then from (12), assuming that $d_T \ll b$, $d_H \ll b$:

$$T-H = \pi(2b)^{1/2}(\tau_\infty - \tau_D) \tag{13}$$

where

$$T = \int_{b-d_T}^{b} \frac{\tau_1(s)}{(b-s)^{1/2}} ds \tag{14}$$

$$H = -\int_{-b}^{-b+d_H} \frac{\tau_1(s)}{(b+s)^{1/2}} ds \tag{15}$$

$$\tau_1(s) = \tau_o(x) - \tau_D \tag{16}$$

T and H will be called the moduli of tearing and healing, respectively. They are assumed to be characteristics of the material, depending only on the velocity V and the superposed pressure. It can be shown that both T and H are positive. They are of exactly the same structure as the modulus of cohesion, introduced by Barenblatt [4] for mode I cracks at static equilibrium.

Since the shear stress τ_{yx} is now known for all x at y=0, the stresses and the displacements at all points x,y can be calculated. Of special interest are the mass velocities. Details are found in [3]. Here the main results will be summarized.

6. THE MAIN RESULTS

The length of the pulse of sliding is related to the moduli T and H and to the "overstress" $(\tau_\infty - \tau_D)$ in the following way:

$$2b = \frac{1}{\pi^2}\left(\frac{T-H}{\tau_\infty - \tau_D}\right)^2 \tag{17}$$

The amount of total slip (at each side) is found to be

$$\Delta u = \frac{f(\beta)}{4\pi(1-k^2)G} \cdot \frac{T^2 - H^2}{\tau_\infty - \tau_D} \tag{18}$$

Since $f(\beta)$ and certainly also $[T(\beta)]^2 - [H(\beta)]^2$ increase with β a minimum amount of dislocation at each side,

$$\Delta u_o = \frac{[T(0)]^2 - [H(0)]^2}{4\pi(1-k^2)G(\tau_\infty - \tau_D)} \tag{19}$$

is required for sustained slip propagation.

The energy dissipation per unit of area of the interface, after the pulse of sliding has passed by, is

$$\frac{dW}{dS} = 2\tau_D \Delta u + \frac{f(\beta)}{2\pi(1-k^2)G}(T^2 - H^2) \tag{20}$$

(In [3] the sign in front of H^2 is erroneous.) One observes that the part of the shear stress that is involved in healing performs a negative work. Thus the healing process is accompanied by energy radiation from the healing region.

By introduction of the tearing/healing parameter

$$q = \frac{T/H + 1}{T/H - 1} \tag{21}$$

that can be intuitively estimated better than T and H separately, the mass velocities in the solid can be calculated for different propagation velocities. An example is shown in Fig. 10. Also the lateral displacement of

the fault can be calculated, see Fig. 11. It is quite
obvious that movements in the direction normal to the
fault are particularly pronounced (Figs. 10 and 11).

The choice of q and β at a mathematical representation
of an earthquake event, rather than the primary quantities T and H, is motivated for the sake of convenience.
In fracture mechanics laboratory tests can be carried
out, enabling rather direct determination of the fracture toughness or some other closely related quantity.
A major earthquake event along the same fault is very
rare. The experimental situation is, therefore, radically different from the laboratory replicability. Direct
experimental determination of T and H seems to be very
difficult, but the propagation velocity of the pulse
of sliding can probably be accurately determined. Therefore a representation, say in the form of curves of
different β, as in Figs. 10 - 14, seems to be convenient.
Moreover, estimates of q could, perhaps, be done rather
safely. One would assume that q is not much larger
than unity. Assumption of H=0.2T gives q=1.5. Since
the energy associated with tearing and healing is proportional to the square of the respective modulus, the
assumption implies that the energy released at healing
is 4 per cent of the energy required at tearing.

Fracture in an engineering structure is dangerous just
because the broken piece loses its function. The main
goal of the design philosophy is, therefore, to prevent
fracture. The danger of an earthquake event consists of
the action of the stress waves on different structures.
Therefore, the main interest is focussed on the stress
waves. This implies that research in fracture mechanics
is concentrated primarily on other phenomena than research in the earthquake field. However, recent interest
in dynamic fractures seems to be directed towards studies

of the stress waves, since, in finite bodies, the wave pattern plays an important part of the arrest possibilities. It has also been found that in several materials (e.g. steel [7]) the crack tip moves with a very irregular velocity. This phenomena certainly has its counterpart at the propagation of the sliding region at an earthquake event. To each fault could be ascribed a certain signature, possible, maybe, to describe in terms of stochastic quantities. Naturally such irregularities may influence the mass velocity very much, even though the displacement-time relation, due to the smoothing effect of integration, should be less affected.

The study of slip propagation, described here, concerns the simple case of a pulse travelling with constant velocity. The study should be continued by consideration of the more realistic case that involves pulse creation, acceleration, deceleration and arrest. Similar studies have been performed in the field of fracture mechanics, but the analytical difficulties are great. However, the tools to be used at such studies would certainly be essentially identical in fracture mechanics and earthquake theory.

SUMMARY

The use of solid mechanics in earth sciences has been briefly discussed, and, specially, comparisons have been made between treatments of tectonic earthquake events and fracture mechanics. To this end a model of slip propagation along an earthquake fault has been discussed in some detail. In the terminology of fracture mechanics this is a mode II deformation, which rarely occurs as fracture in engineering structures, but in spite of this fact many features of earthquake sliding motion parallel those of dynamic mode I fracture.

Therefore, developments in one field should be of interest to the other one. In particular the current research on dynamic crack propagation in finite bodies should be of considerable interest in earthquake physics, and, at the same time, profit from developments made in seismology on stress wave propagation and reflection.

REFERENCES

[1] Broberg, K. B., *On the Behaviour of the Process Region at a Fast Running Crack Tip*, IUTAM Symposium on High Velocity Deformation of Solids, Tokyo 1977. To appear in the Symposium volume.

[2] Rice, J. R. and Simons, D. A., *The Stabilization of Spreading Shear Faults by Coupled Deformation--Diffusion Effects in Fluid-Infiltrated Porous Materials*, J. Geophys. Research, (1976), 81, pp. 5322-5334.

[3] Broberg, K. B., *On Transient Sliding Motion*, Geophys. J. R. astr. Soc. (1978), 52, pp. 397-432.

[4] Barenblatt, G. I., J. appl. Math. Mech. (PMM), (1959), 23, p. 622.

[5] Bergkvist, H., *The Motion of a Brittle Crack*, J. Mech. Phys. Solids (1973), 21, pp. 229-239.

[6] Muskhelishvili, N. I., *Singular Integral Equations*, P. Noordhof, N. V., Groningen, Holland 1953. (Trans. from Russian 2nd Edn, 1946.)

[7] Carlsson, J., *Experimental Studies of Brittle Fracture Propagation*, Trans. Roy. Inst. Technology, Stockholm, Sweden 1962, No. 189.

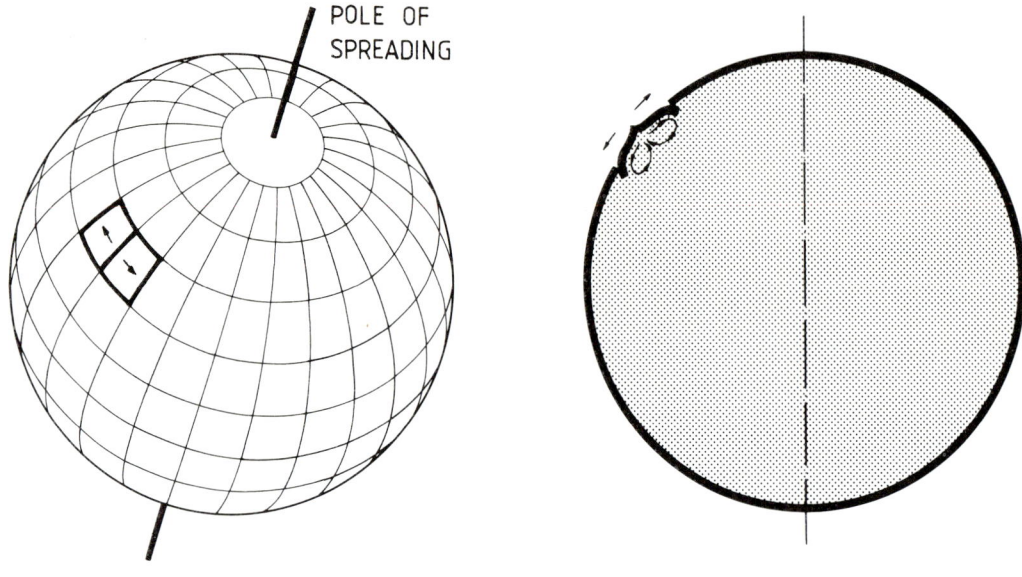

Figure 1. Schematical illustration of a moving tectonic plate. The pole of spreading does not (necessarily) coincide with the pole of rotation.

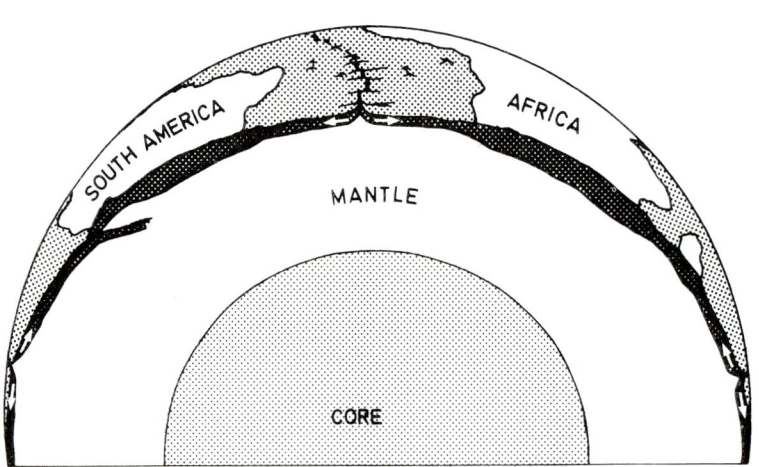

Figure 2. Example of tectonic plate motion. Source: See the end of the paper.

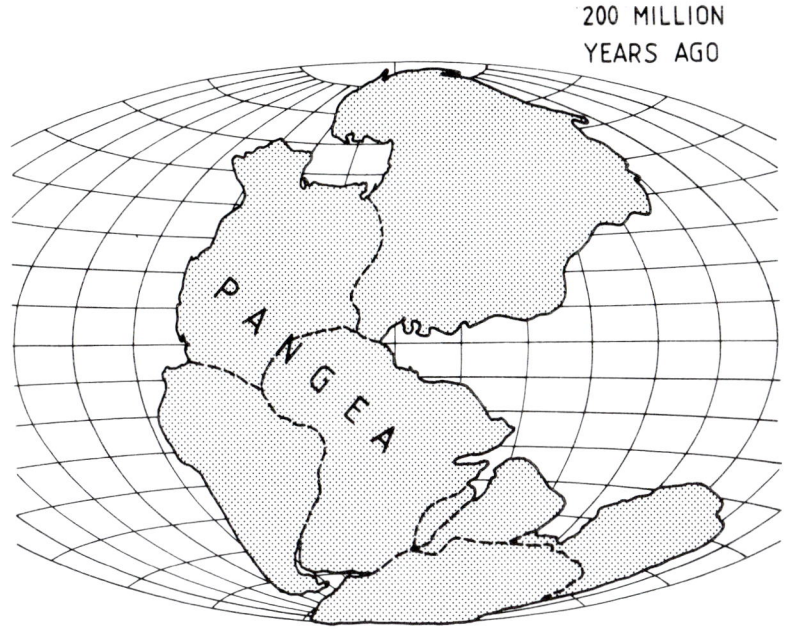

Figure 3. The "supercontinent" Pangea. Source: See the end of the paper.

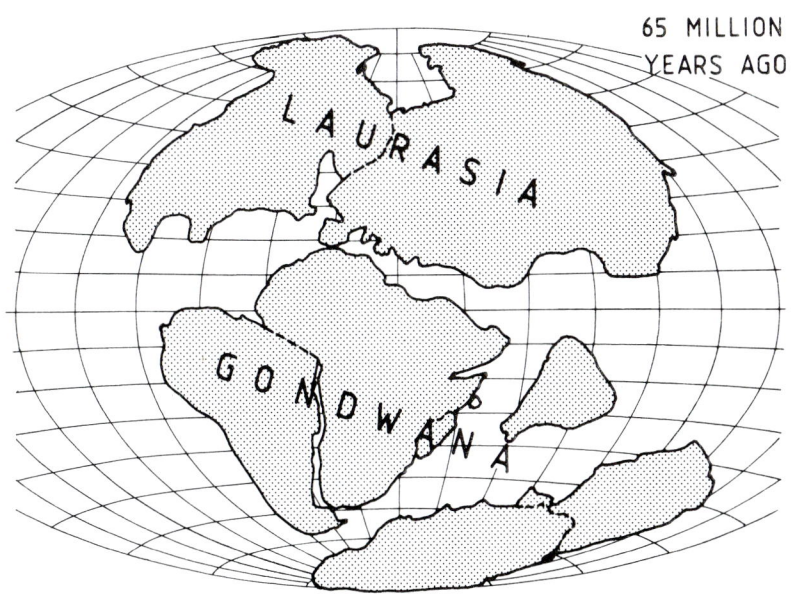

Figure 4. Splitting up of Pangea into Laurasia and Gondwanaland and formation of the present continents. Source: See the end of the paper.

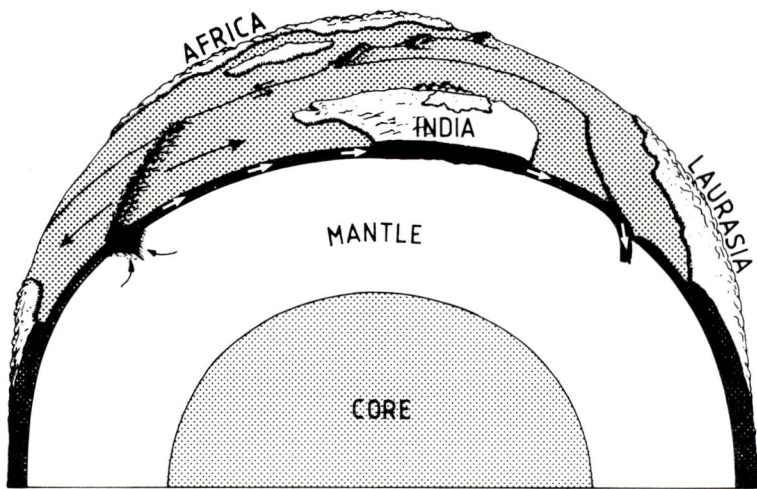

Figure 5. The Indian continent was transported on a plate from the Antarctic region to Asia. At the collision the Himalayan mountain chain was created. Source: See the end of the paper.

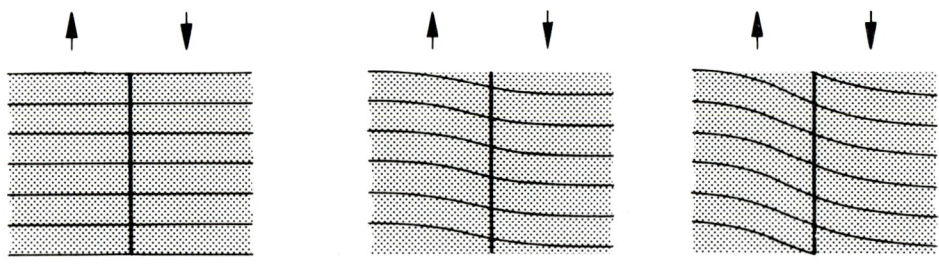

Figure 6. Shear strains created near the interface between two tectonic plates. Arrows show moving directions.

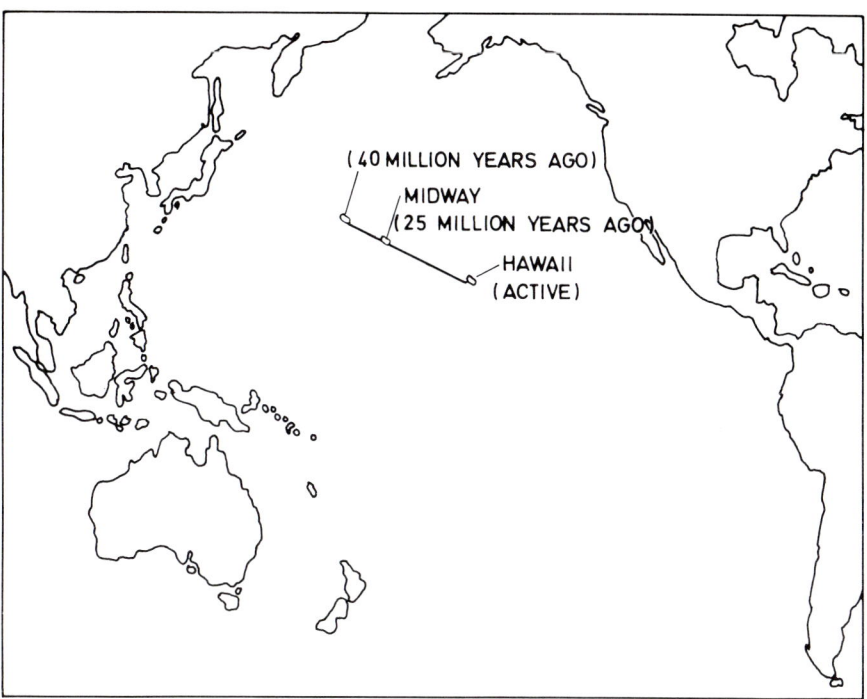

Figure 7. The Hawaiian island chain is created by volcanic activity from a deep-seated hot spot as the pacific plate passes over. Source: See the end of the paper.

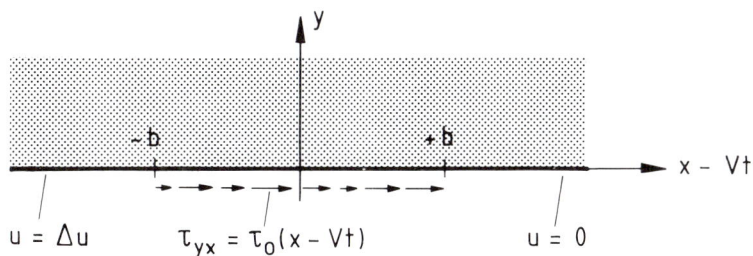

Figure 8. The mixed boundary conditions of the mathematical problem.

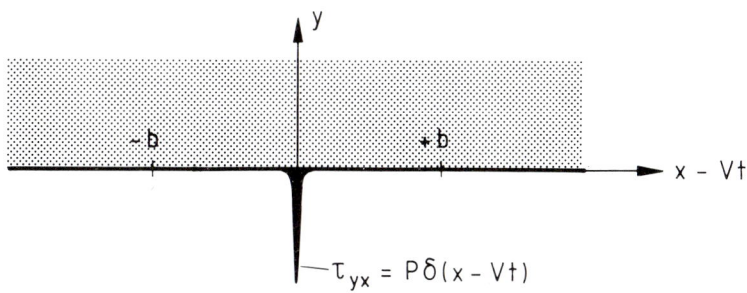

Figure 9. The simple boundary conditions of the problem first treated.

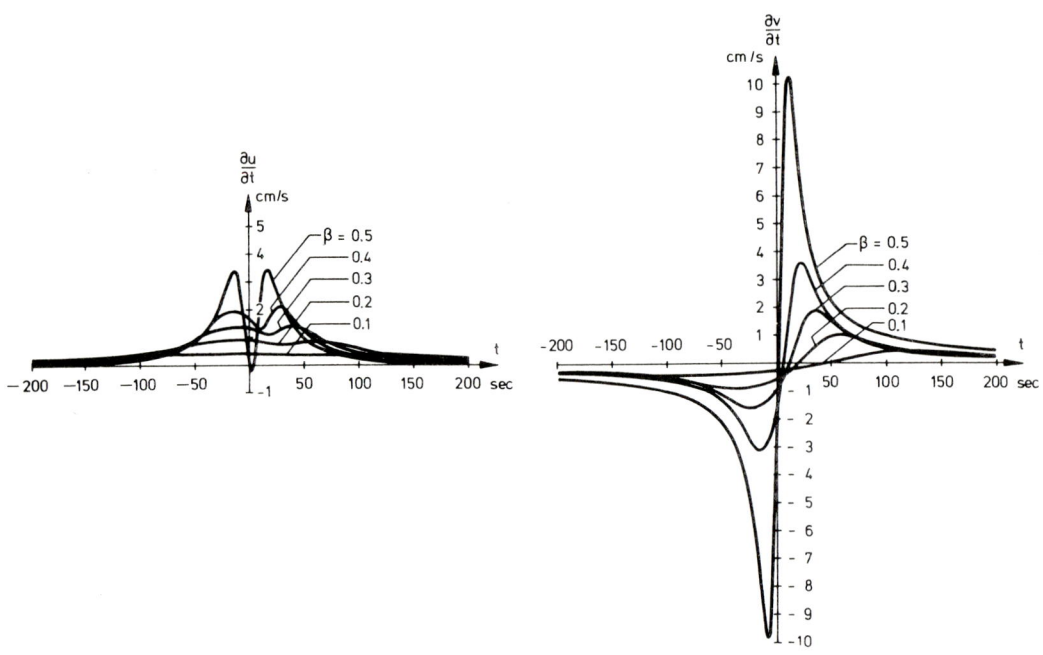

Figure 10. Particle velocity at a point situated at 50 km distance from the fault as function of time. It is supposed that $(\tau_\infty - \tau_D)/G = 2 \times 10^{-5}$, $2\Delta u = 3$ m, $c = 5.2$ km/s, $k^2 = 1/3$ and $q = 1.5$. Time $t = 0$ is (arbitrarily) chosen to the instant at which the centre of the region of sliding is most close.

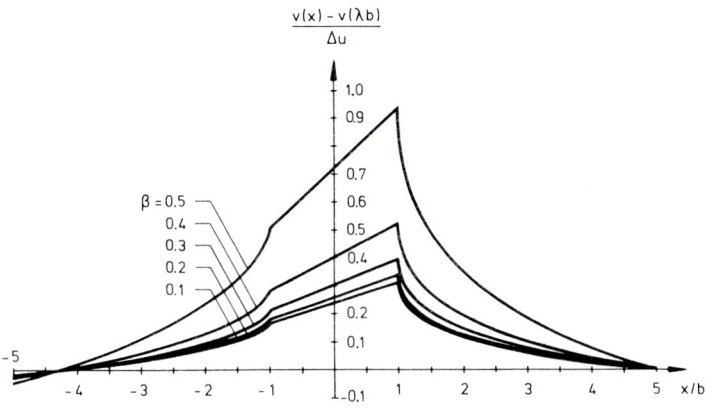

Figure 11. Non-dimensional lateral displacement of the fault. The displacement at $x = 5b$ is (arbitrarily) put equal to zero. x stands for $(x - Vt)$.

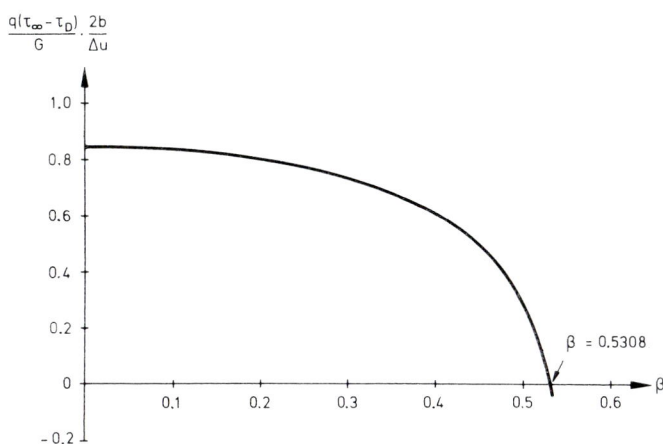

Figure 12. Non-dimensional length of the sliding region as a function of non-dimensional slip propagation velocity β. $k^2=1/3$.

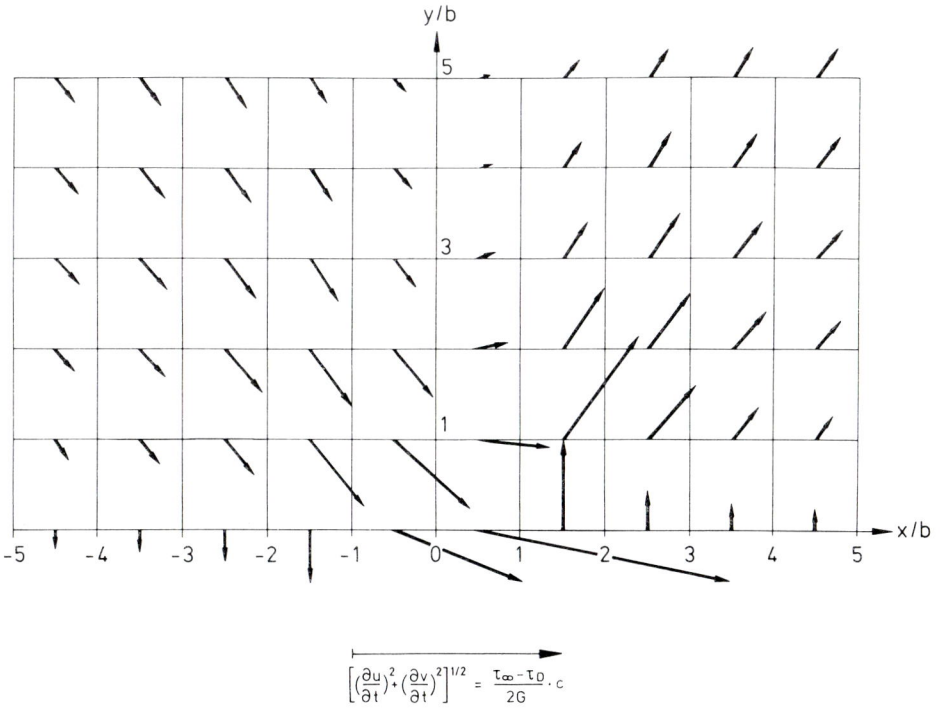

Figure 13. The velocity field in the neighbourhood of the region of sliding. Arrow tails indicate position, arrow lengths the magnitude of the velocity according to the scale shown. $\beta=0.3$, $k^2=1/3$, $q=1.5$. x stands for (x-Vt).

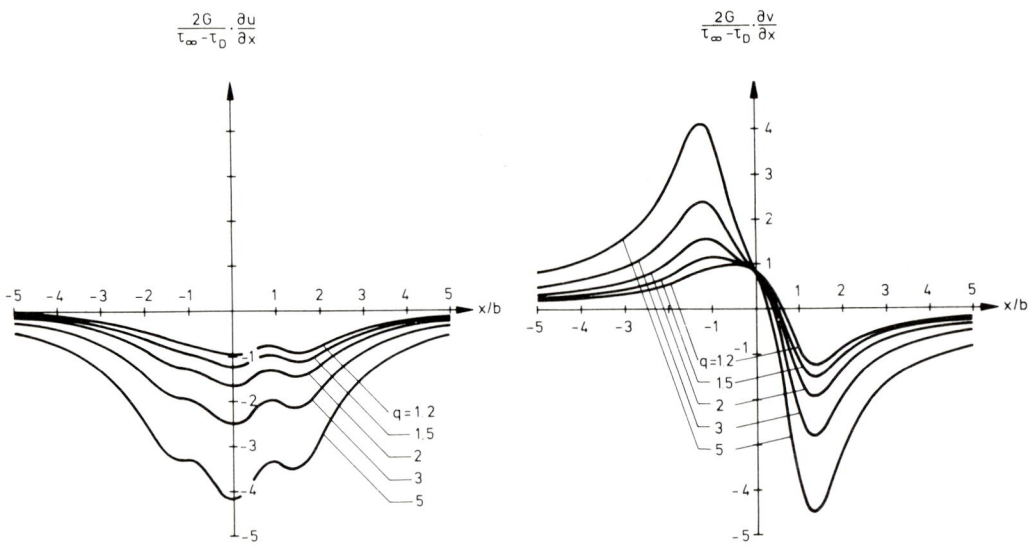

Figure 14. Non-dimensional displacement gradients at y/b=1, for different values of q. $k^2=1/3$, $\beta=0.3$. x stands for (x-Vt).

SOURCES TO SOME ILLUSTRATIONS

Some of the illustrations are drawings made after illustrations in Scientific American:

Figure 2, August 1976, p.48
Figure 3, October 1970, p.34
Figure 4, October 1970, p.36
Figure 5, October 1970, p.33
Figure 7, August 1976, p.50

THE INTER-RELATIONSHIP: EARTH SCIENCES — MATERIAL SCIENCES

P. Feltham

Brunel University, Uxbridge UB8 3PH, England

ABSTRACT

The development of concepts and models appertaining to the mechanisms of plastic deformation in solids is reviewed, with particular reference to thermally activated processes. Experience has shown that simple models of the plastic flow, based on a single intrinsic energy barrier, need to be refined, to allow for the heterogeneous internal-stress field, before they can describe adequately the flow of ductile materials above the "glass-transition" temperature; they cannot account for the ductile behaviour, observed down to temperatures approaching 0 K in some materials, e.g. face-centred cubic metals, in which a brittle-to-ductile transition is not observed. The complexity of the microstructure of most materials calls for models of plastic flow in which the dispersion of energy-barrier heights is an integral part of a description in terms of evolutionary stochastic processes. Techniques of computation and sophisticated experimental methods, particularly the various types of electron microscopy, are providing the technological basis for the formulation of theories in which the macro-deformation is described in terms of observed microprocesses. These conclusions are illustrated by reference to a tentative, stochastic, interpretation of the plastic flow of clay.

1. INTRODUCTION

The present phase of the development of our knowledge of the microprocesses occurring in materials in the course of plastic deformation, and their relation to the observed macro-response, is somewhat reminiscent of the state of understanding of conductivity, i.e. electron transport, in metals, in what may be called the "crisis stage", about 1925 (1). For, just as we have been dealing for many years with the deformation of crystals in terms of isolated dislocations which were "jumping" over isolated energy barriers, and have found this approach to be wanting, so in the mid-twenties the shortcomings of the classical gas-kinetic theory of electrical conductivity advanced in the first decade of this century by Drude, Lorentz and others, in terms of billiard-ball models of classical charged particles, were well known.

The then relatively new quantum-mechanical treatments, involving Fermi-Dirac statistics, were introduced by Sommerfeld in 1928 into his "free-electron gas" model of conduction, and in the same year Bloch refined this by allowing for the periodic potential of the crystal lattice. These developments removed the major difficulties, and facilitated a deeper insight into the processes of conduction.

The late J.D. Bernal, taking a social, evolutionary, perspective of science, realised many years ago that cooperative models, which had proved so successful in the quantum physics of the condensed state, would eventually find broad application also outside quantum physics; it is rather interesting that quite recently Bohlin, Kubát and Rigdahl (2) at the Chalmers University of Technology advanced, what I believe to be the first cooperative interpretation of stress-relaxation in solids, based essentially on the Ising model of ferromagnetism and antiferromagnetism. A further contribution: a co-operative model of the flow-kinetics of dislocations in the creep of metals, has been proposed by Himstedt (3). The correctness of the view that sophisticated models, embodying the cooperative and evolutionary stochastic features would, in the course of time find application in practically all fields of applied science as well, has been vindicated in the last 10 to 20 years; they became realisable firstly when technology, e.g. high-resolution electron microscopy with all its variants, revealed the complex microstructure of solids and, secondly, when modern computers and associated techniques added new dimensions to the scope for data processing.

Thus, just as the availability of electric power, vacuum techniques and other technological advances had paved the way to the laboratory generation of X-rays, and thus opened the door to the elucidation of the atomic structure of crystals, the recent technological developments in instrumentation have provided the basis for revolutionary new advances, not only as a result of the possible resolution of detail, measured in terms of Ångstroms rather than microns, but also in the scope for formulating and solving mathematically, i.e. in the language of science, the equations of the new models.

In relation to the plasticity of crystals it is therefore realistic to emphasise, as Kröner (4) recently has, that it was now necessary to redirect researches towards efforts to reduce the gap between experiment and theory by correlating experimental and theoretical results relating to the micro-scale with the observed plastic marco-response of materials. However, he remarks that one could not hope to derive theories which would allow one to model the process of evolution to the current mechanical state of a material by taking account of the complex mass of details shown for example, in transmission or scanning electron-micrographs. Thus statistically formulated treatments of the microdeformation (4-10) can remain the only tractable path to a realistic description of the mechanical macro-behaviour.

So as to make practical use of this insight in terms of the development of usable theories of the plasticity of complex materials, e.g. natural and synthetic "composites", such as clay or paper, we shall briefly outline the salient features of earlier models, and trace some phases in their evolution. This should provide a functional background for attempts at new syntheses.

2. EARLIER MODELS OF FLOW AND PLASTICITY

Since the introduction of the concept of the initiation of slip in crystals through thermally activated disturbances of the lattice, by Becker in 1925 (11), many attempts have been made to apply "rate theory" in the physics of plastic deformation. In particular we shall refer, firstly, to the work of Eyring and coworkers (12,13) and then to a few other papers where subsequent work in this field was reviewed, e.g. the role of rate theory in interpretations of viscosity (14), and in theories appertaining to the influence of molecular structure on viscosity (15).

In the case of crystalline materials Kauzmann (16) attempted to explain the flow of solid metals from the standpoint of chemical rate theory as early as 1941, and this work left its trace on many subsequent papers on the creep of metals and other solids. More recently Pink (17) surveyed the application of the Eyring rate-theory to the macro-deformation of glassy polymers. He notes, _inter alia_, that the existence of a rather broad mechanical relaxation-peak in the range _below_ the glass transition, i.e. in the "glass", suggests that there exists a spectrum of energy barriers to plastic flow, while the prerequisite for the usual Eyring type of analysis, e.g. in a fairly recent treatment of the behaviour of Kaolin under shear loading (18), is the presence of only one type of intrinsic barrier. Nevertheless, he considers that the defects in interpretations of the experimental data by means of the Eyring rate-equation and its various modifications are not a case against the theory, but rather point to the need for its development to a more advanced level. As may be inferred from the above reference to the glass-transition, this conclusion applies to "solids"; in polymer melts, by contrast, a single-barrier model may not be inappropriate (14,15). Fox et al (19) pointed out some time ago that it provides a physically acceptable basis for the widely observed temperature dependence of the viscosity:

$$\eta = \text{const. exp} (E_v / kT) \tag{1}$$

where E_v is the energy of activation of viscous flow. Even here however Eyring's theory fails to account for the observed increase of the pre-exponential "constant" with temperature, a shortcoming which Bueche (see ref. 15) later succeeded in remedying by assuming that the process of activation involved _cooperative_ motion, i.e. more than one polymer segment, and not just the jump of one chain link into a "large" hole. Thus even in liquids the neglect of the influence of structure on the activation processes represents an appreciable oversimplification. The pronounced temperature dependence of the viscosity, introduced through the Boltzmann term in eq. (1) readily explains the rapid "freezing" of the plastic flow in materials deformed at a constant strain rate at the glass-transition temperature, a phenomenon commonly observed, for example, in polymers (17).

The dependence of the jump frequency of polymer segments on the configurational arrangements, i.e. the free volume associated with a given configuration, can be viewed as resulting in a distribution of the entropy, but not of the heat, E_v, of activation in the Helmholtz free energy of activation for the segments contributing to the viscous or plastic flow (19). Allowance for this effect renders the "constant" in eq.(1) somewhat temperature dependent, though the exponentional term

still remains dominant, by far, with respect to the influence of
temperature on the flow rate. By implication the "glass-transition"
is preserved, and the need to introduce a spectrum of activation
energies, in which the entropy of activation is distributed,
becomes apparent.

3. ACTIVATION-ENERGY SPECTRA IN VISCO-ELASTIC FLOW

A substantial literature on activation-energy spectra as applied
to creep and stress-relaxation exists; a brief introduction will be
found in a review of stress relaxation by De Batist (20). In the case
of linear, visco-elastic materials, Feltham (21) noted that if the
entropy of activation of a polymer segment were taken proportional to
its length while the height of the energy barrier (heat of activation)
were length-invariant, a plausible assumption for non-cross-linked
elastomers, then a normal distribution of chain-lengths would imply a
log-normal distribution of relaxation times. Such a distribution had
already been suggested by Wiechert in 1893 on heuristic grounds, and
appears to be of wide occurrence. With this distribution of dwell-times
he finds that, at constant strain, the stress in the material will
relax from its initial value σ_o according to the relation

$$\frac{\sigma}{\sigma_o} = \frac{1}{2} \left[1 + \mathrm{erf}\left(b \ln \frac{\tau^*}{t} \right) \right], \tag{2}$$

with the convention that if $t \geq \tau^*$ then

$$\mathrm{erf}\left[b \ln (\tau^*/t) \right] = - \mathrm{erf}\left[b \ln (t/\tau^*) \right],$$

where $\tau^*(T)$ is the most probable value in the dwell-time
spectrum, and is equal to t when $\sigma/\sigma_o = \frac{1}{2}$; b characterises the
width of the spectrum and, depending on the material, may depend on
temperature. Fig. 1 shows stress-relaxation results in polysulphide
rubber (H.11); the points represent experimental values, while the
lines comply with eq.(1), with b increasing approximately linearly
from 0.28 at 40°C to 0.60 at 100°C.

The frequent observation of a Gaussian distribution in the logarithm
of the relaxation times has prompted Nowick and Berry (22) already
some time ago to devise nomograms facilitating the analysis of stress-
relaxation data for which this distribution is appropriate. Given the
stress-relaxation function then, for "linear" materials the corresponding
creep function can be derived. Laws and McLaughlin (23) recently
developed a self-consistent method for estimating viscoelastic creep of
composite materials, illustrating the method with two types of composites
of common structure: a dispersion of elastic spheres in a viscoelastic
matrix, and a unidirectional fibre-reinforced composite. They assume
as the starting point for the self-consistent determination of the
parameters in the equation for the creep strain $\gamma(t)$

$$\gamma(t) = \gamma(0) + \sum_{j=1}^{n} g_j (1 - e^{-\lambda_j t}) \tag{3}$$

a spectrum, i.e. a linear superposition of terms in the retardation
times $1/\lambda_j$. The equation is here phenomenological; it represents a

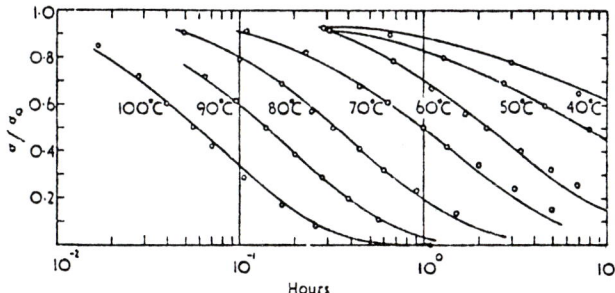

Fig. 1.   Stress relaxation at constant strain of polysulphide rubber (H.11). Full lines represent experimental result; points were evaluated by means of eq. (2) with values of b increasing linearly from 0·28 at 40°C to 0·60 at 100°C. From ref. (21).

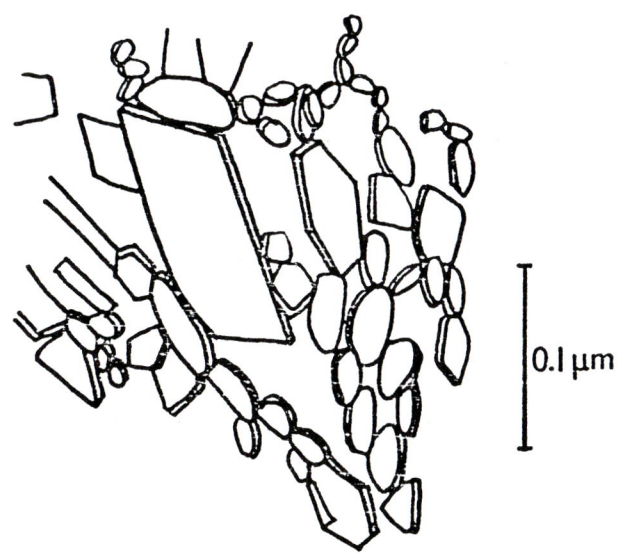

Fig. 2.   A "weak" group of particles forming a link between larger, rigid, aggregates (not shown). Schematic drawing from a micrograph. From ref. (25).

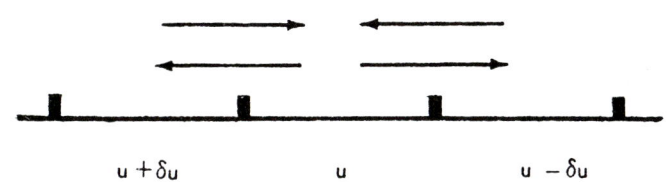

Fig. 3.   Scheme of activated jumps to and from a given activation-energy level.

rheological model of an elastic spring, series-connected with a chain of Kelvin (Voigt) elements, i.e. each consisting of a spring and a dashpot connected in parallel. In the first example, in which spheres of E-glass act as inclusions or reinforcing phase in a cold-setting epoxy polymer, they used 34 retardation times, equally spaced between 1 sec and 10^{22} sec on the logarithmic time axis, for the description of the process. Although retardation times beyond about 3×10^9 sec (about 100 years) are unlikely to contribute significantly to the deformation in practice, except for geological processes, it is clear that without computers the problem would not be tractable. The work complies with Kröner's suggestion that the macro-response should be derived from that of the micro-structure, for the dispersion, the geometrical form of the inclusions, as well as the elastic constants of the inclusion, and the creep compliance of the matrix are the parameters from which the overall creep behaviour can then be theoretically evaluated.

In this treatment the micro-structure is not taken into account at the molecular level, as was attempted in the stress-relaxation experiments interpreted by eq. (2), in which a characteristic energy-barrier was in fact deduced. Rather, the structure is specified in less fundamental terms, adequate for engineering purposes, i.e. the elastic compliances of the "filler" and of the matrix. The method may prove of substantial practical value in the earth sciences, i.e. in modelling the time-dependent mechanical response of natural composites, be they rocks or clays, provided the deformation is sufficiently small to remain "linear", i.e. visco-elastic.

Axelrad and co-workers (8-10) have developed a theory of some mathematical complexity in which the observed macro-response of inhomogenous solids, e.g. fibrous materials such as paper, is derived from a synthesis based on the behaviour of "meso-domains", i.e. representative volumes of the material sufficiently large to be regarded as homogenous; however the viscoelastic response of these domains is derived by means of a stochastic model. In this model the distributions of the deformations in the micro-domains, in which the structural elements of the material are resolved, e.g. dislocations in a metal grain, fibre links in a paper-pulp, are assumed to be random, i.e. Gaussian, and to remain so in the course of creep under an applied constant stress. The macroscopic creep-response must be derived by an averaging procedure over a meso-domain, i.e. generally referred to as a "representative volume" (23).

The transition from the structural detail via a statistical treatment to the macro-behaviour has its classical parallel in the Gibbsian ensemble, i.e. the transition from the molecule-space to gas-space or, more concretely, from Maxwell's law of the distribution of the velocities of molecules in a gas to the evaluation of the gas pressure.

The above examples show that materials science has developed to the extent that it is possible to formulate mathematically the problem of the time dependent deformation of materials which are structurally complex, including composites. The treatments referred to apply, however, only to linear viscoelastic behaviour, i.e. when the constitutive relations for any phase of the system can be expressed in terms of a Riemann-Stieltje "Boltzmann superposition" integral

$$\varepsilon(t) = \int_{-\infty}^{t} M(t - \tau) \, d\sigma(\tau), \qquad (4)$$

in which the creep compliance tensor $M(t)$ is not a function either
of stress σ or of strain ε (9,23). It is therefore appropriate to
many composites in which the phases are viscoelastic. In the case
of the creep of clays it may be applicable only under rather restric-
tive conditions, at relatively low stresses. This is apparent if one
considers the simple case of the application of an increasing deviatoric
stress at time t=0, with the stress equal to zero for all $t < 0$. Then,
using a constant stress-rate $d\sigma/d\tau$ sufficiently high to enable one
to assume that $M(t-\tau)$ will remain essentially constant during the
period of deformation, one has

$$\varepsilon(t) = M(0) \int_0^t \frac{d\sigma}{d\tau} d\tau \tag{5}$$

or $\quad\quad \varepsilon(t) = M(0) . \sigma(t) , \tag{6}$

i.e. we obtain a linear stress/strain relation. This may be appropriate
in the case of heavily overconsolidated clays (24), so that eq. (4),
and by implication eq. (6), impose boundary conditions which restrict
the use of "viscoelastic" theory to a specific class of systems.

4. PLASTIC DEFORMATION

In so far as we are here concerned with materials science in its
general bearing on the deformation of materials of interest in soil
science, I want to refer here specifically to an apposite, recent paper
by Pusch (25) relating to a complex material: the deformation of
illitic clay - a void/particle composite - in the light of electron
microscopy of the microstructure of sheared specimens.

Referring to Fig. 2, a drawing based on a micrograph, taken from his
paper, it is evident, as he points out, that the local shear stresses
in the microdomains represented by the aggregates, resulting from applied
external forces, will be relatively low in large, dense, aggregates while,
by contrast, it will be high in the smaller groupings of particles wedged
between them. He questions the applicability of Eyring's rate theory
to that type of structure. In such applications of Eyring's theory,
e.g. to fine-grained consolidated Kaolin at room temperature (18),
Eyring's "molecules" are replaced (in an idealised card-house structure)
by a uniform distribution of particle-contact zones, which represent
the flow units. The inter-particle contact area per unit volume, and
the stresses in these areas of contact are determined by an averaging
procedure from a knowledge of the initial void-ratio and the mean Stokes'
diameter of the disc-shaped particles. Clearly, such a "homogenised"
model-structure automatically obliterates the stochastic features of
the deformation which the observations of the actual structure imply.
In fact results of attempts to use Eyring's theory in the interpretation
of the flow of "composites" such as soils (26) have not led to definite
conclusions concerning the activation energy of the rate-controlling
process and, hence, the nature of the barriers or bonds involved.

Thus, as in the transition from the classical "billiard-ball" model of electrical conductivity to the quantum-statistical interpretation, we now search for a development from the simple molecular rate-theory to a more discerning model, which should allow for the micromechanic and evolutionary features observed to occur in the course of deformation in materials of that type.

In our illustrative discussion of the bearing of the micromechanics of deformation on the observed creep or stress/strain relation we shall consider as an example, appropriate here, observations relating to clay, as evident from the work of Pusch already referred to (25), and from other evidence obtained by scanning electron-microscopy, e.g. in compacted specimens of dispersed or floculated Kaolin or Illite (27). It is clear from these, as pointed out by Pusch (25), that at sub-failure stress levels the stress system at any point in the material is governed primarily by the local aggregate-and-pore geometry. As changes occur in this micro-topology in the course of plastic flow, changes in the micro-stress pattern will take place simultaneously. One is thus led to the conception of a spatially and temporally varying micro-stress field and, with thermally activated slip processes, this implies that the dwell-times of the slip units will be distributed, i.e. one has a spectrum of activation energies. Pusch (28) proposed the application of a stochastic model of creep, and examined the specific micro-mechanical features which such a theory must take into account; the aim is thus in harmony with Kröner's view, referred to above, of deriving the description of the macro-response from the micromechanics via statistical models.

5. SOME ASPECTS OF THE MICROMECHANICS OF CLAY PLASTICITY

Before proceeding to a review of the principal features of Pusch's interpretation of the plastic flow of clays, based primarily on his two recent papers (25,28), we have to note that clays subjected to stresses will deform not only due to plastic shears in the particles or in the contact zones between adjacent ones, but also as a result of consolidation due to volume changes associated from the loss of pore-water, permeating through the structure. This hydrodynamic contribution to the deformation (29,30), i.e. consolidation which may be coupled to the shear component, is excluded from the plastic shear-deformation considered in the stochastic model used by Pusch. The creep accounted for by the model would therefore be best observed in consolidated specimens when the clay behaves plastically, e.g. in the remoulded state, below the Atterberg "liquid limit".

Referring to Fig. 2, we see that groups of small particles form links between larger ones. If a shear stress is applied to a specimen containing such groupings, it is not hard to imagine that relatively high local stresses will develop where small, sharp, particles are pressed against larger ones; the system is somewhat akin to an indentor about to penetrate a flat test-piece. By contrast, where larger, blunter, particles are pressed against similar ones, relatively low local stresses will prevail. If the stress is high enough for irreversible flow to take place, it will start to set in preferentially at the former

locations. However, as the tips of the sharp particles blunten in
the course of flow, just as a piece of moist chalk would be seen to
do when pressed against a blackboard, rapid local flow and an
accompanying relaxation of local stresses would take place at such
points. Larger, blunter, particles would be less prone to deformation.
Thus we see that the first effect of the deformation beyond the elastic
limit is a plastic "consolidation" at points of high local effective
stress. If the externally applied stress were not increased, but
maintained steady, then one can visualise that deformation in such
regions would continue, even though more slowly than at the instant
of the application of the external stress, and the material would
creep locally. As a consequence of the creep, the local stresses
would relax, i.e. tips would blunten, but resulting from the variations
in the local displacements, higher stresses would be thrown onto
neighbouring sites.

Assuming that elementary slip is controlled by a stress-dependent
activation energy, the above process implies, on the one hand, a
gradual <u>increase</u> of activation energies in the asperities as they
become blunted; on the other hand, the increasing stresses sustained
by the neighbouring particles would there lead to a <u>lowering</u> of the
energy-barriers to slip, thus facilitating flow in previously inactive
domains. These newly activated domains will now take on the role of
"tips", and a cycle of local relaxation and flow will take place as
discussed above. The process of plastic consolidation will continue
with, in general, a gradual "blue shift" of the activation energy
spectrum, and a concomitant decrease in the overall rate of creep.
Pusch (28) discussed the potential application of a creep model of this
type, and related the micromechanics of deformation to the specific
features and shear-induced micro-structural changes observable in
clays; we shall briefly discuss the model and show that, with appropriately modified boundary conditions, it is equally applicable to other
materials, differing in their plastic response substantially from clay,
e.g. in which a "glass-transition" is observed.

6. STOCHASTIC MODEL OF CREEP

For the present we shall regard a test specimen not smaller than a
representative volume of the material as a unit, i.e. we shall not
focus on the individual particles or aggregates, but consider an
"energy-barrier space" (7). In that space potential slip-units have
allocated to them dwell-times determined by a spectrum of activation
energies, u, ranging from a minimum u_1 to a maximum u_2 (Fig. 3). The
dwell-time θ of a slip unit held up at a barrier of magnitude u,
would then be given by

$$\theta = (1/\nu) \exp(u/kT), \qquad (7)$$

where ν is taken to be a vibrational frequency of the order of
10^{11} sec^{-1}, independent of u. We sub-divide the interval $u_1 - u_2$ into
discrete steps of magnitude δu, and shall denote the number of
available slip-units in such an energy interval by $n(u).\delta u$. Referring
to Fig. 3, if following a jump over a barrier of height u the proba-

bilities that a slip unit should become held up at a new barrier either higher or lower than u by the amount δu are taken to be equal, then $\partial n/\partial t$ at any u-level is determined by the four fluxes indicated by the arrows. Elementary considerations then show that the continuity equation of the process is

$$\frac{\partial n}{\partial t} = D \frac{\partial^2 [n \exp(-u/kT)]}{\partial u^2}, \quad D = \tfrac{1}{2} \nu (\delta u)^2, \tag{8}$$

solutions of which, with appropriate boundary conditions, yield the energy-barrier distribution $n(u,t)$. If each activated jump is assumed to make the same contribution to the overall shear, then the creep rate is proportional to the integral of $n(u,t) \cdot \exp(-u/kT)$ over the u-spectrum. With relatively simple boundary conditions (5-7) one can then deduce some of the well-known creep laws, e.g. the logarithmic (7) and the t^m-type with m $\approx$ 1/3 (31), observed with clay at low values of the normalised deviatoric stress (32), as well as the sigmoidal curves found, for example, in covalent semiconductor crystals (5,6) at relatviely high temperatures.

If the continuity equation is to hold at the end-points of the spectrum, then at u_1 one must have a "generating barrier", at which slip units are formed at a rate compensating for the missing inflow to and outflow from the empty interval $u_1 - \delta u$. Similarly, at u_2 one has an absorbing barrier. In practice this may mean that one neglects jumps for which $u > u_2$, the dwell-times of which may be incommensurately long compared with the creep times of relevance in the experiment.

The picture of the processes thus shows that at points in the material jumps will take place from a range of barriers, some to higher u-values, some to lower ones. New slip units are brought into operation at the low-u end of the spectrum, i.e. in general at points of high local stress, such as the "tips" referred to. On the whole, the spectrum undergoes a blue-shift in the course of time, and the flow stops. The model is thus physically appropriate. We shall illustrate the scope of the model by two examples. In the first instance we select a material, e.g. a germanium crystal in which a single intrinsic barrier appears to control creep above the "glass" transition range, i.e. above about 450°C (5); in GaSb crystals the corresponding temperature is about 370°C (6). The u-range will thus be centred around the intrinsic value of the single type of barrier u^*, and a fairly sharp "glass"-transition would be expected on lowering the temperature below a critical value, as is in fact observed. This can be seen by taking as a somewhat rough criterion for the transition that, with $u = u^*$ in eq. (7), the jump rate should be neither too low nor too high to lead to "observable" creep, i.e. one may take θ = 1 second. With $\nu \approx 10^{11}$ sec^{-1}, the transition temperature T^* should then be given by

$$u^* \approx 25\, kT^*. \tag{9}$$

In germanium, where the Peierls potential is the intrinsic barrier which governs the mobility of dislocations and, hence, creep, $u^* \approx 1\cdot 6$ eV, and this value does, indeed yield the observed $T^* \approx 450°C$. Thus below about 400°C creep is hard to induce in germanium, the material is rather brittle, glass-like.

By contrast, we may consider a face-centred-cubic metal, such as copper which, when pure, remains soft, i.e. ductile even at temperatures close of 0 K. The lattice resistance to the movement of dislocations, unlike in germanium, is low, and the obstacles to slip are formed in the course of work-hardening. Thus the initial loading of a specimen, in a creep test, may result in an "instantaneous", i.e. rather rapid plastic deformation until a level of work-hardening is attained where the magnitude of the most probable barrier to slip would comply with eq. (9), with T now equal to the temperature at which the experiment is being carried out. In this case u* is therefore determined by the lattice defects arising from work-hardening; a glass-type transition is not observed. The activation energy of creep, which would be derived from a "thermal cycling" experiment would yield a value close to 25 kT: about 0.6 eV at room temperature (1).

The plastic behaviour of overconsolidated soils is sometimes likened to that of work-hardening materials, e.g. metals (33), and might provide suitable material for studies of the scope of stochastic models in the interpretation of creep in soils. The model is free from the difficulties of the simple Eyring treatments we have referred to and, as we have seen, takes into account the microstructural processes, such as the "creep-consolidation" in illitic clay subject to shear.

It seems to me that, at least from the point of view of availability of "high resolution" experimental techniques and of the mathematical and computational facilities nowadays at hand, materials science is providing the framework for a systematic study of the deformation behaviour of heterogeneous materials. The rate of advance however, depends on the availability of resources, not least human, and on their efficient use. Concerning stochastic models, we are probably as yet "at the beginning", and considerable refinement will, foreseeably, take place in the near future. So, perhaps it behoves us to conclude with Voltaire's admonition: "We must not say "let us begin by inventing principles whereby we may be able to explain everything", rather we must say "let us make an exact analysis of the matter", and then we should try and see with much diffidence whether it fits in with any principle."

REFERENCES

(1) Feltham, P., Collective dislocation models in crystal plasticity. Rev. Deform. Beh. Materials 1 (1976) 1-30.

(2) Bohlin, L., Kubát, J. and Rigdahl, M., A cooperative model for stress relaxation in solids. Proc. VIIth Int. Congr. Rheology, Gothenburg, Sweden 1976 pp. 312-313.

(3) Himstedt, N., Transient effects during tensile tests in a model of correlated dislocation motion. Acta. Metallurgica 26(1978) 351-356.

(4) Kröner, E., Recent progress in the understanding of the mechanical behaviour of single and polycrystalline metals and alloys. Proc. IVth Int. Conf. Strength of Metals and Alloys, Nancy, France 1976 pp 937-957.

(5) Bekirovic, M. and Feltham, P., *Creep and stress relaxation of germanium single crystals.* J. Mater. Sci. 9 (1974) 383-388.

(6) Ashraf, M., *Creep and stress relaxation of gallium antimonide in the range 250-600°C.* Ph.D. Thesis, Brunel University, Uxbridge, England, 1976.

(7) Feltham, P., *A stochastic model of crystal plasticity.* J. Physics D. 6 (1973) 2048-2056.

(8) Axelrad, D.R., *Rheology of discrete media.* Proc. VIIth Int. Congr. Rheology, Gothenburg, Sweden 1976 pp. 43-49.

(9) Provan, J.W. and Axelrad, D.R., *Probabilistic micromechanics of solids.* Rev. Deform. Beh. Materials 2 (1977) 174-209.

(10) Axelrad, D.R., *Micromechanics of solids.* Elsevier, New York 1978.

(11) Becker, R., *Über die Plastizität amorpher und kristalliner fester Körper.* Phys. Zeit. 26 (1925) 919-924.

(12) Glasstone, E., Laidler, K.J. and Eyring, H., *The Theory of Rate Processes.* McGraw Hill, New York 1947.

(13) Ree, T. and Eyring, H., *The relaxation theory of transport phenomena.* Rheology (II), Academic Press, New York 1958 pp. 83-144.

(14) Bondi, A., *Theories of viscosity.* Rheology (I), Academic Press, New York 1956 pp. 321-350.

(15) Bondi, A., *Viscosity and molecular structure.* Rheology (IV), Academic Press, New York 1967 pp. 1-84.

(16) Kauzman, W., *Flow of Solids from the standpoint of chemical rate theory.* Trans. Amer. Inst. Mech. Eng. 143 (1941) 57-63.

(17) Pink, E., *Applications of the Eyring rate theory to the macro-deformation of glassy polymers.* Rev. Deform. Beh. Materials 2 (1977) 37-80.

(18) Foster, R.H., *Behaviour of Kaolin fabric under shear loading at low stress.* Stress-Strain Behaviour of Soils, Foulis, Henley-on-Thames, England 1972 pp. 81-88.

(19) Fox, T.G., Gratch, S. and Loshack, S., *Viscosity relationships for polymers in bulk and in concentrated solutions.* Rheology (I), Academic Press, New York 1956 pp. 431-493.

(20) DeBatist, R., *Stress relaxation: a review of principles and applications.* Rev. Deform. Beh. Materials 1 (1976) 71-116.

(21) Feltham, P., *On the representation of rheological results with special reference to creep and relaxation.* Brit. J. Appl. Phys. 6 (1955) 26-31.

(22) Norwick, A.S. and Berry, B.S., Lognormal distribution function for describing anelastic and other relaxation processes. I.B.M. Journal of Research and Development 5 (1961) 297-320.

(23) Laws, N. and McLaughlin, R., Self-consistent estimates for the viscoelastic creep compliances of composite materials. Proc. Roy. Soc. London A359 (1978) 251-273.

(24) Wroth, C.P., Some aspects of the elastic behaviour of over-consolidated clay. Stress-Strain Behaviour of Soils, Foulis, Henley-on-Thames, England 1972 pp. 347-361.

(25) Pusch, R., Shear deformation of clay microstructure. Proc. VIIth Int. Congr. Rheology, Gothenburg, Sweden 1976 pp. 270-271.

(26) Mitchell, J.K., Campanella, R.G. and Singh, A., Soil creep as a rate process. J. Soil Mech. Fdn. Div., Amer. Soc. Civil Eng. 94 (1968) 231-239.

(27) Barden, L. Examples of clay structure and its influence on engineering behaviour. Stress-Strain Behaviour of Soils. Foulis Henley-on-Thames, England 1972 pp. 195-205.

(28) Pusch, R., Creep mechanisms in clay. Private Communication. See also Proc. Conf. on Mechanisms of Deformation and Fracture. Luleå, Sweden 1978.

(29) Biot, M.A., General theory of three-dimensional consolidation. J. Appl. Phys., 12 (1941) 155, 426, 578.

(30) Schiffman, R.L., Chen, A.T. and Jordan J.C., An analysis of consolidation theories. J. Soil Mech., Fdn. Div., Amer. Soc. Civil Eng. 95 (1969) 285-294.

(31) Feltham, P., A simple stochastic model of low-temperature creep and stress-relaxation in solids. Proc. VIIth Int. Congr. Rheology Gothenburg, Sweden 1976 pp. 166-167.

(32) Vaid, V.P., and Campanella, R.G., Time-dependent behaviour of undisturbed clay. Proc. Amer. Soc. Civ. Eng., J. Geotechn. Eng. Div., 103 (1977) 693-699.

(33) Pendle, M.J., A model for the behaviour of overconsolidated soil. Geotechnique 28 (1978) 1-9.

THE INTER-RELATIONSHIP: MATERIAL SCIENCES—SOLID MECHANICS

J. F. Knott

Department of Metallurgy and Materials Science, University of Cambridge, Pembroke Street, Cambridge CB2 3QZ, England

ABSTRACT

The paper reviews macroscopic and microscopic aspects of deformation and fracture in materials, showing that a number of concepts from solid mechanics have application for different scales of physical size. Attention is paid particularly to slip-line fields, to fracture mechanics, to dislocations and cracks, and to elastic wave propagation.

1. INTRODUCTION

A complete understanding of why a material possesses a particular strength or resistance to fracture can, in the final resort, be gained only from a knowledge of inter-ionic potentials and the energy levels of bonding electrons. However, the application of solid mechanics to flow and fracture phenomena in materials enables a number of useful predictions to be made, based on measurements of a few macroscopic parameters, such as elastic modulus, yield strength and fracture toughness. In the present paper, attention is paid to the application of a number of concepts from solid mechanics, which are of particular value in trying to understand the behaviour of structural engineering materials. Apart from a brief reference to creep crack growth, no treatment of rates of deformation is given, but a good, general discussion of this topic is to be found in the work of Ashby (1).

The concepts find application in different size-scales of interest and these may be divided roughly into three regimes: the macroscopic, or "engineering" scale; the microscopic, where the treatment mainly involves the stresses around line defects known as dislocations; and the microstructural, where the microstructure itself is treated as a small engineering structure: for many materials, this involves features whose sizes lie in the range 0.1μm to 0.1mm.

2. ELASTIC PROPERTIES

Structural metals are polycrystalline aggregates of grains, usually well-bonded at their boundaries. The crystal symmetry is high (cubic, or hexagonal) but single crystals, even of cubic metals, generally exhibit some degree of anisotropy in elastic behaviour (2,3). Elastic anisotropy in cubic metals is expressed by the ratio $2(S_{11}-S_{12})/S_{44}$ where S_{11}, S_{12}, S_{44} are appropriate elastic compliances. This ratio is unity for an isotropic material. It has a value of unity for tungsten (b.c.c.), 1.2 for aluminium (f.c.c.), 2.7 for iron (b.c.c.) and 3.2 for copper (f.c.c.). However, the random orientations of grains in a large polycrystalline mass lead to near-isotropy in macroscopic elastic constants and it is possible to specify elastic properties in terms of two independent parameters: either Young's modulus, E, and Poisson's ratio, ν; or Lamé's constants, λ and μ (the shear modulus). Relationships between these parameters are given by:

$$\mu = \frac{E}{2(1+\nu)} \; ; \quad \lambda = \frac{\nu E}{(1+\nu)(1-2\nu)} \tag{1}$$

These simplifications are used with great success in conventional engineering design, but clearly require modification if the symmetry is lower than that in the cubic system, e.g. some titanium or magnesium alloys; or if the orientation of grains is not random, as in rolled and annealed "textured" sheets, in directionally-solidified turbine blades and perhaps in some weld deposits. The problems are accentuated in composite materials, where properties are highly directional.

3. PLASTIC DEFORMATION

In addition to ensuring elastic stability, the design engineer needs to prevent failure of his structure by general yield. He therefore requires to know the conditions under which his material begins to undergo plastic deformation and how these may be affected by the stress state. If a design, based on yield stress, is employed, there should not be any need to consider macroscopic tensile plastic instability as a mode of failure in structures, but such instability, in different stress states, is of importance in metal forming (particularly, in sheet pressing, deep-drawing and stretching operations). Reference may be made to forming limit diagrams (5,6).

Yielding in a single crystal begins at a critical value of shear stress, resolved on the slip-plane (usually, the close-packed plane) in the slip direction (the close-packed direction). In a randomly oriented (isotropic) polycrystalline aggregate, it is found experimentally that the onset of yielding at ambient temperature is unaffected by (moderate) hydrostatic tension or pressure. This means that the macroscopic yield criterion involves only a critical value of the deviatoric component of the stress tensor, independent of the hydrostatic component. The two most commonly used yield criteria are those of Tresca and Von Mises:-

Tresca: $\quad \sigma_1 - \sigma_3 = 2k = \sigma_Y \tag{2}$

Von Mises: $(\sigma_1-\sigma_2)^2 + (\sigma_2-\sigma_3)^2 + (\sigma_3-\sigma_1)^2 = 6k^2 = 2\sigma_Y^2 \tag{3}$

where σ_1, σ_2, σ_3 are the principal stresses ($\sigma_1 > \sigma_2 > \sigma_3$), k is the yield stress in pure shear and σ_Y is the yield stress in uniaxial tension. Neither criterion seems to be related simply to the critical

resolved shear stress criterion for a single crystal, but the value
of the uniaxial yield stress for a polycrystalline specimen has been
expressed in terms of the critical resolved shear stress, using the
"Taylor factor" which takes account of the number of available slip
systems (7,8).

The macroscopic yield criteria for metals are therefore simpler than
are those for materials such as soils or rocks, where the value of
the compressive stress normal to the shear plane affects the critical
value of shear stress. A similar effect is found in some polymers,
where the slip processes involve bond rotation or unfolding, which is
easier to operate if the compressive stress normal to the slip plane
is small. Non-random orientations in a polycrystalline metal (texture)
give rise to anisotropy which affects the ease of producing flow in
different directions (as in the "earing" of deep-drawn cups formed from
oriented cold rolled sheet) and this can be represented by an
asymmetrical yield locus in stress space and by a modification of the
Von Mises criterion (9). Occasionally, use is made of texture to
improve mechanical properties in fabricated objects, but these effects
are often adventitious.

The ability to spread plastic deformation throughout a polycrystalline
aggregate is a feature of importance. To produce a uniform shape
change throughout the piece as a whole, it must be possible to
accommodate completely the arbitrary shape change of a given grain, by
plastic deformation in the grains adjacent to it. The symmetrical
strain tensor has six components (three tensile and three pure shear),
but only five of these are independent if the volume is to remain
constant, as it must, if yielding is a pure slip process. It is then
found that a material must possess five independent slip systems to
accommodate an arbitrary plastic shape change. These are readily
available in cubic metals and, in hexagonal metals, slip on non-basal
planes or deformation twinning can help to provide sufficient systems.
The positive and negative charge distributions in ionic solids place
severe restrictions on the number of independent systems, however, so
that most polycrystalline ceramic materials are unable to bear high
tensile loads, because they fall apart at the grain boundaries. In
materials such as rocks, even compressive deformations may be treated
by "clastic theory" (which regards the material as a set of disjointed
units) rather than by continuum mechanics (10).

4. SLIP-LINE FIELDS

A common idealisation of a material's flow properties is to assume
non-hardening, rigid/plastic behaviour. Strictly, this assumption is
viable for two extremes. One is hot working, where the large shape
changes mean that the plastic strains are very much larger than the
elastic strains, and where continuous recovery and recrystallisation
prevent work-hardening. The other extreme involves the deformation
of a heavily cold-worked material. There is no further hardening if
the work-hardening capacity has been exhausted, but it is not always
easy to find deformation systems in which the elastic strain can be
ignored, or in which tensile instability does not occur. Rigid/plastic
behaviour is often assumed, even for plasticity contained within an
elastic field, such as a local enclave around a notch, or below a

hardness indenter. Limit analysis can often give good upper and lower bound solutions to rigid/plastic problems, but the most satisfactory solutions are those given by slip-line fields (1,4,9). Slip-lines simultaneously show the directions of maximum shear stress and the directions of (tangential) velocity discontinuities. They lie at 45° to the maximum and minimum principal stresses and are orthogonal to each other. In $(X_1 X_2)$ plane strain, $\varepsilon_3 = 0$ and it can then be shown that $\sigma_3 = \frac{1}{2}(\sigma_1 + \sigma_2) = \frac{1}{3}(\sigma_1 + \sigma_2 + \sigma_3) = \sigma_H$, so that the slip-lines are at 45° to the two in-plane principal stresses, σ_1 and σ_2. The general slip-line pattern can often be determined from symmetry and from the fact that the lines must meet free surfaces at 45° (the stress normal to a free surface is a principal stress).

Some slip-line fields of interest are shown in fig.1. The first, fig.1a), is for simple plane-strain tension, where $\sigma_1 = \sigma$, $\sigma_2 = 0$, $\sigma_3 = \sigma/2$, and the lines are at 45° to the X_1 and X_2 axes. Fig.1b) shows the effect of circular symmetry. Considering the circular arc as part of the section of an internally pressurised cylinder, it is clear that the principal stresses are tangential (the "hoop" stress) and radial. The slip-lines then follow logarithmic spirals, as shown. Curvature in a slip-line implies a change in the hydrostatic component of stress ($\sigma_H = \sigma_3$ in plane strain): following Hencky's relationships(9):

$$\begin{aligned} p + 2k\phi &= \text{const.} \quad \text{along an } \alpha\text{-line} \\ p - 2k\phi &= \text{const.} \quad \text{along a } \beta\text{-line} \end{aligned} \qquad (4)$$

where $p(= -\sigma_H)$ is the pressure and ϕ is the anticlockwise rotation of the α-line from the X_1 axis. It may be deduced that the tensile stress in the X_2 direction at a radial distance x from the circumference is given by:

$$\sigma_2 = 2k \left\{ 1 + \ln\left(1 + \frac{x}{a}\right) \right\} \qquad (5)$$

The stress therefore rises, to a maximum value, at the plastic/elastic interface <u>below</u> the notch. (A simple physical explanation for this is given in (4)). The model is used commonly also for notched specimens under axial load, and, less properly, for problems involving circular symmetry, such as necked geometries and hardness tests, using ball indenters (9). Use of the technique for problems of similar geometry in the earth sciences, e.g. deformation around rock tunnels, is complicated by lack of homogeneity and isotropy and by the fact that the value of the rock crushing strength is affected by the hydrostatic stress (10).

Indentation by a flat punch provides another basic pattern, as indicated in fig.1c) where the flow of material leads to some "piling-up" around the punch. This "Prandtl field" gives an indentation pressure of

$$p = 2k\left(1 + \frac{\pi}{2}\right) \simeq 3\sigma_Y \qquad (6)$$

using Von Mises' criterion. It provides also a good model for a double cracked plane strain testpiece, if the configuration is mirrored about the centre line and the sign of the applied stress is reversed; fig.1d). If the indenter or notch has an included angle θ, the pressure or

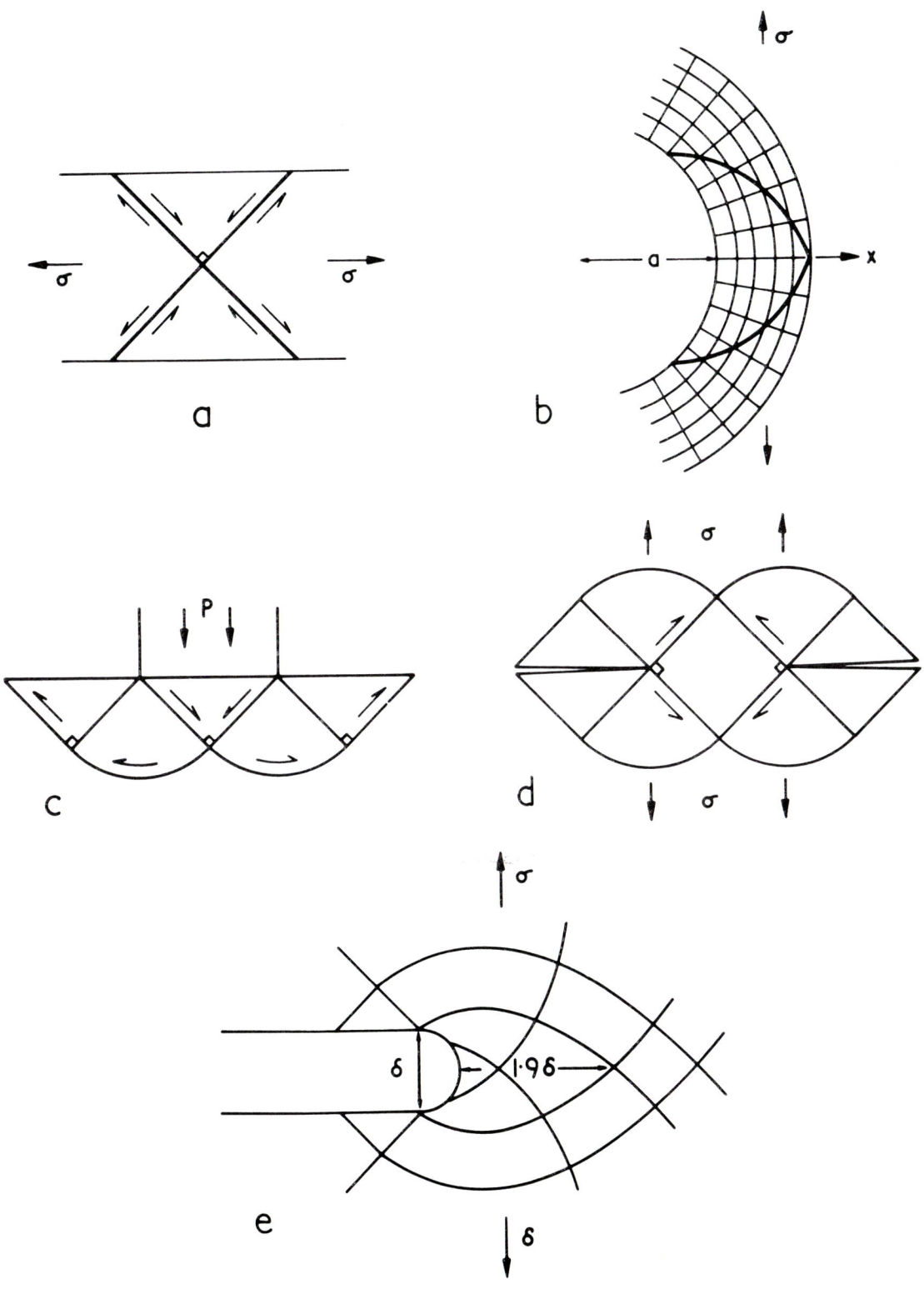

Fig.1. Some Slip-line Fields of Particular Interest to Deformation and Fracture Studies (for description see text).

tensile stress is given by:

$$p, \sigma = 2k \left(1 + \frac{\pi}{2} - \frac{\theta}{2}\right) \qquad (7)$$

Rice (11) has combined the features of figs.1b) and 1d) to provide a model for a blunted crack, fig.1e). This matches the Prandtl field away from the crack tip, but conforms to the circular notch field over a distance of 1.9δ, where δ is the crack opening. The predictions of the model are that there is a discrete crack opening displacement, δ; that the peak stress is to be found at a distance of 1.9δ from the tip; and that the region of high hydrostatic tension is to be found at distances $>1.9\delta$ beyond the tip. All these predictions have implications with respect to fracture processes in pre-cracked pieces.

The slip-line fields in figs.1a)-e) have particular relevance to the theme of this paper, but the general use of the method extends to other applications, such as metal working (extrusion, rolling etc.) (9,12) and plastic collapse calculations in testpieces and structures. Even though finite element stress analysis results are now often available, slip-line fields still provide useful bounds to solutions and give a good idea of the way in which the material flows. The results of two finite elements analyses are shown in figs.2 and 3, giving the increase in tensile stress ahead of a blunt notch with increasing load (13), and the distribution of stress and displacement around an initially sharp crack, under conditions of "small-scale yielding" (14). This matches the slip-line field solution (fig.1e) for non-hardening material.

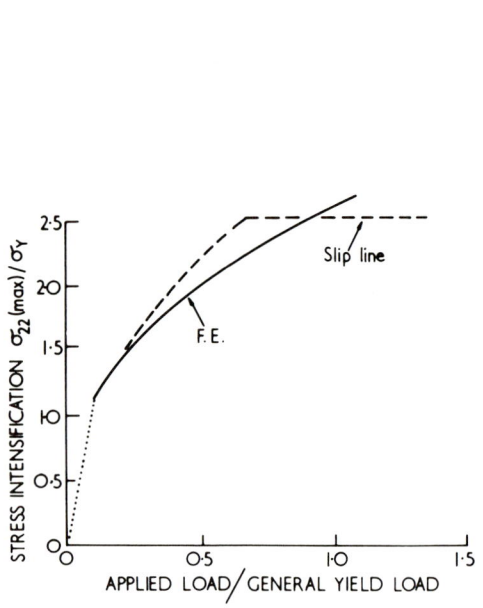

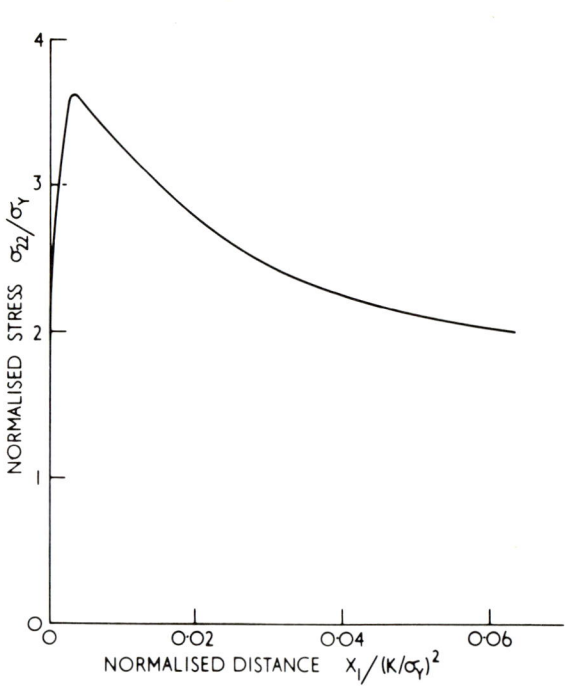

Fig.2. Stress Intensification ahead of a blunt $45°$ Notch as a Function of Applied Load (see ref.(13)).

Fig.3. Stress Distribution ahead of a Sharp Crack under conditions of Small-Scale Yielding (see ref.(14)).

5. FRACTURE MECHANICS

If a structure contains sharp, crack-like defects, such as quenching cracks, or cracks around welded joints, it is possible for the cracks to grow, either catastrophically ("brittle" or "fast" fracture) or by "sub-critical" mechanisms (fatigue, stress-corrosion, creep), at an applied stress less than the design stress, which is based on plastic collapse considerations. In his classic work on brittle fracture in glass, Griffith (15) assumed that the cracks were sufficiently sharp for the stresses at their tips to exceed the strength of the atomic bonds and was able to set up a purely energetic condition for catastrophic crack propagation, by balancing the potential energy release against the work of fracture. For a central, through-thickness crack, of length 2a, lying normal to an applied stress, σ, in an infinite body, Griffith's analysis gives the fracture stress, σ_F, as:

$$\sigma_F = \left(\frac{2E\gamma}{\alpha\pi a}\right)^{\frac{1}{2}} \tag{8}$$

where α is unity in plane stress ($\sigma_3 = 0$) and is equal to $(1 - \nu^2)$ in plane strain ($\epsilon_3 = 0$); 2γ is the elastic work of fracture per unit thickness and is equal to the area under the atomic force:displacement curve. An alternative way of writing equation (8) is to define a parameter, termed the <u>potential energy release rate</u> (per unit thickness), $G = (1/B)(dW/da)$; where dW is the potential energy release for a crack extension, da, and B is the thickness, and then to express the fracture condition as the attainment of a critical value of G (4). In the above case, we would simply write:

$$G_{crit} = 2\gamma \tag{9}$$

The relationship of G to applied stress, crack length, and testpiece geometry may be calculated by elastic stress analysis or may be measured experimentally, using compliance changes in pre-cracked testpieces (4).

The most powerful approach (16) is to consider the energy change when a crack is given a virtual extension, δa ($\delta a \to 0$). The tensile stress ahead of a crack tip along the line of crack extension has a high value and work is done as this stress drops to zero by the crack extension over δa (see (4)). Consider the central, through-thickness crack of length 2a in an infinite body. Let it extend to distances $+a$ and $-a$ in the $+X_1$ and $-X_1$ directions respectively, and let it be subjected to an applied stress, σ, in the X_2 direction. The configuration is symmetrical, so that only half the field, $x_1 > 0$, need be treated. Then, along the line $x_2 = 0$, the stress σ_{22}, for $x_1 > a$, is given by

$$\sigma_{22} = \frac{\sigma x_1}{(x_1^2 - a^2)^{\frac{1}{2}}} \tag{10}$$

which tends to infinity as $x_1 \to a$. For $x_1 < a$, σ_{22} is zero. For short distances ahead of the crack tip, $r = (x_1 - a)$, where $r \ll a$, equation (10) may be written in the form:

$$\sigma_{22} = K/(2\pi r)^{\frac{1}{2}} + \ldots \tag{11}$$

where $K = \sigma(\pi a)^{\frac{1}{2}}$. It is found, for different loading systems, (shear, internal pressurization, point loads of various forms etc.) and different geometries (edge cracks, circular cracks, finite boundaries etc.) that the form of equation (11) holds close to the crack tip, although the factor K takes different values, depending on the configuration under consideration: generally, K is proportional to $\sigma f(a/W)$, where W is the testpiece width. The value of K gives the strength, or intensity, of the singular stress field ($\propto r^{-\frac{1}{2}}$) ahead of a crack tip and K is termed the <u>stress intensity factor.</u>

In a similar manner, the displacement, u_{22}, along the line $x_2 = 0$, within the crack ($x_1 < a$), may be expressed in terms of K/E, over distances small compared with a. For a crack virtually extended to a length $(a + \delta a)$, the displacement is given by:

$$u_{22} = 2\alpha \sqrt{\frac{2}{\pi}} \frac{K}{E} (\delta a - r)^{\frac{1}{2}} \qquad (12)$$

where $\alpha = 1$ in plane stress, or $(1 - \nu^2)$ in plane strain. Then, the change in energy per unit thickness for a virtual extension of δa is given by:

$$G = \int_0^{\delta a} \sigma_{22} u_{22} \, dr = \frac{2\alpha K^2}{\pi E} \int_0^{\delta a} \left(\frac{\delta a - r}{r}\right)^{\frac{1}{2}} dr \qquad (13)$$

which may be evaluated to give:

$$G = \alpha \frac{K^2}{E} \qquad (14)$$

This is a fundamental relationship, for all elastic crack tip geometries under tension, which reduces the problem of calculating the energy release rate, $G = (1/B)(dW/da)$, to one of stress analysis (calculation of K). In terms of Griffith's fracture criterion, the critical value of G: $G_{crit} = 2\gamma$ (equation 9) may be equated to a critical value of K through equation 14. Hence, $K_{crit} = (EG_{crit}/\alpha)^{\frac{1}{2}} = (2E\gamma/\alpha)^{\frac{1}{2}}$. Substituting the value for K, corresponding to the central crack in an infinite body, $K = \sigma(\pi a)^{\frac{1}{2}}$ (following equation 11), we may derive the fracture stress, $\sigma_F = K_{crit}(\pi a)^{-\frac{1}{2}}$ as:

$$\sigma_F = \left(\frac{2E\gamma}{\alpha \pi a}\right)^{\frac{1}{2}}$$

in agreement with equation (8). This is, of course, a calculation for a single configuration, and, in general, the relationship $K = \text{const.} \, \sigma f(a/W)$ would be used to calculate the fracture stress.

The critical value, K_{crit}, is known as a material's <u>fracture toughness</u> and is of great importance because it characterises <u>the onset of</u> catastrophic failure in precracked geometries and hence provides a link between fracture stress and defect size. Reproducible values of K_{crit}, denoted by K_{IC}, are found in thick (plane strain) specimens, where total instability is coincident with the initiation of fracture at the crack tip. The toughness, K_{IC}, is, however, a critical stress intensity-factor. To interpret the value of K_{IC} for a material in terms of local fracture criteria, such as a critical stress or strain, it is necessary to incorporate a length parameter, which is a function of the material's microstructure. Several examples will be given in later sections.

6. BRITTLE AND DUCTILE BEHAVIOUR

It is found that, in some materials, cracks do propagate in the ideally elastic manner envisaged by Griffith, but this is not general. The situation has been described as a competition between the relative ease of fracturing the crack tip bond in tension and that of blunting the crack by local plastic flow [17]. In a layer silicate such as mica, the cleavage plane is only weakly bonded and it is easy to propagate a crack before slip can be induced on any other planes: the fracture is microscopically brittle. Similar behaviour may be observed in hexagonal metals, for example, in (0001) cleavage of single crystals of zinc at low temperatures [18]. Microscopic brittle behaviour is also observed if it is particularly difficult to propagate slip through the lattice, as in diamond, in some ionic solids, or in some b.c.c. transition metals (W, Mo). At the opposite extreme of material behaviour lie pure f.c.c. metals: for these, it is easy to blunt the crack by slip, at an applied stress lower than that required to break the crack tip bond by tensile force, unless the bonding has been weakened by chemical interaction with the environment, e.g. the brittle behaviour of aluminium in a mercury environment [19]. It is, however, possible for ductile rupture in hardened f.c.c. alloys to be associated with extensions so low that macroscopically the material appears to be brittle i.e. it fractures at a load less than the general yield load. From the calculations, iron appears to be a borderline material, and it is not easy to predict whether it is microscopically brittle or ductile [17,20].

7. CRACK PROPAGATION IN STRUCTURAL MATERIALS

Engineering materials which have to bear tensile loads require high toughness, and substantial local plastic deformation at a crack tip occurs before the crack propagates. This deformation affects the linear elastic assumptions made in the calculations of G or K, but, provided that the plastic zone is small, compared with all other dimensions, the errors in using elastic analyses are not large. In plane stress, the effect of a (non work-hardened) plastic zone, d_Y, on the stress analysis for a crack of length $2a$, may be derived [4,21,22] by matching the K-value, produced by loading a crack of length $2c$ ($c > a$) with a uniform applied stress σ, over the lengths, $d_Y = |c-a|$, with point loads equal to the yield stress, σ_Y (see fig.4). The results give:

$$\frac{a}{c} = \cos\left(\frac{\pi \sigma}{2\sigma_Y}\right) \tag{15}$$

or, at very low stress levels, i.e. $\sigma/\sigma_Y \ll 1$,

$$d_Y = \pi K^2/8\sigma_Y^2 \simeq 0.4 (K/\sigma_Y)^2 \tag{16}$$

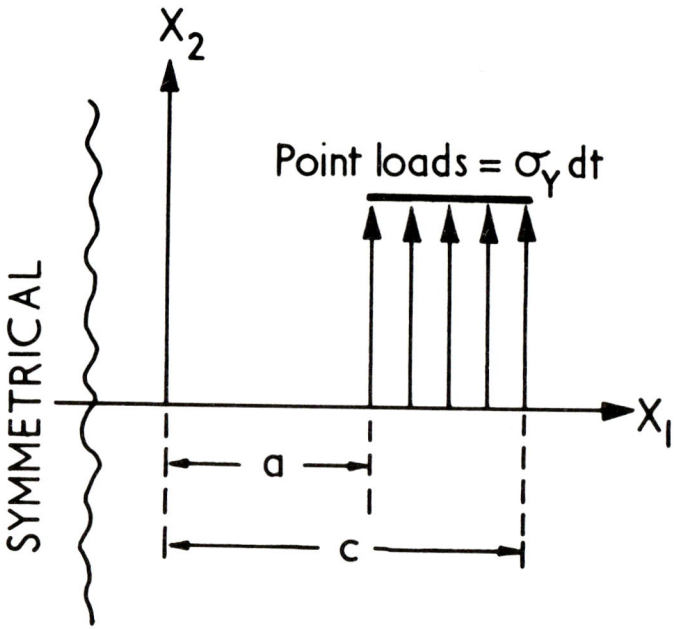

Fig.4. Model used to represent Plastic Zone of Length $|c-a|$ (see refs. (4,21,22)).

It is found that the elliptical opening of an elastic crack is replaced by discrete openings at the positions of the original crack tip ($\pm$ a) given by

$$\delta = \frac{8}{\pi} \frac{\sigma_Y}{E} \, a \, \ln \, \sec \left(\frac{\pi\sigma}{2\sigma_Y}\right) \qquad (17)$$

or, at low stresses,

$$\sigma_Y \delta = K^2/E = G \qquad (18)$$

The displacement, δ, is known as the <u>crack opening displacement</u> (C.O.D.) (23,24) and the equating of G to $\sigma_Y \delta$ in equation (18) may be thought of in terms of a virtual work calculation where the stress σ_Y ahead of the crack tip does work by moving through the displacement δ(4).

In plane strain, finite element calculations (14,25) matched to local crack tip slip-line fields (e.g. fig.1e)) yield fairly similar results. Under conditions of "small-scale yielding", the yield zone has the form of two lobes, with maximum extent:

$$R_{IY} = 0.16 \, (K/\sigma_Y)^2 \qquad (19)$$

at $\pm 70.5°$ to the X_1 axis and minimum extent

$$r_{IY} = 0.04 \, (K/\sigma_Y)^2 \qquad (20)$$

along the X_1 axis. The value of C.O.D. is given by

$$\delta \simeq 0.5 K^2/\sigma_Y E \qquad (21)$$

As plasticity becomes more extensive, it becomes convenient to characterise the crack tip stress- and displacement-fields by means of a path-independent integral, J, which incorporates both the strain energy stored within any contour Γ, and any work done by surface tractions acting on the contour (26). At low stresses, J becomes the potential energy release rate, G. For a given material and geometry, J and δ are inherently related to one another : at low stresses in an infinite body of non work-hardening material:

$$\delta = 0.55\text{-}0.65 \ J/\sigma_Y \qquad (22)$$

For higher stresses in a deeply cracked bend geometry, the relationship between J and δ is shown in fig.5 (after (27)).

It should be emphasised that when plasticity is significant, Griffith's fracture criterion cannot be applied. In plane stress, it may be shown formally that the system is in neutral equilibrium (28): under a given stress, a small increase in crack length leads to an increase in C.O.D., but not to elastic energy release. The parameters δ or J should therefore be regarded as characterising parameters, which summarise the integrated stress and strain fields around the crack tip. Critical values of δ or J associated with initiation of further fracture ahead of a crack, or with a given amount of crack growth, can be derived from microscopic observations, but these merely tell us how much input is required to achieve the given value and do not bear the energy release interpretation inherent to K_{crit} or G_{crit} values. One approach that could be followed to set up a fracture condition, when yielding is extensive, is to derive a critical value of δ or J from a knowledge of the micro-mechanisms of fracture and then use an expression such as equation 17) or a finite element calculation, to find the value of σ associated with this value, in any particular configuration. A similar approach (29) has recently been

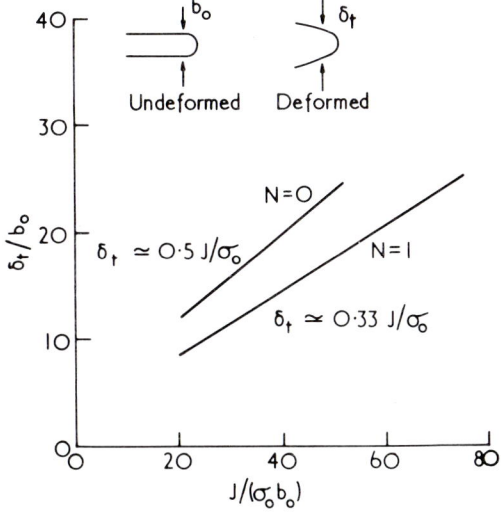

Fig.5. Relationship of C.O.D. to J in a Deeply-Cracked Bend Testpiece (see ref. (27)).

incorporated in a design procedure (30): a "true" toughness, K_{true} is defined as $(E\sigma_y\delta)^{\frac{1}{2}}$ (see equation 18)) and the relationship between this and the apparent value, $K_{app} = \sigma(\pi a)^{\frac{1}{2}}$, is given by substitution in equation 17). This procedure, however, is still one of characterisation, rather than energy balance.

A method attempting to calculate energy release in plane strain (31,32) provides an interesting limitation to continuum mechanics fracture theory. The plastic zone around a crack is often computed by finite element techniques, but, despite the fact that these treat the material as a framework rather than a continuum, they are not essentially different in what they tell us about the stress and strain distributions (the centres of the discretely computed values should fit onto a smooth curve). However, to calculate energy release in the plastic zone, a finite element link is broken and the new value of energy stored in the system is calculated: hence, the energy change, $\Delta W/\Delta a$, for a discrete amount of crack advance, rather than an infinitesimal change $(\delta a \to 0)$, may be computed. At this stage, the value of Δa is governed simply by the arbitrarily chosen spacing of the finite element network, but, in some fracture processes, it might be possible to identify microstructural features which govern discontinuous crack growth, so that the increments could be given physical meanings. It should be noted that the unknown is a length parameter.

8. SUB-CRITICAL CRACK GROWTH

Defects which do not propagate catastrophically on the first application of load may nevertheless grow as a function of time: by creep or stress-corrosion mechanisms if the load is maintained, or by fatigue or corrosion-fatigue if it fluctuates. For each of these mechanisms, attempts have been made to relate the crack growth rate to the value of K or ΔK ($\Delta K = K_{max} - K_{min}$, where these are the maximum and minimum values respectively in a fatigue cycle). The success of such relationships depends on the amount of plasticity involved and on the extent to which purely mechanical effects control the rate of growth. For fatigue in dry air, the crack growth rate per cycle, da/dN, is given in simple form by the Paris relationship (33):

$$\frac{da}{dN} = A \Delta K^m \qquad (23)$$

where A and m are constants. In more detail, it is found that there is a "threshold" value, ΔK_{th}, below which growth does not occur, and that the crack shows further acceleration as K_{max} approaches the material's fracture toughness. Between these extremes, m is found to take values in the range 2-4 for ductile materials: higher values are associated with additional "static modes" of fracture (34).

Continuum models of fatigue crack growth (35,36) are essentially two-dimensional in nature and relate the crack advance per cycle to some fraction of the reversed crack opening, $\Delta\delta$. By analogy with equation 18) this would suggest a dependence of da/dN on ΔK^2, i.e. m = 2. In three dimensions, however, the crack may tunnel forward preferentially at some points along its front: this effect could give rise to apparently higher values of m for the average growth rate (37). In coarse-grained, pure f.c.c. alloys which exhibit planar slip (because they have low stacking fault energy or because they are zone-hardened) fatigue cracks grow in plane strain by slip on two intersecting {111} planes, leading to facets which are macroscopically {001}, although in fact composed of small ridges of appropriate {111} form (38). There are marked tilt and twist misorientations between facets in adjacent grains along the line of the crack front, so that it seems unlikely that the crack grows uniformly at low ΔK levels. Experiments on textured material would clearly be informative.

At higher ΔK values, the assumption of uniform growth appears to be more reasonable, but interference by "static modes" may then occur (34,39). These may be brittle (cleavage or intergranular) or may involve void growth, which is of particular importance in systems which contain a distribution of closely-spaced non-metallic inclusions, such as weld metals (37). Modifications of equation 23) involving a $(K_{IC}-K_{max})$ term have been proposed (40) and attempts to treat the threshold have also been made, although here the behaviour is influenced very strongly by the environment.

The application of fracture mechanics to creep crack growth is at present still a matter of debate, but there do seem to be situations, in rather brittle materials, where behaviour is dominated by the conditions in the crack tip region and the growth rate, da/dt, is given by:

$$da/dt = C K^q \qquad (24)$$

where C and q are constants (41). The meanings of q values, which may be high, are not clear, but a suggestion has been made (42) that they may be approximated to values of the stress exponent n in the creep law

$$\dot{\varepsilon} = B \sigma^n \qquad (25)$$

where $\dot{\varepsilon}$ is the creep rate and B is a constant. Correlations have also been made with the rate of change of C.O.D. and J integral (see (41)).

In modes of growth involving corrosion, solid mechanics can at best provide a model for the mechanical component of crack growth. Often, the rate controlling step is the rate of anodic dissolution or hydrogen uptake at the crack tip (43), but these are in turn affected by the rate at which clean surface is exposed for attack: this presents the main problem in trying to deduce the overall growth rate by combining mechanical growth rates and corrosion rates.

9. DISLOCATIONS AND PLASTIC DEFORMATION

A field in which the principles of solid mechanics have been much applied is that of the stress and strain distributions around line defects known as dislocations (44). The continuum theory was developed initially by Volterra (45), but the concepts were applied to materials by Taylor (46), to try to explain the large difference between the ideal shear strength of a crystalline solid (approx. $\mu/10$) and those observed in practice (between approx. $\mu/1000$ and $\mu/100$ for common engineering alloys and perhaps $\mu/10,000$ for some metal single crystals). By introducing a line defect, bonds can be broken and re-made only a line at a time, rather than simultaneously across the whole slip plane. A dislocation is defined by its line vector, $\underline{l}$, and its Burger's vector:- sighting along the line vector a right-hand circuit is made from atom to atom; the circuit is then repeated in perfect lattice and the closure failure vector is the Burger's vector, $\underline{b}$. A screw dislocation has $\underline{b}//\underline{l}$: an edge has $\underline{b} \perp \underline{l}$. The stress system around a screw is, to a first approximation, a simple shear stress of magnitude

$$\sigma_{\theta z} = \frac{\mu b}{2\pi r} \qquad (26)$$

in terms of cylindrical coordinates, r, z, θ. The stress field around an edge has both tension and shear components of the form:

$$\sigma_{rr} = \sigma_{\theta\theta} = -\frac{\mu b}{2\pi(1-\nu)r} \cos\theta \qquad \sigma_{zz} = \nu(\sigma_{rr}+\sigma_{\theta\theta})$$

$$\sigma_{r\theta} = \sigma_{\theta r} = -\frac{\mu b}{2\pi(1-\nu)r} \sin\theta \qquad (27)$$

where the line $\theta = 0$ lies along the "extra half plane" and normal to $\underline{l}$. As for a crack, these stresses tend to infinity as $r \to 0$, and a "core", of radius approx. $\underline{b}/0.2\pi$, is generally held to be excluded from the region in which linear elasticity applies.

Due to the fact that the movement of dislocations involves the breaking and re-making of bonds, the ease of movement is a function of the lattice type and of the strength and localisation of inter-atomic bonding. Movement is particularly difficult in strongly covalently bonded materials, such as diamond, and is rather easy in f.c.c. metals: the implications of this, with respect to brittle and ductile behaviour, have been treated earlier. In a continuum analysis, this difficulty of movement is treated as lattice "friction"(the Peierl's-Nabarro stress), using the analogy of two rough planes sliding over one another. If the lattice contains a high density of dislocations as a result of cold working or phase transformations, it becomes more difficult to move new dislocations to produce further slip, because work must be done against the stress fields of the dislocations already there. If dislocations cut one another, extra work must be done to create jogs.

Two further barriers to dislocation movement are grain boundaries and second phase particles. The Hall-Petch model for the effect of grain size on the lower yield stress of mild steel (47) envisages an array of dislocations in a yielded grain, of diameter d, "piled up" against the boundary with an unyielded grain, and producing high local shear stress in this second grain. The "pile-up", under applied shear stress τ_{app}, is modelled as a freely slipping shear crack of length $2a = d$, under stress $(\tau_{app} - \tau_i)$, where τ_i is the "friction stress". The (shear) stress ahead of the shear crack is given (cf. equation 11)) by:

$$\tau = (\tau_{app} - \tau_i)(d/4r)^{\frac{1}{2}} \qquad (28)$$

The argument is then that yielding is triggered off in the unyielded grain, when the value of τ at a critical distance r* (the average distance to the nearest dislocation source) attains a critical level, τ* (the stress required to operate the source). Rearrangement of equation 28) then gives:

$$\tau_{app} = \tau_i + (4\tau^* r^*)^{\frac{1}{2}} d^{-\frac{1}{2}} \qquad (29)$$

As statistical averages over a large number of grain, the terms τ* and r* are taken as constants. If the appropriate Taylor factor(7) is applied, to convert shear stress to the applied stress at the lower yield point in the polycrystalline aggregate, we obtain the usual form:

$$\sigma_Y = \sigma_i + k_Y d^{-\frac{1}{2}} \qquad (30)$$

It should be noted that this model takes a stress intensity (albeit in shear) and then postulates a microstructure-controlled distance, r*, at which the critical stress, τ*, must be attained.

The hardening of a matrix by second-phase particles depends on their size, structure, spacing and coherence. Zones in underaged alloys can be cut through by dislocations: this process becomes easier once the first dislocation has cut the particle, and so planar slip tends to predominate in these alloys (48). In quenched and tempered steels, dislocations bow between hard carbide particles at an applied shear stress

$$\tau_{app} (-\tau_i) = \mu b/l_o \qquad (31)$$

where l_o is the initial particle spacing. As strain continues, successive dislocation loops are left around the particles, reducing the effective value of l_o and hence raising the stress required for further flow. The loops around the particles are rather like pile-ups and, as the number of loops increases, so the stress on the particle increases, to a level such that particle fracture or interface decohesion can occur. The cracking of carbide particles is of importance in the nucleation of cleavage cracks and interface decohesion is often a critical stage in the ductile fracture of alloy steels.

10. DISLOCATIONS AND CRACKS

In the Hall-Petch yielding model, an array of dislocations was represented by a crack. It is possible also to represent cracks by arrays of dislocations, ordered in such a manner as to replace the r^{-1} singularity by the required $r^{-\frac{1}{2}}$ singularity. In such representations, it is often convenient to smear out the discrete displacement changes pertaining to a finite number of dislocations, by means of a continuous distribution, containing $f(x_1)dx$, dislocations, each of Burger's vector $\underline{b} > 0$ in the distance dx_1 (49). The method has been applied particularly to the case of a crack with yielded zones at its ends, as fig.4 (50). In antiplane strain (mode III), where all the dislocations are of screw character, with $\underline{l} = l\underline{x}_3$, $\underline{b} = b\underline{x}_3$; the function must be such that, under an applied shear stress σ_{32}, the friction stress is zero for $|x_1| < a$, and is equal to τ_i (= τ_Y) for $a < |x_1| < c$; there are no dislocations present for $|x_1| > c$. When a suitable distribution function has been found, it is possible to determine the length of the yield zone:

$$\frac{a}{c} = \cos\left(\frac{\pi\sigma_{32}}{2\tau_Y}\right) \tag{32}$$

and the relative crack tip (shear) displacement:

$$s = \frac{4}{\pi} \frac{\tau_Y}{\mu} a \ln \sec\left(\frac{\pi\sigma_{23}}{2\tau_Y}\right) \tag{33}$$

These equations are of the same form as 15) and 17) and the dislocation approach can be used also for plane stress tension (mode I) or plane strain shear (mode II) and for arrays of cracks (51,52).

Dislocations were introduced to explain why slip could occur in materials at stresses much lower than the ideal shear stress and are defined in terms of a closure failure when the circuit around the boundary between slipped and unslipped material is repeated in undislocated reference material. The treatment of dislocation arrays, particularly by the continuous distribution approach, tends to ignore any details of the crystallographic lattice and to treat the matrix as a continuum. Types of "dislocation loop" have been discussed in the context of flow in glassy polymers (53) (which are amorphous) and geological faulting (54). In this latter case, analogies are drawn between faults (thrust faults, strike-slip faults, dip-slip faults) and conservative dislocations, and between fissures (sills and dykes) and non-conservative dislocations. It is not clear, however, whether such "dislocation loops" possess the r^{-1} singularity, and have fully re-made bonds within the loop, as would a crystal dislocation loop, or whether they are more of the nature of shear cracks, with an $r^{-\frac{1}{2}}$ singularity, and no cohesion within the loop.

11. CRACKS AND "END REGIONS"

In their discussion of slip surfaces in over-consolidated clay, Palmer and Rice (55) treat the propagation of a "shear band", across which there is a residual shear strength, τ_r, greater than zero, but less than the peak shear strength of the lattice, τ_p. (In this sense, the model is not unlike that of a dislocation pile-up, section 9). The relationship between shear resistance, τ, and relative shear displacement in a slip surface is taken to be of the form in fig.6: τ decreases from τ_p to τ_r as δ increases, perhaps because the higher strains permit more diffusion of water, which weakens the bonding. The total work of fracture is thus $\int(\tau-\tau_r) d\delta$, or, if a "characteristic displacement", $\bar{\delta}$, is defined, such that:

$$(\tau_p - \tau_r)\bar{\delta} = \int (\tau-\tau_r) d\delta \tag{34}$$

it may be written as $(\tau_p-\tau_r)\bar{\delta}$. In one limit, the shear band is envisaged as being situated in an elastic matrix, although an "end region" exists ahead of each shear band tip, rather like the yield zone ahead of a tensile crack, under conditions of "small-scale yielding". The Griffith equation in plane strain shear (mode II) may be written (as equation 18)), modified by a factor of $(1-\nu^2)$),

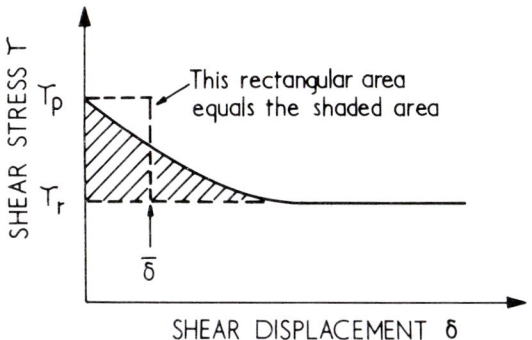

Fig.6. Assumed Stress-Displacement Relationship across "Shear-Band" in Over-Consolidated Clay (see ref.(55)).

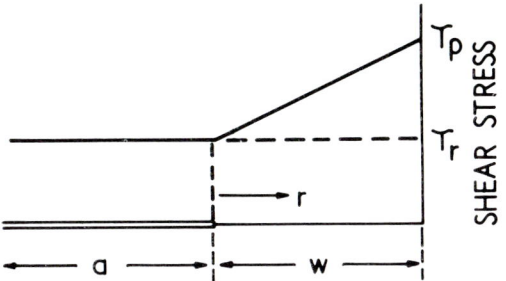

Fig.7. Variation of Shear Resistance with Position along the "End Region" (see ref.(23)).

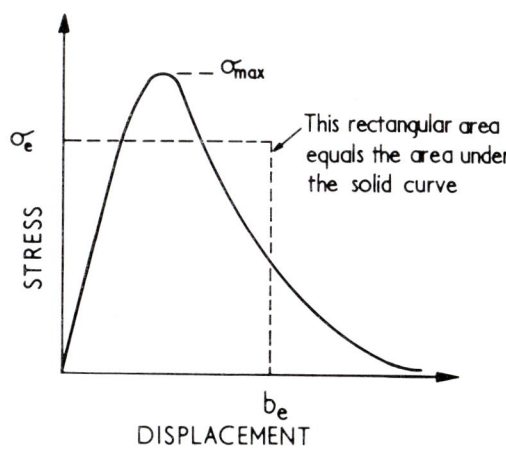

Fig.8. Idealisation of Atomic Stress-Displacement Curve (see ref. (57)).

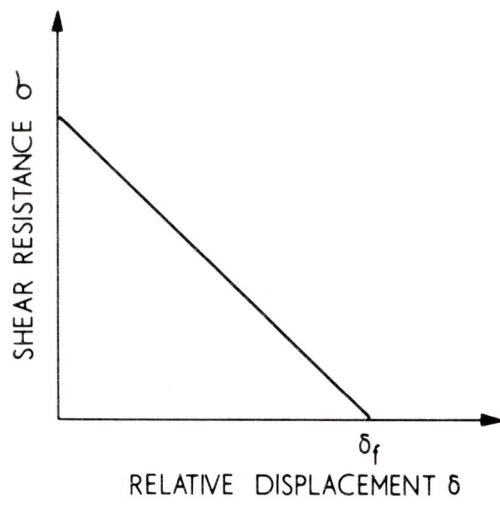

Fig.9. Stress-Displacement Curve for a Softened Flow Zone (see ref.(58)).

$$\frac{K^2(1-\nu)}{2\mu} = \frac{K^2}{E}(1-\nu^2) = (\tau_p - \tau_r)\overline{\delta} \tag{35}$$

or, if the applied stress at infinity is τ_∞ and the length of the shear band is $2l$, $K = (\tau_\infty - \tau_r)(\pi l)^{\frac{1}{2}}$ (cf. equation 28).

$$\tau_\infty - \tau_r = \left(\frac{2E}{\pi(1-\nu^2)\,l}(\tau_p - \tau_r)\overline{\delta}\right)^{\frac{1}{2}} \tag{36}$$

The length of "end region", w, associated with a given value of $\overline{\delta}$ may be estimated, using the same method as that for a plastic zone, (quoted in equations 15) and 16) and derived in ref. (4)). The main difference is that the shear resistance in the "end region" varies with position (see fig.7), whereas that in the yield zone is constant. To derive the result obtained by Palmer and Rice (55), use may be made of equation 16) together with the assumption that the shear resistance increases in a linear fashion over the length, w, from τ_r at the end of the shear band to τ_p at the boundary between the "end region" and the elastic surroundings. At any point, the shear resistance is $(\tau - \tau_r) = (\tau_p - \tau_r)(r/w)$. Let $T = (\tau_p - \tau_r)$, so that $\tau' = (\tau - \tau_r)$ varies from 0 to T along the "end region". If the shear resistance were a constant value T throughout the zone, the value of w (= d_y in equation 16)) would be equal to $\pi^2 K^2/8\,T^2$. The integrated effect of T may be equated to that of a rectangle of area Tw acting through its centre of gravity, at w/2, i.e. $Tw^2/2$. In contrast the linear variation of τ' from 0 to T is equivalent to the triangular area $Tw^2/2$ acting through its c.o.g., at 2w/3 from the end of the shear band, i.e. $Tw^2/3$. The total shear resistance is therefore only two-thirds as great, so that w is now given by $\pi^2 K^2/8(2T/3)^2$. Giving K^2 its critical value from 35) we have:

$$w = \frac{9\pi}{16(1-\nu)} \frac{\mu}{(\tau_p - \tau_r)} \overline{\delta} \tag{37}$$

Note that this length is a basic material property, independent of applied stress or crack length. Values of μ/τ_p of approximately 50 have been quoted (56) so that, assuming $\tau_p \simeq 2\tau_r$ and that $\nu = 0.25$ (a soil), the relationship becomes $w \simeq 250\overline{\delta}$. Palmer and Rice suggest values for $\overline{\delta}$ in the region of 2-10mm which gives w as 0.5-2.5 m. They use this result to suggest that this particular type of instability, whilst not observed in laboratory shear boxes, may yet be important in the field, where the overall dimensions of interest can be very much greater than the size of the "end region".

Particular attention is given to this model, because similar approaches have been used at the microscopic and microstructural levels. Cottrell (57) has examined the atomic stress/displacement curve (fig.8) and approximates this to rectilinear form, in terms of an equivalent stress, σ_e, and displacement, b_e, in a manner similar to that of equation 34):-

$$\sigma_e b_e = \int_0^\infty \sigma(b)\,db = 2\gamma \tag{38}$$

where b is the displacement from the rest position, b_o, and 2γ is the elastic work of fracture (as, for example, in equations 8) or 9)). Values of $\sigma_e = 0.2E$ and $b_e = 0.5 b_o$ give a figure for γ of $0.05 Eb_o$, which is in fair agreement with experimental surface tension results. The size of the "end region", w_1, may be calculated from equation 16) by substituting σ_e for σ_Y and using equations 14) and 9), with $\alpha = (1-\nu^2)$ and $2\gamma = \sigma_e b_e$ to give the critical value to substitute for K:

$$w_1 = \frac{\pi}{8(1-\nu^2)} \frac{E}{\sigma_e} b_e \qquad (39)$$

The analogy with equation 37) is clear, but because the stress in the end region is so high, $\sigma_e = 0.2E$, the value of w_1 is only just greater than a lattice spacing ($w_1 = 1.25 b_o$). For a "softer" force law, values of $\sigma_e = 0.1E$, $b_e = b_o$, might be appropriate and the length of the end region would then be $5b_o$. In detail, of course, the stress must first rise and then fall as the displacement increases (fig.8), as, in reality, does the shear stress/displacement curve for a soil: fig.6 is an idealisation.

Similar behaviour is observed during the ductile fracture of high strength steels, where the stress first rises as a result of work-hardening, but then may drop catastrophically if flow localises (see section 13). Smith et al. (58) model this situation in a manner that is completely analogous to that used for the soil. Their stress-displacement curve in the softened flow zone is shown in fig.9), assuming a critical fracture displacement. Again, this may be put in rectilinear form, with effective (shear) flow stress, τ_f, and effective displacement, δ_f. The "end region", w_2, is calculated to be of order

$$w_2 \simeq \frac{\mu}{\tau_f} \delta_f \qquad (40)$$

cf. equations 37) and 39). Taking $\mu = 80$ GNm^{-2} and a value for τ_f of 800 MNm^{-2}, we see that $w_2 \simeq 100 \delta_f$. Thus a value of $\delta_f = 1\mu m$ (perhaps a strain of order unity around a carbide particle) implies a value of w_2 of some 100μm, which is similar to the spacings of non-metallic inclusions in steels. Smith et al. also consider the effect of micro-cracks in the flow zone.

These three situations give good examples of the same type of solid mechanics model applied to the three size-scales: macroscopic, microscopic and microstructural. In each case, an autonomous "end region" exists, to accommodate the critical displacement, and this "end region" must be very much smaller than all testpiece dimensions for Griffith-type propagation to occur, just as plastic zone sizes must be small compared with dimensions in valid toughness tests. The local fracture criteria have, however, been critical displacements: $\bar{\delta}$, b_e or δ_f. In the following section, consideration is given to a plane strain fracture mode in which the fracture criterion, on the microstructural scale, is a tensile stress and in which the necessary length parameter is a function of the microstructure.

12. TRANSGRANULAR CLEAVAGE FRACTURE IN MILD STEEL

The typical microstructure of a low carbon (mild) steel in the normalised or annealed condition consists of a set of randomly oriented ferrite (b.c.c. iron) grains, with platelet cementite (iron carbide, Fe_3C) at the grain boundaries. As indicated earlier, cracks are produced in the carbides at low plastic strains, by pile-ups or by fibre loading from the plastically strained matrix. A successful plane strain model of the fracture process has been developed by Smith (59), who examines the condition for propagation of a crack nucleus in a grain-boundary carbide under the joint action of pile-up stress and applied tensile stress. This becomes:

$$\sigma_F^2 \left(\frac{C_o}{d}\right) + \tau_{eff}^2 \left[1 + \frac{4}{\pi}\left(\frac{C_o}{d}\right)^{\frac{1}{2}} \frac{\tau_i}{\tau_{eff}}\right]^2 \geq \frac{4E\gamma_p}{\pi(1-\nu^2)d} \qquad (40)$$

where $\tau_{eff} = (\tau_{app} - \tau_i)$, C_o is the carbide thickness and γ_p is the effective surface energy of the ferrite grain. If the second term is ignored, the equation reduces to the Griffith condition (equation 8)), with $C_o = 2a$. It should be noted that brittle particles give a low value of γ for the initiation process compared with the value appropriate to propagation into the ferrite: if γ remained constant, fracture would be initiation (shear stress) controlled rather than propagation (tensile stress) controlled. This situation has recently been observed in pure zinc, which contained no brittle particles (60).

If notched specimens of mild steels are fractured at low temperatures, it is possible to calculate the local stresses required to produce cleavage fracture, using slip-line field analyses (e.g. equation 7)) or finite element analysis (e.g. fig.2). These fracture stresses can be related quite consistently to the microstructural parameters, using equation 37), if γ_p is taken as $14 Jm^{-2}$. The same value may be used for the fracture stresses of spheroidal carbide microstructures, if the stress analysis is modified to take account of "penny-shaped" crack nuclei (61,62). Possible reasons for the discrepancy between the figure of $14 Jm^{-2}$ for γ_p and the value of $2 Jm^{-2}$ for the surface energy of iron have been discussed elsewhere (62).

Given a critical tensile stress, σ_F, required to produce cleavage fracture it is necessary to postulate a "critical distance" before a toughness value, K_{IC}, can be calculated, Ritchie et al. (63) compared K_{IC} values and σ_F values in mild steel, using a finite element analysis for the stress distribution ahead of a sharp crack, (fig.3) and arguing that the plastic zone grows to a size such that a stress equal to σ_F is attained at a constant critical distance, d_c, ahead of the tip. Their value for d_c was two grain diameters, but subsequent, more detailed studies have shown that d_c varies with grain size. In essence, d_c is a measure of the probability of finding a carbide, ahead of the tip, sufficiently coarse to give a nucleus which will propagate at a given stress, σ_F. It is found that the ratio C_o/d is almost linear in mild steels (61), and it can be reasoned that it is relatively more difficult to find the carbides in fine-grained steels, so that d_c/d increases.

13. FIBROUS FRACTURE AT A CRACK TIP

Another fracture mode, in which a distance parameter shows clearly, is that of fibrous fracture. Here, in the simplest case, a crack advances from inclusion to inclusion by void growth and internal necking, so that the C.O.D. at the original crack tip is a simple multiple of the inclusion spacing (62). Rice and Johnson's(11) predictions for the variation of δ/X_o with X_o/R_o, where X_o and R_o are respectively the inclusion spacing and inclusion radius, are shown in fig.10, together with experimental results for a number of steels(see 65). Several of the points lie close to the predicted line and, for these, it is therefore possible to calculate the value of K at the initiation of fracture, using equation 21), directly from the microstructural parameter, X_o.

A number of experimental points in fig.10 lie well below the predictions. This is because the internal necking process between crack tip and void is truncated by a fast shear process or because the necking is replaced completely by "zig-zag" fracture. Such behaviour occurs when flow localisation is favoured (62,66). One example is in a precipitation hardened matrix containing widely spaced non-metallic inclusions (a clean, quenched and tempered steel). As the matrix is strained, dislocations loop around carbides until, eventually, such high local stress levels are reached that the interfaces decohere. At this point, the resistance to flow drops catastrophically and a fast shear fracture ensues. The shear nature of the fracture is supported by the observations of Spretnak on notched bar and torsional ductilities (67). In a recent experiment (68) carbide/matrix interfaces have been embrittled by impurity element segregation and a 50% decrease in the value of δ_i for this ductile separation has been obtained. It is not clear whether a heavily work-hardened dislocation structure in a precipitate-free matrix can become spontaneously unstable and allow flow to localise, but the large numbers of point defects produced by neutron irradiation would certainly appear to permit such flow concentration to occur.

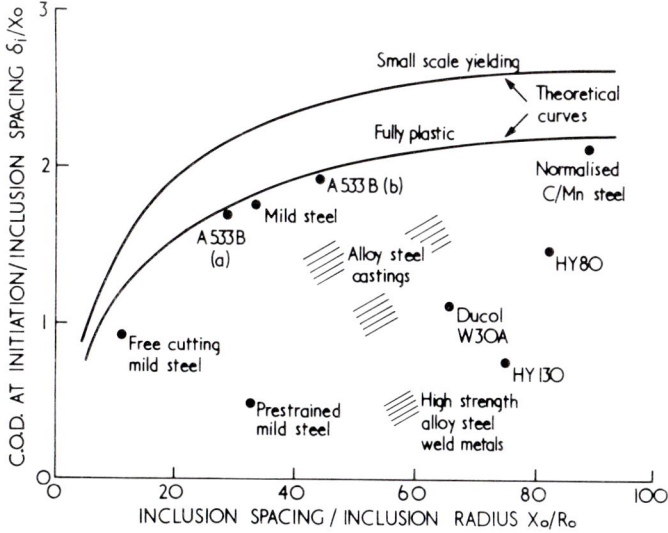

Fig.10. Variation of Crack Tip Ductility with Inclusion Spacing (see ref.(62)).

Fibrous initiation does not, of course, automatically produce total instability, even in a large structure. If continuum theory is employed, the point of instability in a non work-hardening elastic/plastic material cannot be predicted, but energy changes, $\Delta W/\Delta a$, for discrete crack advances in a finite element model have been evaluated (31,32). There, the crack steps were link spacings in the finite element network, but, for a fibrous growth process, individual steps could perhaps be identified with inclusion spacings (32). Initial results of this sort are promising: low values of δ and $\Delta\delta/\Delta a$ rapidly produce instability; high values lead to the conclusion that instability is impossible in anything but unrealistically large bodies. Another possibility is that for the zig-zag fast shear process, the inclusion spacing may correspond to w_2, the size of the "end zone" in Smith's localisation model (58) (see section 10). Here, if the local carbide fractures can occur for values of w_2 less than the inclusion spacing, it would appear that the value of K_{IC} should not depend on the inclusion content.

14. ELASTIC WAVES

In any paper dealing with crack propagation in engineering materials, it is important to point out the difficulties encountered in detecting and measuring defects in structures. This is particularly appropriate when the interaction with solid mechanics is being emphasised, because the increasing use of ultrasonics as an NDT technique means that the propagation of elastic waves in material must be properly understood, to appreciate effects of reflection, diffraction and mode change. Moreover, in a meeting which seeks to provide interdisciplinary links with earth sciences, it is pertinent to draw attention also to the techniques of acoustic emission, which is the direct analogue of seismology, in its attempts to locate and characterize sources which emit elastic waves.

If the internal stresses in a body are not in static equilibrium, Newton's second law gives the initial condition for an elastic wave to be generated:

$$\frac{\partial \sigma_{ij}}{\partial x_j} = \rho \frac{d^2 u_i}{dt^2} \qquad (41)$$

where i and j take values 1, 2 or 3 and ρ is the material's density. To convert this to an equation in the single variable, u, (displacement), the stresses σ_{ij} are converted into strains, via Lamé's constants (equation 1) for isotropic materials or the crystal constants, (C_{ijkl}) for other systems, and the strains are then written as differentials of displacements: $\varepsilon_{ij} = \frac{1}{2}(\partial u_i/\partial x_j + \partial u_j/\partial x_i)$. For the simple isotropic case, the solution of the wave equation gives, for the velocity of a longitudinal disturbance:

$$C_L = \{(\lambda + 2\mu)/\rho\}^{\frac{1}{2}} \qquad (42)$$

and for a shear disturbance:

$$C_s = (\mu/\rho)^{\frac{1}{2}} \qquad (43)$$

Usually, $\lambda = 2\mu$; $C_L = 2C_s$. Acoustic impedance, Z, is defined as $Z = \rho C$, and for a longitudinal wave which is incident normal to an interface of acoustic mismatch ($Z_2 < Z_1$) it may be shown that the reflected intensity I_r is related to the incident intensity, I_o, through the expression(69)

$$I_r = \frac{(Z_2 - Z_1)^2}{(Z_1 + Z_2)^2} I_o \qquad (44)$$

For a metal/air interface ($Z_2 << Z_1$), this means that almost all the transmitted intensity is reflected by the back face of a material or by an air-filled crack within it, and a "pulse-echo" back reflection technique is used with success to detect cracks. Sizing is accomplished by traversing the ultrasonic transceiver (usually a PZT probe) until the reflected signal drops by a standard amount (say, 6dB): this marks the "end" of the crack. In practice, diffraction by the probe or defect can cause confusion in sizing, and texture in the part being examined can affect the assumptions of isotropy and the amount of attenuation observed. These problems are under active investigation (70,71).

In acoustic emission, the transucers are passive monitors, often arranged in a standard array on a pressure vessel or structure undergoing a proof test. Acoustic sources may then be located by recording the time of arrival at different transducers and employing triangulation techniques (72). The method is able to locate sources within an area whose dimensions are of the order of the plate thickness. Characterisation of the source from the signals received is not, however, well-developed. Low frequency vibrations, pressure surges and extraneous noise can be mainly eliminated by comparing the signals received at different transducers in the array, but the fundamental question of whether plastic deformation or cracking has caused the acoustic event remains at present unresolved. Amplitude characterization has been used with some success for brittle cracking processes, and even for discontinuous, fast shear fracture, in laboratory testpieces (73), but the problems in applying this in practice are great. The emissions travel mainly as surface waves, so that attenuation is extremely sensitive to surface condition (scratches, rust) and to acoustic mismatch (air, pressurizing fluid). Any laboratory calibration, ascribing a particular amplitude to a particular event, therefore requires modification for every different service configuration.

One possible solution is to use frequency analysis to characterise emissions. Although this does not appear to be feasible using commercial "broad-band" PZT transducers, because they drown the frequency response with a large number of peaked transucers resonances(74), encouraging results have been found, using broad-band capacitive transducers (75) (which, unfortunately, at the present state of development, are not suitable for use in service). These suggest that the frequency content of waves emitted by fast-moving brittle cracks is different from that emitted by dislocations, during normal plastic flow. The differences between cracks and dislocation arrays are not always clear, however, as we have seen in previous sections. In some situations, dislocations move slowly, because the friction stress is high, whilst a crack, such as a cleavage crack, moves rapidly. In other situations, such as "fast shear" decohesion, the difference between dislocation movement, once a critical stress has been exceeded, and shear crack propagation, is reduced almost to a matter of semantics and little difference in characteristic "voice prints" could be expected.

15. CONCLUSIONS

This paper has presented a brief survey of some important features of elastic and plastic deformation in materials and has discussed factors controlling fracture toughness. To relate K_{IC} values to local crack-tip fracture criteria, it is necessary to postulate some material-controlled distance parameter which, in different fields of interest, may be as large as one or two metres or as small as a lattice spacing. In many structural materials, it is unnecessary to seek the refinement provided by atomic bond treatments: fracture stresses or displacements characterise fracture over distances equal to a few carbide spacings or inclusion spacings.

In accordance with the interdisciplinary theme, it is worth drawing attention first to the similarities of the solid mechanics models in earth sciences and materials science , whilst appreciating the difficulties of dealing with anisotropic behaviour and a pressure-dependent yield criterion. Secondly, the use of the term "dislocation" might be noted: in materials science, this is a line defect: in earth sciences, it may often be a shear crack. The difference lies in the possibility of catastrophic deformation or fracture: this is not a feature of (line defect) dislocation movement through crystals, but can occur readily for shear cracks, where there are no "re-made" bonds within the faulted area. This area, and the terminology "dislocation", warrants further attention.

From the materials science point of view, possibly the most fruitful field for interdisciplinary activity is that of using elastic waves; both to detect defects and to characterize sources. The latter area is comparatively new in the materials field and here materials science should reap enormous benefit from solid mechanics, which can provide models of possible source spectra, and from the earth sciences, which have accumulated a store of experience in detection techniques, source location, signal handling and source identification, applied to seismic studies. It is hoped that the present paper has shown that there may be benefit to solid mechanics and earth sciences in a knowledge of the mechanical behaviour of structural materials. On the one hand, it may excite more realistic models of flow and fracture: on the other, it may enable earth scientists to observe, in the failure of rather simple and well-behaved microstructures, the operation of failure mechanisms, which may be of relevance to their own, rather more complicated, geological materials.

16. ACKNOWLEDGEMENTS

Thanks, as always, are due to Professor R.W.K. Honeycombe, Goldsmith's Professor of Metallurgy, for his continuing support of fracture studies at Cambridge. At various times, useful discussions have been held with Dr. A. Kfouri, Dr. G.C.P. King, Dr. A.C. Palmer, Dr. A.C. Pickard, Prof. J.R. Rice, Dr. N.W. Ringshall, Dr. R.O. Ritchie, and Dr.C.P. Wroth, all of whom are, or were, attached to Churchill College. The Collegiate system, operating in a science-based community, is one in which subject barriers can be broken down, and in which an inter-disciplinary view can emerge.

REFERENCES

(1) Ashby M.F., Microstructure and Design of Alloys. Proc.3rd ICSMA; Institute Metals and Iron and Steel Inst. 1973 2 p.8.

(2) Dieter G., Mechanical Metallurgy. McGraw Hill - New York - 2nd Ed. 1976.

(3) Nye J.F., Physical Properties of Crystals. Oxford University Press, 1972.

(4) Knott J.F., Fundamentals of Fracture Mechanics. Butterworths, London 1973.

(5) Embury J.D. and LeRoy G.H., Fracture 1977, Proc. ICF4, Ed. D.M.R. Taplin, Pergamon 1978 p.15.

(6) Wilson D.V., Effect of Second-Phase Particles on The Mechanical Properties of Steel, Iron and Steel Inst., 1971, p.28.

(7) Taylor G.I., Jnl. Inst. Metals, 1938, 62 p.307.

(8) Smith E. and Worthington P.J., Phil. Mag. 1964, 9, no.98, p.211.

(9) Hill R., The Mathematical Theory of Plasticity, Oxford University Press 1950.

(10) Jaeger C., Rock Mechanics and Engineering, Cambridge University Press 1972.

(11) Rice J.R. and Johnson M.A., Inelastic Behaviour of Solids (ed. M.F.Kanninen et al.) McGraw Hill, New York, 1970, p.641.

(12) Johnson W. and Mellor P.B., Engineering Plasticity, Van Nostrand Reinhold Company London 1973.

(13) Griffiths J.R. and Owen Jnl. Mech. and Phys. Solids, 1971, 19 p.419.

(14) Tracey D.M., Ph.D. Thesis, Brown University, Providence, R.I. 1973.

(15) Griffith A.A., Phil. Trans. Roy. Soc., 1920, A221, p.163.

(16) Irwin G., Trans. ASME Jnl. Appl. Mech. 1957, 24, p.361.

(17) Kelly A. Tyson W.R. and Cottrell A.H., Phil. Mag. 1967, 15, p.567.

(18) Maitland A. and Chadwick G.A., Phil. Mag. 1969, 19, p.645.

(19) Kamdar M.H. (review) Fracture 1977, Proc. ICF4 Ed. D.M.R. Taplin Pergamon 1978, p.387.

(20) Rice J.R. and Thomson R., Phil. Mag. 1974, 29, p.73.

(21) Dugdale D.S. Jnl. Mech. Phys. Solids, 1960, 8, p.100.

(22) Burdekin F.M. and Stone D.E.W., Jnl. Strain Anal., 1966, 1, p.145.

(23) Cottrell A.H., Steels for Reactor Pressure Circuits, Iron and Steel Inst. Spec. Rep. no.69, 1961, p.281.

(24) Wells A.A., Crack Propagation Symposium Proceedings, Cranfield, 1961, 1, p.210.

(25) McMeeking R.M., Jnl. Mech. Phys. Solids 1977, 25, p.357.

(26) Rice J.R. Trans. A.S.M.E. Jnl. Appl. Mech. 1968, p.379.

(27) McMeeking R.M. and Parks D.M. ASTM Symp. on Elastic-Plastic Fracture, Nov. 16-18, 1977, Atlanta.

(28) Rice J.R., Proc. 1st Intl. Congress on Fracture ed. T. Yokobori et al. Jap. Soc. for Strength and Fracture, Tokyo, 1966, 1, p.309.

(29) Heald P.T., Spink G.M. and Worthington P.J., Mater. Sci. Eng. 1972, 10, p.129.

(30) Milne I., Loosemore K. and Harrison R.P., Tolerance of Flaws in Pressurized Components, Inst. Mech. Engrs. 1978, p.317.

(31) Kfouri A.P. and Rice J.R., Fracture 1977, Proc. ICF4 Ed. D.M.R.Taplin Pergamon 1978, p. 43.

(32) Rice J.R. and Sorensen E.P., to be published in Jnl. Mech. Phys. Solids.

(33) Paris P.C. and Erdogan F., Jnl. Basic Eng. (Trans. ASME, D), 1963, 85, p.528.

(34) Ritchie R.O. and Knott J.F., Acta Met., 1973, 21, p.639.

(35) Laird C. and Smith G.C., Phil. Mag.1963, 8, p.1945.

(36) Tomkins B. and Biggs W.D., Jnl. Mat.Sci. 1969, 4, p.544.

(37) Pickard A.C., Ritchie R.O. and Knott J.F., Metals Technology June 1975, p.253.

(38) Garrett G.G. and Knott J.F., Acta Met. 1975, 23, p.841.

(39) Ritchie R.O. and Knott J.F., Mater. Sci. Eng. 1974, 14, p.7.

(40) McEvily A.T., Metal Science 1977, 11, p.274.

(41) Ellison E.G. and Harper M.P., Jnl. Strain Anal. 1978, 13, p.35.

(42) Siverns M.J. and Price A.T., Int. Jnl. Fracture, 1973, 9, p.199.

(43) Parkins R., Metal Science, 1977, 11, p.405.

(44) Cottrell A.H., Dislocations and Plastic Flow in Crystals, Oxford, 1953.

(45) Volterra V., Paris, Ann. École. Norm. (Ser 3) 1907, 24, p.401.

(46) Taylor G.I., Proc. Roy. Soc. 1934, A145, p.362.

(47) Petch N.J., Jnl. Iron and Steel Inst. 1953, 173, p.25.

(48) Kelly A. and Nicholson R.B., Prog. Mater. Sci. 1963, 10, p.350.

(49) Weertman J. and Weertman J.R., Elementary Dislocation Theory, Macmillan 1964, p.128.

(50) Bilby B.A., Cottrell A.H. and Swinden K.H., Proc. Roy. Soc. 1963, A272, p.304.

(51) Smith E., ibid 1964 A282, p.422.

(52) Smith E., ibid 1965 A285, p.46.

(53) Bowden P.B. and Raha S., Phil. Mag. 1974, 29, p.149.

(54) King G.C.P., Phil. Trans. Roy. Soc., 1978, A288, p.197.

(55) Palmer A.C. and Rice J.R., Proc. Roy. Soc. 1973, A332, p.527.

(56) Wroth C.P., Stress-Strain Behaviour of Soils, ed.R.H.G. Parry, Foulis, Henley-on-Thames, 1972, p.347.

(57) Cottrell A.H., First Tewksbury Symposium on Fracture, Univ. Press Melbourne 1963, p.1.

(58) Smith E., Cook T, and Rau C., Fracture 1977 Proc. ICF4, Ed. D.M.R. Taplin Pergamon 1978, p.215.

(59) Smith E., Proc. Conf. Physical Basis of Yield and Fracture, Inst. Phys. and Phys. Soc. 1966, p.36.

(60) Curry D.A., King J.E. and Knott J.F., Metal Science, 1978, 12, p.247.

(61) Curry D.A. and Knott J.F., Metal Science, to be published.

(62) Knott J.F., *Fracture 1977*, Proc. ICF4 Ed. D.M.R. Taplin, Pergamon 1978, p.61.

(63) Ritchie R.O., Knott J.F. and Rice J.R., *Jnl. Mech. Phys. Solids* 1973, **21**, p.395.

(64) Curry D.A. and Knott J.F., *Metal Science* 1976, **10**, p.1.

(65) Green G. and Knott J.F., *Trans. ASME Series H Jnl. Eng. Mat. Tech.* 1976, **98**, p.37.

(66) Clayton J.Q. and Knott J.F., *Metal Science* 1976, **10**, p.63.

(67) Griffis C.A. and Spretnak J.W., *Trans. Iron & Steel Inst. Japan* 1969, **9**, p.372.

(68) King J.E., private communication.

(69) Lamb J., *Principles and Practice of Non-Destructive Testing* (ed. J.H. Lamble), Heywood and Co. Ltd., (London) 1962, p.118.

(70) Lumb R.F., *Tolerance of Flaws in Pressurized Components*, Inst. Mech. Engrs. 1978, p.71.

(71) Coffey J.M., ibid p.63.

(72) See *ASTM STP 571*, 1975.

(73) Clark G. and Knott J.F., *Metal Science* 1977, **11**, p.531.

(74) Ringshall N.W., *Ph.D. Thesis*, University of Cambridge, 1978.

(75) Scruby C.B., Collingwood J.C. and Wadley H.N.G., *AERE Report* 8915, 1977.

Session I

DISCUSSION

E. Suoninen: What about other possibilities of heat transfer from the interior of the Earth (besides conduction)?

B. Broberg: Convection also plays a role. In fact, difficulties with nonconvecting models are that they lead to calculated temperatures of the upper mantle above the melting point, which evidently is incorrect since seismic studies show that the mantle transmits shear stresses. Note, however, that the reason why I talked about how slowly heat is transferred through distances of the size of the Earth radius only intended to show implications of scale effects. In reality, the actual heat flux through the surface of the Earth is only to a very small extent due to heat transfer from the inner core. The main contribution is due to heat transfer from sources (nuclear reactions) in the Earth's crust and in the upper mantle (contributing to about 90% of the total heat generation in the interior of the Earth). To a very small extent various dissipative mechanisms also produce heat in the Earth's interior.

G. Swan: In the subject of rock mechanics we speak of joints and cracks, so distinguishing the scale of defects, and even geometry. Fracture mechanics speaks presently about cracks and stress singularities on an extremely small scale. How do you think or envisage that scale effects effect fracture in earthquakes?

B. Broberg: Along an interface where sliding potentially could take place - either it be large or small scale - regions of clutch and of potential immediate sliding alternate, probably. Across regions of clutch very large shear stresses can be transmitted. An assumption that the shear stress is rather small at regions of potential sliding then implies that several "ends" exist, where stress singularities are obtained at a linearized mathematical treatment. One finds that such a model gives rise to non-unique solutions, seemingly typical for the phenomenon of friction itself, not only for mode II events in rocks. Suitable choices (mathematically by the scientist or physically by nature) could lead to stress intensity factors much smaller than those obtained from the classical dry friction model. The inadequacy of this classical model becomes clearly demonstrated in attempts to use it for describing static phenomena related to tectonic faults. In the dynamic case (earthquakes) regions of clutch should disappear (at high sliding velocities) making again the classical dry friction model - and thereby a close parallel to a crack model - feasible.

R. Pusch: Could a creep-failure criterion be introduced in your theory? Can we gain from metal sciences in this respect?

P. Feltham: In many materials, e.g. pure copper and Nimonic 80A a critical level of damage, leading to fracture, accumulates in the coarse of creep. The time to fracture is then found to be inversely proportional to the "steady" creep rate; the constant of proportionality is numerically close to the strain at fracture. The attainment of critical strain level as failure criterion could, in such cases, be used to complement stockastic models, even if the damage accumulation is not explicitly included in the model used.

B. Broberg: The metaphor in your lecture about the eye that nowadays sees more than the brain is able to comprehend - couldn't that also make questionable our inclination to preoccupy ourselves with trying to explain what the eye sees? Can the eye provide information for the whole picture - doesn't it give glimpses from specific angles of view about component parts, while it is their very interconnectivity (and complexity) we really need to understand (not only get pieces of knowledge about).

P. Feltham: In 1934 G.I. Taylor could devise a simple dislocation model of work-hardening. When the process was observed by transmission electron microscopy, about 20 years later, it revealed great complexity, which cannot be encompassed by such a conceptionally simple model. In this sense the information content - what we see - is too great to be encompassed in detailed models. This point is made clearly by Kröner; I refer to it in my paper.

I.J. Smalley: J.B.S. Haldane proposed that no research student should be allowed to carry out research in the subject in which he graduated; have you any comments on this proposal and do you think that it is still applicable?

P. Feltham: It may not be practicable to comply with this suggestion. Haldane no doubt wished to point to the desirability for young people to acquire a scientific attitude to their work and life in general, rather than becoming "one-dimensional" specialists.

J. Hult: You told us how the Greeks thought of Hercules carrying the Earth, so that it would not fall "down" in the Universe. Then we learnt how, eventually, the obvious next question came up: "But, what supports Hercules"? One would have thought that their answer: "Hercules stands on the hard shell of a huge tortoise" would soon have led to an equally obvious next question... How can we understand that they were satisfied with the tortoise answer?

P. Feltham: The question appears not to have been pursued. Newton and Galileo were not yet born, so there was no answer. Generally mankind does not pursue questions for which the answers are not either at hand, or close at hand.

C. Ferguson: Is there a practical problem that co-operative stochastic models of complex polycrystalline materials - even if the technical and mathematical difficulties could be overcome - would become so "fuzzy" (by virtue of the numerous interacting stockastic elements) as to be of little predictive value?

P. Feltham: The more fundamental an explanation is, the more reliance can be placed on it. There are, clearly, practical limits on how fundamental a model can be made at any given stage of understanding of the process. The "resolving power" of the model vill be inherent in the assumptions on which it is based.

J.F. Knott: It seems to me that the point is really one of __scale__. For engineering purposes, one often makes use of fairly macroscopic parameters, such as yield stress or flow stress. A detailed derivation of these __ab initio__ from the observations of individual dislocation movements in commercial materials, using the electron microscope, is an extremely difficult problem, and, because we understand the general principles, we can, with complete confidence, replace a vast amount of stochastic calculation by a simple mechanical measurement. What we are lacking in other situations are suitable criterions for other types of event. In the case of cleavage fracture in steels, the critical cleavage fracture stress seems to be of general use, and is fairly well understood in terms of microstructure and deformation processes. Shear decohesion in high strength materials may be controlled simply by a critical value of shear strain, but the importance of the stress normal to the shear band is not clear. In soils, Professor Pusch asked what failure criterion might apply and this, I think, is a critical point.

On the question of the eye distracting the brain because it now provides so much information, the following quotation may be appropriate:

> "A horde of crazy prophets,
> That, by staring at a crystal,
> Can fill it with more fancies
> Than there's herring in the sea".

SESSION 2
Experimental Studies

EFFECT OF GRAIN SIZE ON THE STATIC AND DYNAMIC FRACTURE BEHAVIOR OF α-TITANIUM

D. A. Shockey, K. C. Dao and R. L. Jones

SRI International, Menlo Park, California 94025, U.S.A.

ABSTRACT

The feasibility of establishing quantitative relationships between microstructural features and dynamic fracture behavior was investigated by performing plate impact experiments on specimens of alpha titanium differing predominantly in grain size, and determining values of parameters describing microfracture nucleation and growth. The dynamic fracture strengths for materials having average grain daimeters of 36 μm and 1.5 μm were essentially identical, but threshold nucleation rates, nucleation stress sensitivities, and microfracture growth viscosities differed by factors of 2 to 3. The coarse-grained material exhibited superior dynamic fracture resistance even though its quasi-static yield and ultimate tensile strengths were significantly lower than those of the fine-grained material. An explanation for this reversal in behavior is given in terms of the microstructure and the microfracture kinetics.

1. INTRODUCTION AND BACKGROUND

Dynamic fracture experiments on a wide range of materials have shown that the accumulation of damage (1) depends on the amplitude and duration of the stress pulse, (2) can be treated as a rate process involving nucleation and growth of microfractures, and (3) can be described by simple rate equations containing empirically determinable parameters [1]. The observed laws governing microfracture nucleation rate $\dot{N}$ and growth rate $\dot{R}$ are as follows

$$\dot{N} = \dot{N}_0 \exp \frac{\sigma - \sigma_{no}}{\sigma_1} \qquad (1)$$

$$\dot{R} = \frac{\sigma - \sigma_{go}}{4\eta} R \qquad (2)$$

where σ is the applied tensile stress and R is the characteristic dimension of the microfracture. The remaining parameters are material-specific quantities and hence act as material properties governing various aspects of dynamic fracture. σ_{no} and σ_{go} may be considered strength properties, indicating the threshold stresses required for microfracture nucleation and growth, respectively. $\dot{N}_0$ is the threshold nucleation rate and is related to the density of nucleation sites (inclusions, grain boundaries, inherent flaws, or whatever the weakest defect may be) in the material. The stress-sensitivity property $1/\sigma_1$ may be related to the size distribution of the nucleating defect. The viscosity term η controls the rate of crack growth.

Equations 1 and 2 appear to describe the fracture kinetics of many materials [2]; differences in the dynamic response of the various materials are evidenced by differences in the numerical values of the parameters. The parameters appear to be sensitive to temperature and to microstructure. The temperature sensitivity has been demonstrated by the results of experiments on beryllium [3] at elevated temperatures, whereas the microstructural sensitivity has been demonstrated by results on three grades of beryllium, differing in BeO content, grain size and texture. However, no systematic study of temperature or microstructure dependence has been previously undertaken. This paper reports the results of a research program [4] conducted to determine the dependence of these dynamic fracture parameters on the grain size of unalloyed α-titanium.

2. PREPARATION AND CHARACTERIZATION OF MATERIALS

Quasistatic tensile test data on commercially-pure α-titanium show substantial increases in yield and fracture strength with decreasing grain size [5, 6]. Moreover, in this material grain size can be varied over a wide range independently of other microstructural features by simple heat treatments.

Forged bars of unalloyed α-titanium* were heavily (80%) cold-worked by swaging into 25-mm dia. bars. Sections of the swaged bar 150 mm long were subjected to isothermal anneals at various temperatures for various durations to produce materials with a range of grain sizes, Table 1. Characterization studies revealed few significant differences among materials with different grain sizes. Basal pole distributions were independent of grain size, but an iron-rich phase detected at grain boundaries in the coarse-grained materials was not detected in the fine-grained materials.

3. QUASISTATIC FRACTURE EXPERIMENTS

Round bar tensile tests at a nominal strain rate of about $3 \times 10^{-4} s^{-1}$ indicated a Hall-Petch relationship [8, 9] between the yield strength σ_y and the grain diameter d (Figure 1),

$$\sigma_y = \sigma_i + kd^{-1/2}$$

where σ_i and k have values of 324 MPa and 7.25 MPa-m$^{1/2}$, respectively A similar variation with grain diameter was observed for the tensile strengths, but significant deviations occurred at coarse grain sizes. The effect of grain size on toughness was indicated by work-to-fracture†

---

*Conforming to ASTM Specification B265, Grade 3.

†The area under the load-displacement curve normalized by the minimum specimen cross section.

Table 1. GRAIN SIZES AND MECHANICAL PROPERTIES OF COLD-SWAGED TITANIUM BARS AFTER VARIOUS ANNEALING TREATMENTS

Heat Treatment	Grain Size (μm)	0.25% Offset Yield Strength (MPa)	Nominal Tensile Strength (MPa)	Uniform Elongation (%)	True Fracture Stress (MPa)	Reduction of Area (%)	Normalized Work to Fracture (MPa-m)
1 hr at 610°C	7	-	-	-	-	-	-
2 hr at 650°C	9	-	-	-	-	-	-
16 hr at 650°C	13	-	-	-	-	-	-
1 hr at 820°C	26	-	-	-	-	-	-
64 hr at 820°C	36	363	470	16.2	787	53	0.159
5 min at 610°C	6	416	553	15.3	897	55	0.228
1.25 min at 610°C	2.0	490	460	15.5	-	-	0.242
2 min at 585°C*	2.5	470	620	15.5	1170	60	0.244
1 min at 585°C*	1.0	565	687	12.0	1070	49	0.219
1.5 min at 585°C*	1.5	522	464	15.3	1055	51	0.228

*Specimen preheated to 275°C.

measurements on notched tensile bars. Figure 1 shows that work-to-fracture tends to maximize at a grain size of 2-3 μm, in agreement with previous work [5].

Scanning electron micrographs of the fracture surface of tensile specimens of different grain size indicated that fracture occurred by the nucleation, ductile growth, and coalescence of microvoids. The absence of second-phase particles in the dimples of all three materials suggests that included particles played no significant role in fracture nucleation. The sizes of the dimples on the fracture surfaces reflect the sizes of the microvoids at coalescence and are expected to be related to the spacing of the void nucleation sites. The dimple size was similar to the grain size in materials with 36-μm and 6-μm average grain diameters, suggesting void nucleation at grain boundaries. The dimples in the fine-grained material, however, were considerably larger than the grains, and may be related to the grain size of the as-swaged material.

Etched cross sections through the necks of fractured tensile bars of coarse-grained material (Figure 2) revealed populations of isolated microvoids, most of which are roughly circular and associated with grain boundaries, particularly grain boundary triple points. We found no evidence that the iron-rich grain boundary phase influenced the fracture process of coarse-grained material. We could not positively associate microvoids in fine-grained material with any microstructural feature, but the voids seemed to be located more frequently at the remnants of boundaries of elongated grains formed by swaging, than at boundaries of the recrystallized grains. This is consistent with the observation that fracture surface dimples are larger than the recrystallized grain size.

4. DYNAMIC FRACTURE EXPERIMENTS

The fracture behavior of α-titanium under high-amplitude, short duration stress pulses was studied using a well-established plate impact technique (Figure 3). Specimen plates of materials having different grain sizes

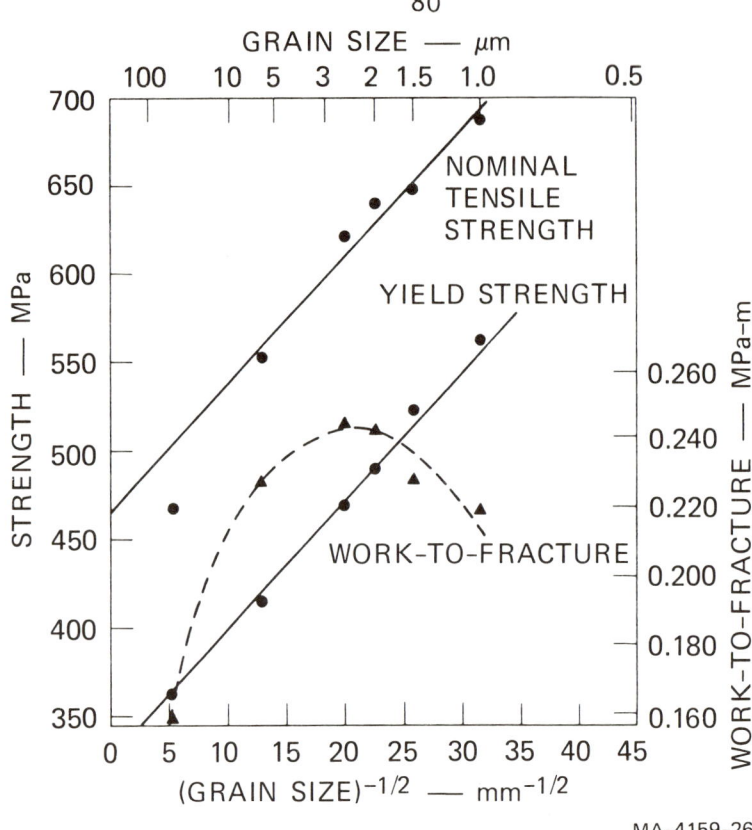

Figure 1. Variation of 0.25% Offset Yield Strength, Nominal Tensile Strength, and Work-to-Fracture With Grain Size for α-Titanium

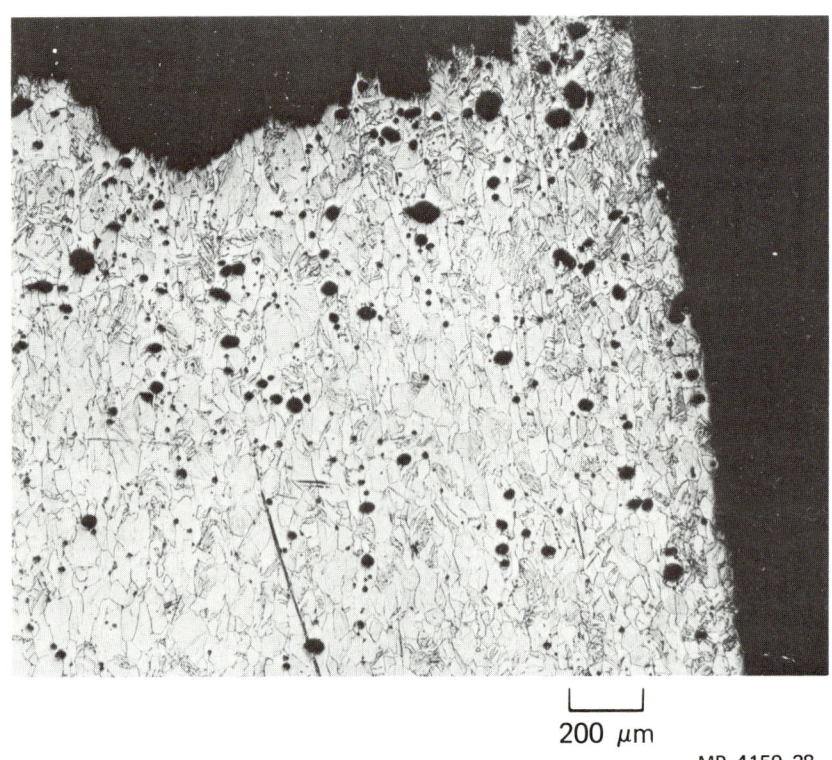

Figure 2. Etched Cross Section Showing Microfractures in the Neck Region of a Tensile Specimen of 36-μm-Grain-Size Material

were positioned at the muzzle of a light gas gun and were impacted uniformly and simultaneously over their planar areas by a titanium flyer plate mounted on the leading edge of a cylindrical projectile.

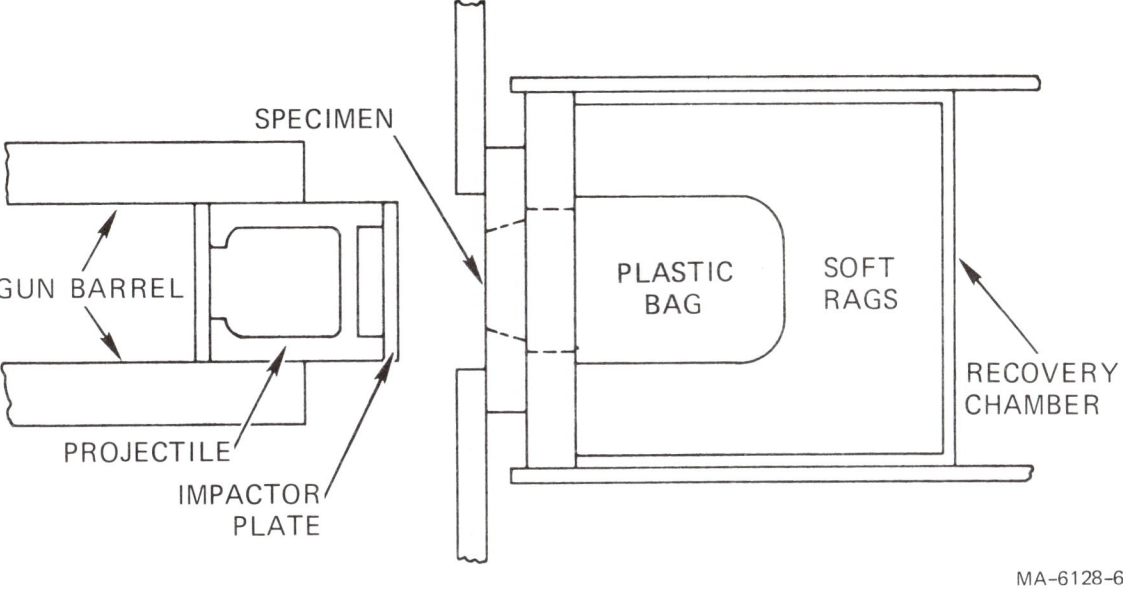

Figure 3. Arrangement for Dynamic Fracture Experiments

A planar impact produces a uniaxial strain compressive wave in both the specimen and flyer plate. Flyer plate thicknesses were made some fraction of the specimen plate thickness, so that the compressive waves reflecting from the free surfaces would meet within the specimen plate and produce a nominally rectangular tensile pulse that endured typically for tenths of microseconds. Pulse amplitude and duration were precisely controlled by impact velocity and plate thicknesses, respectively.

Fifteen plate impact experiments were carried out on material with average grain diameters of 36 and 1.5 μm. The goal of these experiments was to produce incipient levels of internal fracture damage under load pulses of different shapes. Peak tensile amplitudes ranged from 1480 to 5080 MPa and durations ranged from 0.031 to 1.08 μs. Impacted specimen plates were recovered from the catcher chamber, sectioned on a diameter, polished, and examined with an optical microscope to ascertain the extent of fracture damage, which appeared as a population of microfractures (Figure 4a). In general, damage was heaviest near the midplane where the tensile stress first appeared and endured longest; the number and size of the microfractures diminished gradually with distance from the midplane. Evidence of intense local plastic deformation by slip and by twinning was often found in the vicinity of microfractures.

As in the quasi-static tests, grain boundaries and grain boundary triple points were the most common nucleation sites. Individual microfractures ranged in shape from equiaxed voids, Figure 4a, to planar grain boundary separations, Figure 4b, in several specimens particularly those made from the finer-grained materials,

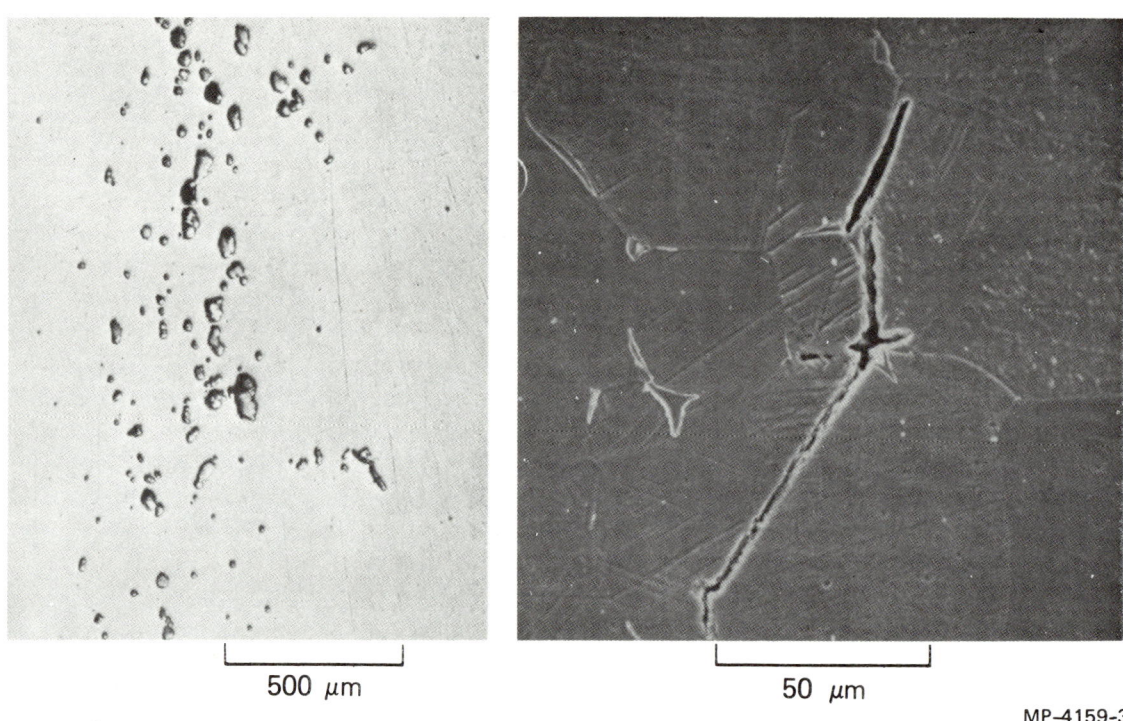

500 μm 50 μm

Figure 4. (a) Distribution of Microvoids Revealed on Polished Cross Section of Impacted Plate of Coarse-Grained Titanium.
(b) Grain Boundary Failure Typifying Microfracture Morphology in a Companion Impacted Plate of Coarse-Grained Material.

microfracture occurred in clusters. No consistent trend with grain size was apparent in microfracture morphology. Similar variability was found by Christman et al. [9], although other commercially pure metals such as OFHC copper, 1145 aluminum, and Armco iron exhibited invariant microfracture morphologies and no tendency to cluster [10].

To quantify the fracture damage, we counted and measured individual microfractures visible on polished cross sections at 200x. These surface data were then converted by a statistical transformation to obtain the size distributions of microfractures per unit specimen volume.

The stress histories of the specimens were computed with a one-dimensional, finite-difference Lagrangian wave propagation code called PUFF. This code computes the density, stress, and energy of one-dimensional material cells at each wave-propagation time step. Required inputs to the code are plate dimensions, impact velocity, and material constitutive relations.

Elastic-perfectly plastic constitutive relations estimated from our quasi-static data and dynamic data taken by other authors [9] were used to calculate the stress history that each of the 15 specimens would have experienced had no fracture damage occurred to affect

the stress pulse.* Peak values of the amplitudes and durations of the
tensile stresses were used to obtain first estimates of the parameters
in the dynamic fracture equations 1 and 2 in the following way. The
total number of microfractures in a zone of a specimen were obtained
from the measured size distributions and divided by the nominal computed
stress duration to obtain a gross nucleation rate and plotted versus
peak tensile stress. The characteristics of the resulting curve were
taken as first-order estimates of the microfracture nucleation parameters. Approximate growth parameters were obtained similarly by
plotting microfracture sizes versus the product of computed peak
tensile stress and duration (that is, the impulse) and comparing the
resulting curve with Eq. 2. The scatter in both plots was considerable.
The stress history calculations for selected specimens were then
repeated, but this time fracture was allowed to occur according to
Eqs. 1 and 2, and the effect of the developing damage on the stress
history was accounted for. The computed size distributions were compared with the measured distributions to ascertain how much the gross
estimates for the parameters in the dynamic fracture model should be
changed. A third or fourth calculation using revised fracture parameters
brought the computed and observed damage into somewhat better agreement,
but the scatter in the data precluded an unambiguous correlation.
The final values are given in Table 2.

Table 2. DYNAMIC FRACTURE PARAMETERS FOR α-TITANIUM IN TWO GRAIN
SIZES

	Coarse Grained (36 μm)	Fine Grained (1.5 μm)
σ_{no}	1800 MPa	1750 MPa
$\dot{N}_0$	3.8×10^{12} cm^{-3}s^{-1}	1.7×10^{12} cm^{-3}s^{-1}
$1/\sigma_1$	3.3×10^{-2} MPa^{-1}	0.91×10^{-2} MPa^{-1}
σ_{go}	1800 MPa	1750 MPa
η	310 poise	110 poise

6. DISCUSSION OF RESULTS

The values in Table 2 are only approximate, but can be compared to
indicate which parameters are sensitive to grain size, the degree of
sensitivity, and the direction of change.

The threshold stress for fracture nucleation σ_{no} was found to be about
1800 MPa for both the coarse- and fine-grained material, and the
threshold stress for microfracture growth σ_{go} was found to be equal
to that for nucleation.† These threshold stresses are 3.5 to 5 times

---

*As fracture damage develops, the material becomes more compliant
and the stresses relax considerably.

†If microfractures nucleate at sizes well below the resolution limits
of our optical microscope, considerable growth will be required
before they reach a detectable size. Thus, the "nucleation conditions"

the quasi-static yield strength of the two materials, (Table 1), a
difference consistent with results obtained for other materials [2, 9].
Surprisingly, the dynamic threshold strengths were not affected by
grain size; the quasi-static findings (Figure 1) led us to anticipate
higher dynamic thresholds for the fine-grained material. The results
suggest that the stress level required to cause failure at grain
boundary triple points is independent of grain size at shock loading
rates.

Grain size influenced the other nucleation and growth parameters more
significantly. The threshold nucleation rate $\dot{N}_0$ for coarse-grained
material was approximately twice that for fine-grained material, but
the nucleation stress sensitivity $1/\sigma_1$ was only about 1/3 as large.
This means that at dynamic stresses just slightly above threshold, more
microfractures will be initiated in coarse-grained material, but at
higher stress levels greater numbers of microfractures will be produced
in fine-grained material. This behavior is explained by microfractures
in the fine-grained material nucleating at sizes too small to be
resolved with the optical microscope, whereas microfractures in the
coarse-grained material (where the grain diameter is some 20 times that
of the fine-grained material) may be considerably larger at nucleation
and hence easily seen. Thus, initially there appears to be a higher microfracture concentration in the coarse-grained material, but at higher
stress levels the microfractures grow larger, and more of those present
in the fine-grained material are discernable. Then, because it has a
much higher density of triple points, the fine-grained material exhibits
higher densities of microfractures.

The rate at which microfractures grow is about three times greater in
the fine-grained material, as indicated by the values for the viscosity
parameter η. This may reflect the differences in microfracture morphology discussed earlier. The crack-like defects found in the fine-grained material may be expected to grow more rapidly than the spherical
void-like defects that were characteristic of the coarse-grained
material, because substantially less energy is required to increase
the length of a grain boundary separation than to increase the diameter
of a sphere expanding by plastic flow.

Taken together, these results (Table 2) suggest that although failure
begins at similar stress levels in shock-loaded fine- and coarse-grained
α-titanium, the failure process proceeds more rapidly with increasing
stress in the fine-grained material. The observed superiority of the
coarse-grained material in dynamic load situations is the reverse of
the behavior expected on the basis of the quasistatic failure behavior.
The reversal is attributable to the much higher concentration of
failure initiation sites (grain boundary triple points) in fine-grained
material, combined with the fact that much larger numbers of these

---

†ascribed at the point where microfractures became visible may actually
be growth conditions, which would explain why identical values were
obtained for nucleation and growth threshold stresses.

sites are activated under short-lived, high-level dynamic loads than under the longer lived but lower level loads that are involved in failure at quasi-static strain rates.

ACKNOWLEDGMENTS

The authors gratefully acknowledge the financial support of the Army Research Office, Durham, NC under Contract DAHCO-75-C-0020, and the technical assistance of many colleagues at SRI International.

REFERENCES

[1] Curran, D. R., L. Seaman, and D. A. Shockey, Dynamic fracture of solids. Physics Today 30 (1977) 46ff.

[2] Shockey, D. A., L. Seaman, and D. R. Curran, Dynamic fracture of beryllium under plate impact and correlation with electron beam and underground test results. Final Technical Report AFWL-TR-73-12 to the Air Force Weapons Laboratory, Kirtland AFB, New Mexico (1973).

[3] Shockey, D. A., D. R. Curran, K. C. Dao, and L. Seaman, Feasibility of producing fracture damage with repetitive laser pulses. Final technical report on Contract DAAH01-75-C-1072 to U. S. Army Missile Command, Redstone Arsenal, AL, (1976).

[4] Shockey, D. A., K. C. Dao, and R. L. Jones, Dependence of dynamic fracture behavior on microstructural variables. Final technical report Contract DAHCO-75-C-0020 to the Army Research Office, Durham, North Carolina (1978).

[5] Jones, R. L., and H. Conrad, The effect of grain size on the strength of alpha titanium at room temperature. Trans. TMS-AIME 245 (1969) 779-789.

[6] Jones, R. L., Titanium science and technology, R. I. Jaffee and H. M. Burte, eds. Plenum Press, New York, 1973, pp 1033-1047.

[7] Hall, E. O., Deformation and aging of mild steel. Proc. Roy. Soc. (London) B64 (1951) p. 747-753.

[8] Petch, N. J, Cleavage strength of polycrystals. J. Iron and Steel Inst. 174 (1953) p. 25.

[9] Christman, D. R., T. E. Michaels, W. M. Isbell, and S. G. Babcock, Measurements of dynamic properties of materials--volume IV alpha titanium. Report MSL-70-23 vol IV to Defense Nuclear Agency, Washington, D.C. (1972).

[10] Barbee, T.W., L. Seaman, R. Crewdson, and D. R. Curran, Dynamic fracture criteria for ductile and brittle metals. J. Matls. 7 (1972) 393-401.

THE MECHANISM OF DEFORMATION OF PEARLITE

D. A. Porter, K. E. Easterling and G. D. W. Smith*

Department of Materials Technology, University of Luleå, S-951 87 Luleå, Sweden
*Department of Metallurgy & Science of Materials, University of Oxford, England

ABSTRACT

Pearlite is a two phase structure of iron and cementite (Fe_3C) which forms as colonies with a lamellar morphology. When fully pearlitic wires are drawn the lamellae rotate until they are all oriented along the axis of the wire. Such wires have enormous strengths and are used e.g. for lifts, cranes, ski-lifts, etc. The basis of these wires' great strength is not fully understood and in this work dynamic deformation experiments are carried out in the electron microscope in an attempt to elucidate the deformation mechanims.

1. INTRODUCTION

Pearlite is an important constituent in many steels. It is produced as a result of the eutectoid reaction in the Fe-C system by which fcc austenite containing ca 0.8 wt % C decomposes into alternate lamellae of bcc ferrite (Fe - 0.02 wt % C) and cementite (Fe_3C). A typical pearlite morphology is shown in figure 1. The structure is divided into colonies within which the cementite lamellae have approximately the same orientation. The spacing of the cementite lamellae depends on the temperature at which the austenite transforms and can vary from ~1 μm at the highest transformation temperatures to less than 0.1 μm at the lowest. Since the volume fraction of cementite is approximately constant the thickness of the cementite also varies in proportion to the interlamellar spacing. The behaviour of pearlite during plastic deformation and fracture is very sensitive to the fineness of the microstructure. All of the mechanical properties of pearlite such as the yield and

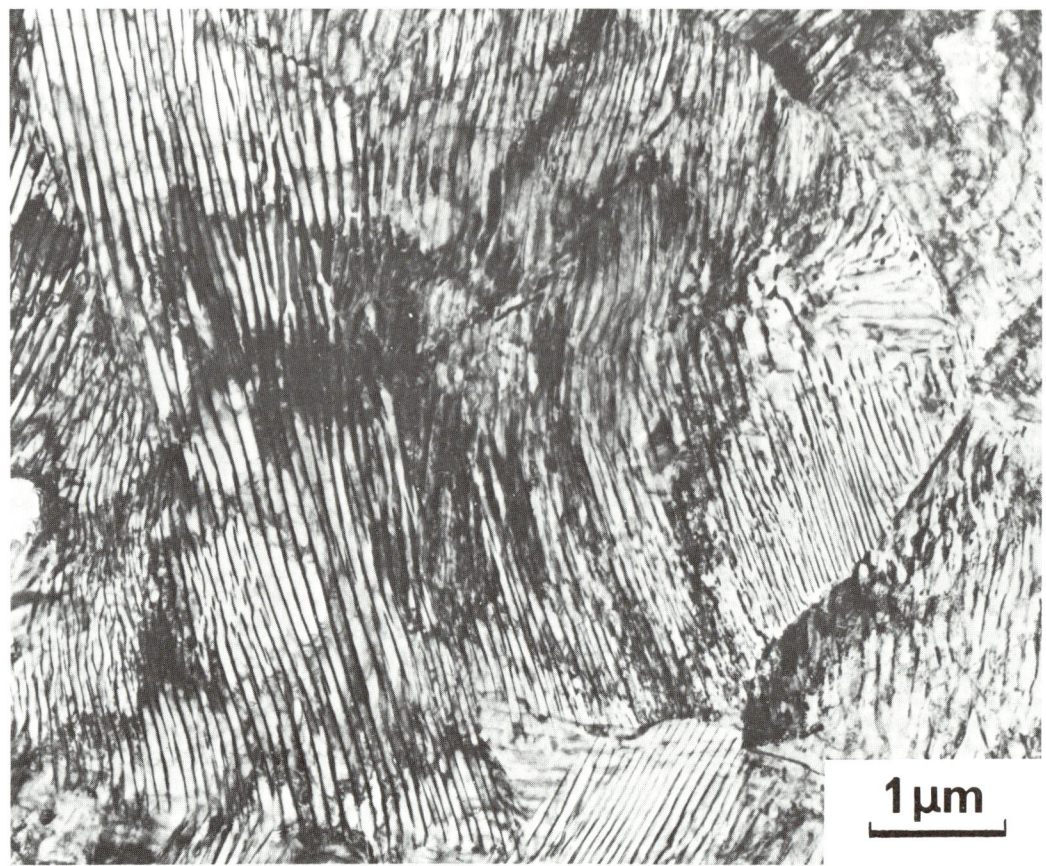

Fig 1. The morphology of undeformed pearlite as revealed by transmission electron microscopy. Cementite lamellae (dark) are interspersed in a matrix of ferrite (light).

fracture stresses as well as the ductility and toughness are improved by decreasing the interlamellar spacing [1 - 5]. However, the reasons for this improvement are not well understood. Therefore it was the purpose of this study to investigate the behaviour of pearlite during tensile testing with particular emphasis on the influence of interlamellar spacing on the micromechanisms of deformation and fracture.

A considerable problem with the study of pearlite deformation is the separation of features resulting from tensile deformation, from those present in the as-grown structure or artefacts arising from specimen preparation. As first pointed out by Butcher & Petit [6], the distinction is sometimes very difficult to make since the structure of undeformed pearlite is quite complex and this has previously lead to wrong conclusions being drawn about deformation mechanisms.

In figure 1 for example it can be seen that the lamellar structure contains many imperfections such as lamella terminations and twists, bends, ripples and changes in thickness of the cementite lamellae. The lamellar structure can also degenerate into ribbons of cementite, perforated lamellae, or even isolated fragments. In a deformed specimen these features would be extremely difficult to separate from any changes caused by deformation. It is really necessary to continously follow the deformation behaviour of individual lamellae or whole pearlite colonies, so that direct comparisons of the same area can be made before and after deformation. Due to the extremely fine spacing of pearlite, this is only feasible in "bulk" samples by the use of a high resolution scanning electron microscope (SEM) fitted with a deformation stage.

2. IN SITU DEFORMATION IN THE SCANNING ELECTRON MICROSCOPE

The basis of in situ deformation experiments in the SEM is that the surface of a bulk tensile specimen can be imaged while the specimen is being deformed. Therefore if the microstructure is revealed by polishing end etching the specimen prior to straining detailed deformation and fracture mechanisms can be unambiguously established by observing the formation of slip markings and cracks as well as the shape changes caused by plastic strain. In the present investigation two scanning electron microscopes have been used: a 50 kV conventional SEM and a 200 kV scanning transmission electron microscope (STEM) operated as a high voltage SEM in the secondary electron mode. Both were equipped with straining stages and video recording facilities so that deformation could be recorded at TV scanning rates. When higher resolution images were required straining could be stopped and slow scan micrographs recorded of features of interest under load. The 50 kV instrument was also equipped with strain gauges providing a similtaneous output of the load/elongation variation. This instrument had a capacity of 2 kN and could therefore be used for deforming quite large specimens, whereas the 200 kV STEM has a 20 N straining stage and no strain gauges. However, the advantage of the high voltage SEM is that the increased gun brightness provides an improvement in resolution for a given signal/noise ratio and therefore allows higher magnifications to be used during the recording of the dynamic sequences at TV scanning rates. This was essential for the study of the fine pearlites which had interlamellar spacings <0.1 µm. The capacity of the straining stage was sufficient for deforming specimens with a cross-section of ~0.2 x 0.5 mm along the gauge length.

Both microscopes were equipped with energy dispersive X-ray analysis facilities so that, in addition to following microstructural changes during deformation, it was also possible to make chemical analyses. This was useful when studying the role of non-metallic inclusions in the specimens.

In all in situ deformation studies there is the problem of how representative the surface observation are of phenomena occurring within the bulk of the material. In the present case the correspondence is believed to be rather good because the cementite is in the form of thin lamellae that extend below the surface and should consequently suffer less stress relaxation due to the free surface. Also, observation of deformed and sectioned bulk samples revealed similar deformation phenomena to those found on the surface.

3. MATERIALS AND SPECIMEN PREPARATION

In order that comparisons could be made between pearlites of different spacings and orientations a range of eutectoidal steels were studied. The materials were supplied from commercial stock by Bridon Wire Limited of Doncaster, England. These included two very fine pearlitic structures in the form of 5.5 mm diameter lead patented rod (designated LP) and a similar rod which had been cold drawn to a 32% reduction in area after lead patenting (CD). Both of these fine pearlites had interlamellar spacings ranging from 40 to 150 nm with a mean value of 90 nm. The pearlite colonies in the lead patented material were randomly oriented whereas in the cold drawn samples the lamellae were mostly aligned parallel to the drawing direction. Two materials containing randomly oriented coarse pearlite were produced from the cold drawn stock by re-austenitizing at $950^{\circ}C$ and either furnace cooling (FC) or isothermally transforming (IT) at $695^{\circ}C$. These pearlites had spacings ranging between 250 and 600 nm with a mean of 400 nm.

The miniature tensile specimens for the 200 kV microscope were made from 2.5 mm thick longitudinal sections of the original bar material so that the rod axis corresponded to the tensile axis. After machining the profile illustrated in figure 2 the specimens were ground and polished to a thickness of 0.15 - 0.20 mm, finishing with 0.1 μm alumina paste, and finally lightly etched in a 2% nital solution.

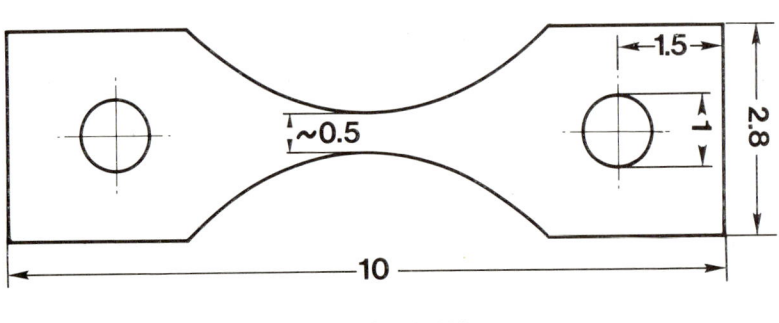

DIMENSIONS IN mm

Fig 2.   The profile of the miniature tensile specimens used in the straining stage of the 200 kV microscope.

4. EXPERIMENTAL RESULTS & DISCUSSION

The results of this investigation which will be reported in detail elsewhere [7] have shown that there are large differences in the deformation and fracture mechanisms between coarse and fine pearlites, especially when the lamellae are aligned parallel to the tensile axis.

The deformation of such colonies in coarse pearlite (Fig. 3) involves slip on oblique narrow slip lines in the ferrite and fracture of the intersecting cementite lamellae due to the action of dislocation pile-up stresses. Figure 3 clearly shows the importance of lamella terminations (at A) in the initiation of slip in the ferrite.

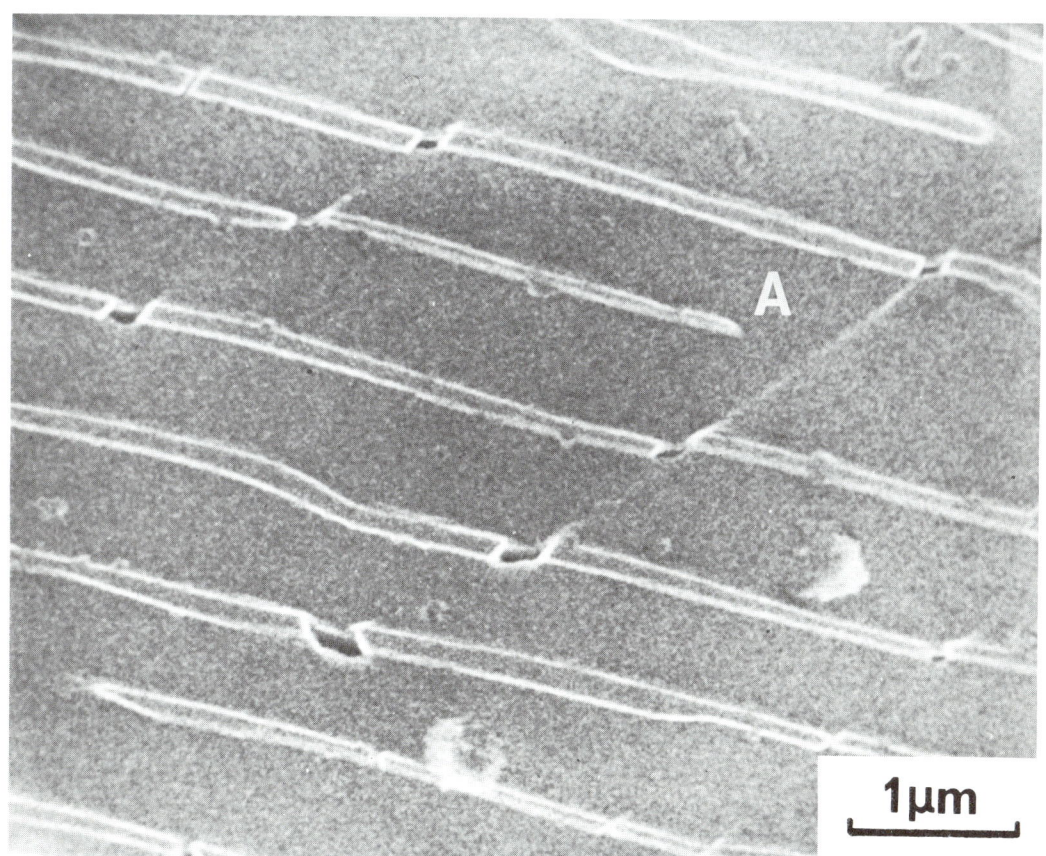

Fig 3. Coarse pearlite showing slip lines in the ferrite and void formation between fractured cementite ends. Tensile axis approximately horizontal.

This is because such faults act as local stress concentrators. Close examination of the intersection of a slip line with a cementite lamella (Fig 4) reveals slip steps at the cementite/ferrite interface indicating that fracture of the cementite is preceded by local plastic flow.

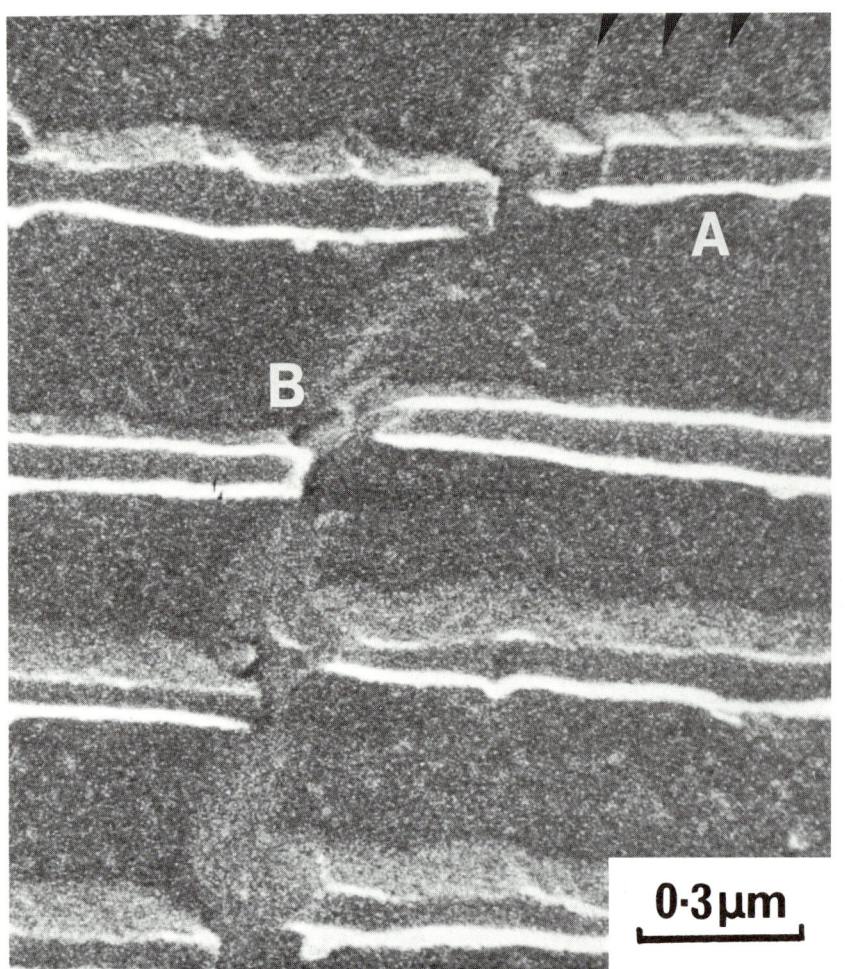

Fig 4. Details of the intersection of a slip line with the cementite lamellae. Slip steps at A. A void at B. Tensile axis approximately horizontal.

The slip step probably increases in size until the local stress concentration is sufficient to propagate a cleavage fracture through the cementite [8]. The lamellae do not, however, thin noticeably during deformation. As the strain increases voids form between the fractured ends and finally link up through the ferrite matrix to produce shear cracks [9].

Cracking in colonies with lamellae transverse to the tensile axis can be seen in the sequence of TV frames in figure 5.

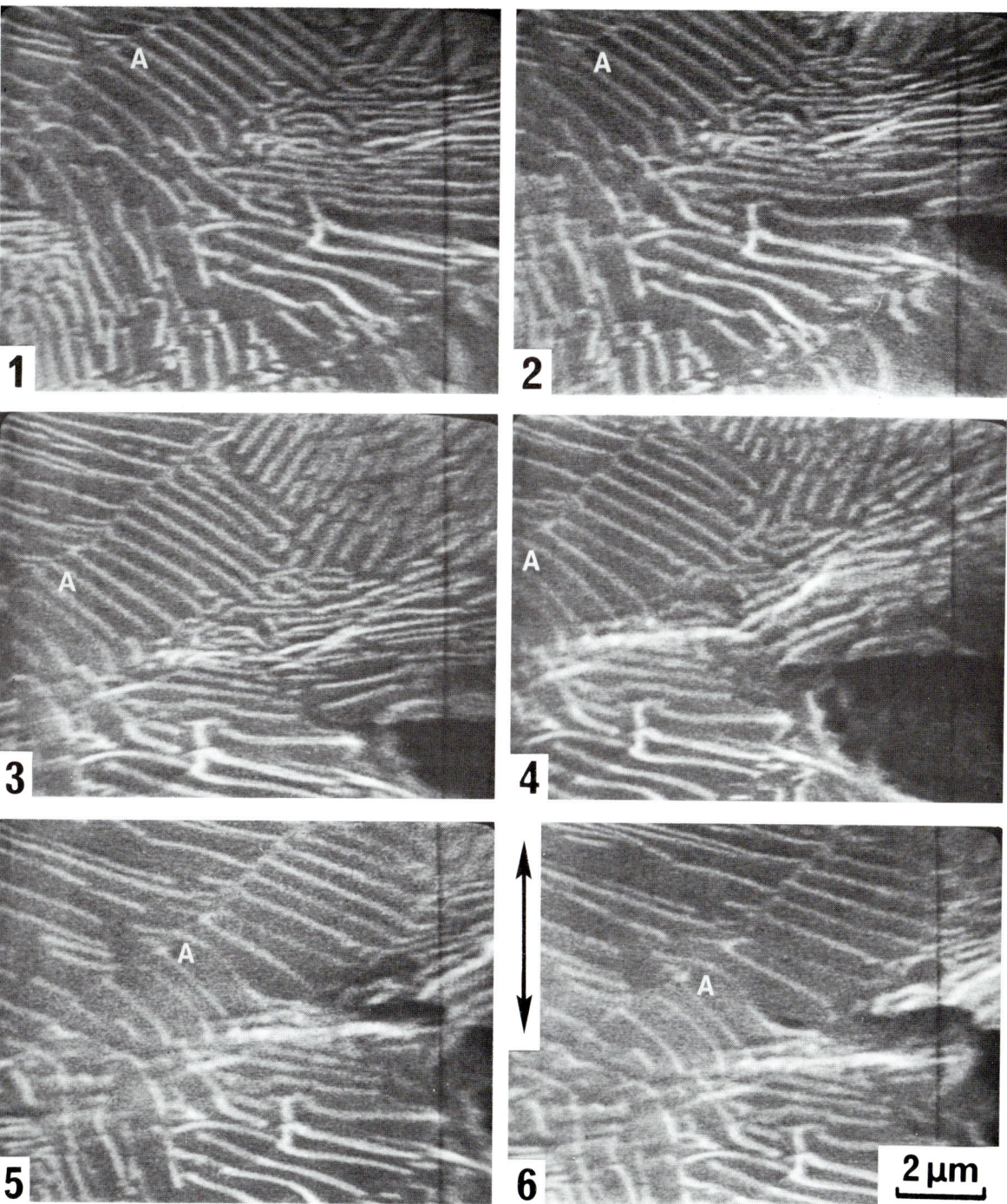

Fig 5. A series of TV frames from a dynamic sequence showing the path of crack propagation in coarse pearlite. Tensile axis indicated in frame 6.

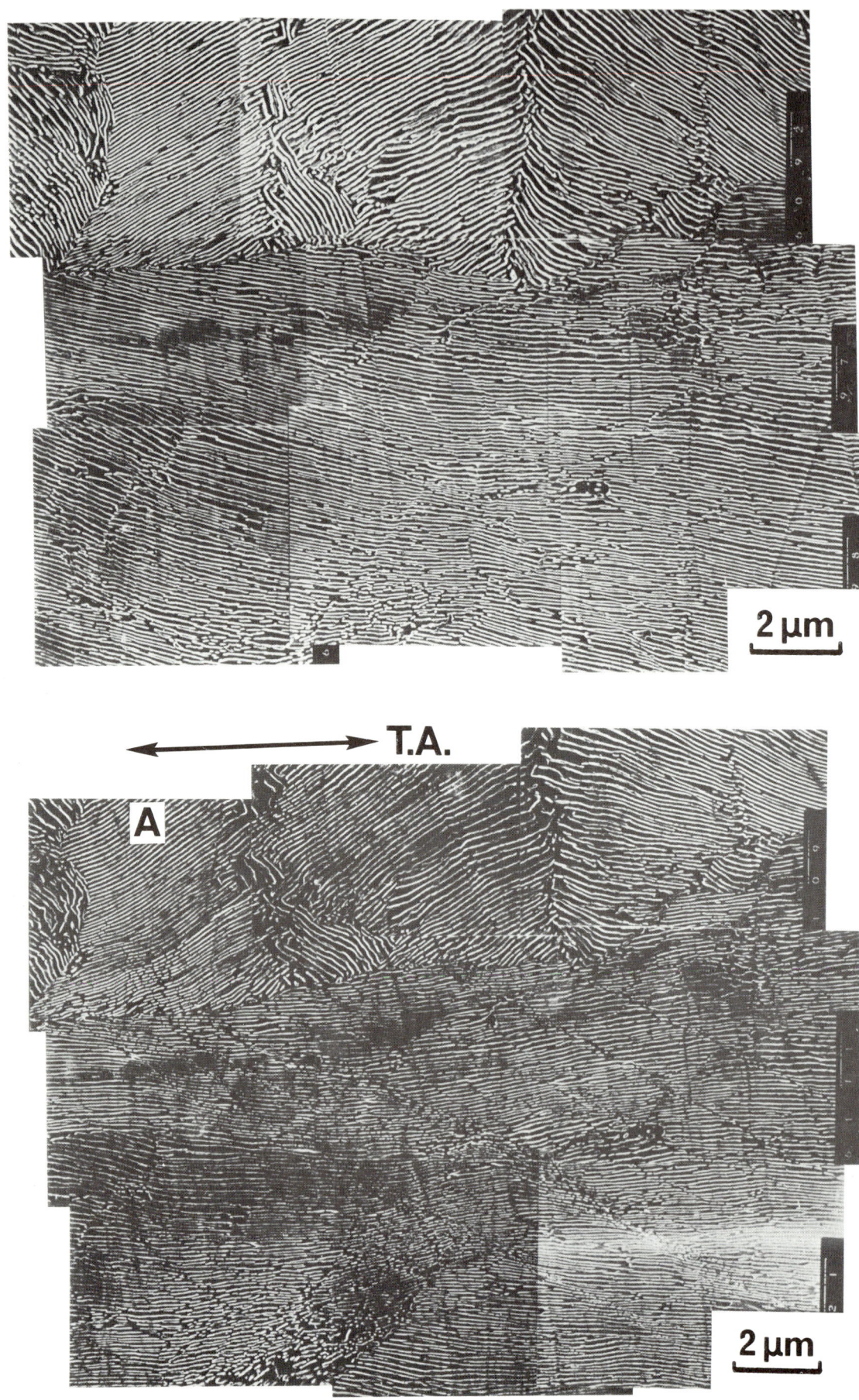

Fig 6. Fine pearlite (CD) (a) before and (b) after ~20% deformation. Tensile axis approximately horizontal.

A growing crack, moving from right to left, is accommodated by the spreading apart of the lamellae in the colony ahead of the crack (at A). In frames 5 and 6 a new crack is seen to form and link up with the main crack.

In contrast to the coarse pearlite the fine pearlite deforms by more homogeneous slip in both phases. Shear cracking does not occur, but rather the cementite is able to neck down in a ductile manner leading to extensive fragmentation at higher strains, but little void formation. Figure 6 illustrates this behaviour in a cold drawn specimen. Figure 6a is the structure before deformation and figure 6b after about 20% elongation. Comparison of these two micrographs shows that almost all cementite lamellae undergo some plastic deformation and there is no evidence of shear crack formation. There has been a great deal of shear parallel to the lamellae causing rotation of the lamellae towards the tensile axis. Quite a large fraction of the total strain in this case is taken up by several bands of enhanced deformation in which the cementite has undergone extensive thinning and fragmentation.

Crack growth in the fine pearlites involves much more plastic deformation than the coarse pearlites. Microvoids eventually form and link up, but only close to the final fracture path. Another source of internal cavities that expand and link up during final fracture are elongated inclusions which were identified from their X-ray spectra as manganese sulphides. These inclusions were present in all the specimens studied and acted as sites for internal cavity formation during deformation by fracture and decohesion as shown in figure 7.

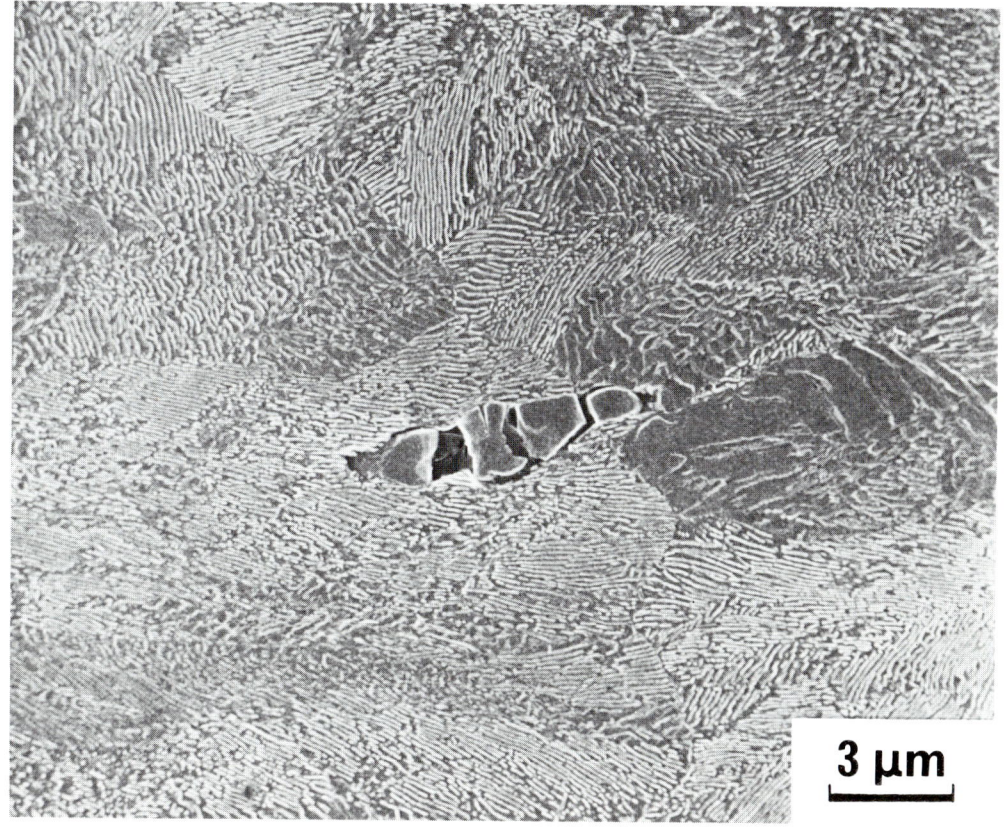

Fig 7. Fracture of a manganese sulphide inclusion in a lead patented specimen.

The results of these experiments provide an explanation for the known increase in the toughness and ductility of pearlite with decreasing interlamellar spacing. The behaviour of colonies containing lamellae aligned parallel to the tensile axis can be rationalised in terms of a model based on the transfer of stress from the weaker ferrite to the stronger cementite. The higher yield stress of the fine pearlite is able to operate many more dislocation sources in the ferrite/cementite interfaces resulting in finer slip in the ferrite than in the coarse pearlite. Consequently the behaviour of cementite in fine pearlite is controlled by fibre loading stresses leading to relatively uniform plastic strain, whereas in coarse pearlite pile-up stresses dominate and localised shear flow and shear cracking develop [7].

REFERENCES

[1]   A.R. Marder and B.L. Bramfitt, Met. Trans., $\underline{7A}$, 365 (1976).

[2]   J.M. Hyzak and I.M. Bernstein, Met. Trans., $\underline{7A}$, 1217 (1976).

[3]   J.D. Embury and R.M. Fisher, Acta Met., $\underline{14}$, 147 (1966).

[4]   T. Takahashi and M. Nagumo, Trans. Japan Inst. Metals, $\underline{11}$, 113 (1970).

[5]   B. Karlsson and G. Lindén, Mater. Sci. Eng., $\underline{17}$, 153 (1975).

[6]   B.R. Butcher and H.R. Petit, J. Iron Steel Inst, $\underline{204}$, 469 (1966).

[7]   D.A. Porter, K.E. Easterling and G.D.W. Smith, Acta Met., to be published.

[8]   A. Inoue, T. Ogura and T. Masumoto, Trans. Japan Inst. Metals, $\underline{17}$, 149 (1976).

[9]   L.E. Miller and G.C. Smith, J. Iron Steel Inst., $\underline{208}$, 998 (1970).

ELECTROMAGNETIC DETECTION OF LOW-CYCLE FATIGUE

L. Jilkén and J. Bäcklund

*Department of Mechanical Engineering, Linköping Institute of Technology,
S-581 83 Linköping, Sweden*

ABSTRACT

An electro-magnetic detection method based on the coupling between the mechanical stress-strain fields and the electro-magnetic field in ferromagnetic bodies has recently been developed at Linköping Institute of Technology. A specially designed detecting unit generates an electro-magnetic flux through the specimen and also registers the change of the flux. Since the unit is the subject of an application for a patent, all details behind the measuring technique can not be given. However, it has proved successful for determining stresses, strains, onset of yielding, initiation of crack growth and damage caused by fatigue.

This paper describes how the metod can be applied to low-cycle fatigue testing of notched specimens. The unit is designed in such a way that it does not touch the narrow part of the specimen. In this way problems of wear that are usually encountered when using axial gauges are completly avoided. The drawback of diametrical gauges, which have a tendency to "climb" along the specimen, is also circumvented by the new method.

1. INTRODUCTION

Low-cycle fatigue can be regarded as a process of *accumulating damage* in the material [1]. The damage is due to repeated yielding and will eventually lead to fracture. The present paper describes how a recently developed measuring technique based on the coupling between the *electromagnetic* and *mechanical stress fields* [2-4] can be used to trace low-cycle fatigue in ferromagnetic solids.

A coupling between the mechanical stress field, the electromagnetic field and the temperature field exsists in ferromagnetic solids. The research work reported in this paper has been focused on the coupling between the first two of these, i e on *magnetomechanical coupling effects*. These phenomena have previously been utilized within the elastic range of the material, then termed *magnetoelastic* coupling, in various kinds of load cells.

In order to register and trace the elastic and plastic deformations in solids due to applied loads, one wishes to be able to obtain signals from some kind of measuring cells distributed throughout the entire volume of the solid. In *ferromagnetic materials* the atoms can be considered to be these cells, provided that they can be properly controlled and their information adequately interpredted. This paper shows how the "*magnetic vectors" of the atoms* can be used for obtaining information on the accumulation of *fatigue* damage in ferromagnetic materials.

2. PATENTED MEASURING EQUIPMENT AND TECHNIQUE

A principal sketch of the measuring equipment is shown in Fig 1. The test *specimen* (1), which can be acted upon by tensile, compressive or twisting forces (2), forms a magnetic circuit with an iron *yoke* (3) equipped with a primary (4) and a secondary (5) *coil*. The driving unit (6) provides the primary (driving) coil

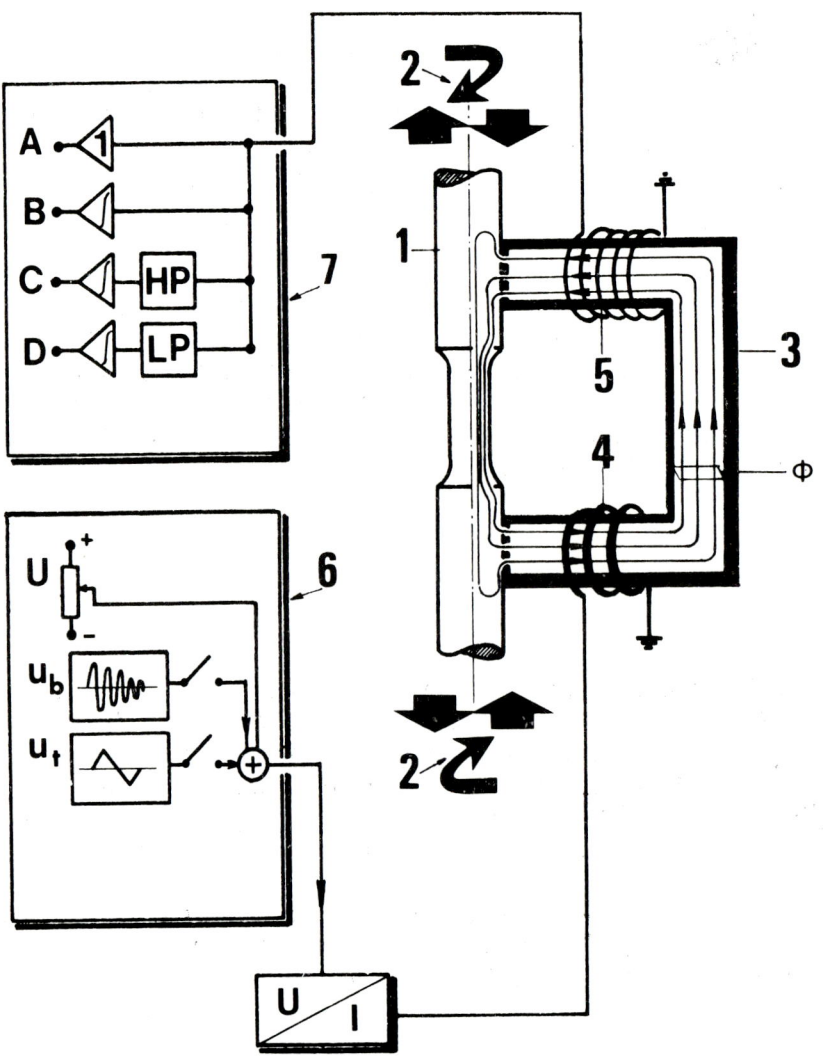

Fig 1 Outline of measuring equipment arrangement

with either a constant direct current, an exponentially decreasing current or a triangular current. Finally, the *detecting unit* (7) takes care of the signal from the secondary coil and integrates it, which gives the *magnetic flux*. The variations in the flux are then used to predict various mechanical phenomena.

As a matter of fact the single yoke unit in Fig 1 is not used at tests or measurements on round bars. Instead, a large number of thin yokes (1) are placed circumferentially around the specimen (2), Fig 2, such that a state of axial symmetry is better approached. The primary (3) and secondary (4) coils are enclosed in a casing and protected by a cover (5).

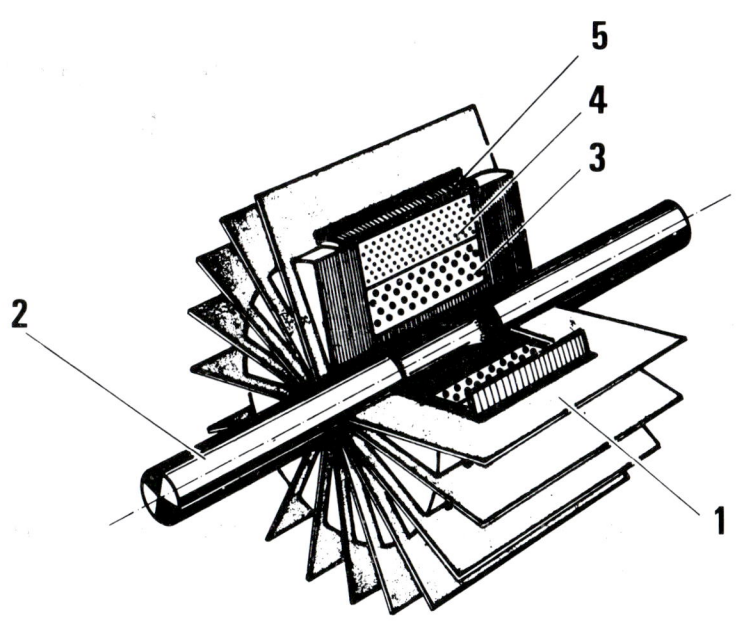

Fig 2 Circumferentially placed thin yokes around a test specimen

The key to a successful use of the unit is the combination of a *constant driving current* in the primary coil, a *constant length* of the magnetic circuit and magnetic *preneutralization* of the material, such that nonlinear effects and hysteresis, respectively, are reduced to a minimum. Each part of the specimen to be tested has to be brought to the *anhysteresis curve* of the material, Fig 3, in order to reflect an applied mechanical stress in a stable way. If some parts are not brought to the anhysteresis curve but fall on the hysteresis curve, the signal is very difficult to interpret, Fig 4.

3. LOW-CYCLE FATIGUE

Fig 5 shows a typical record from a test on a 10 mm diameter bar made of steel SIS 1650. The load P vs elongation δ response is given in Fig 5a whereas the change -Δφ in the electromagnetic flux vs δ is illustrated in Fig 5b. As will be shown subsequently, the change in the flux is a measure of the damage caused in the material.

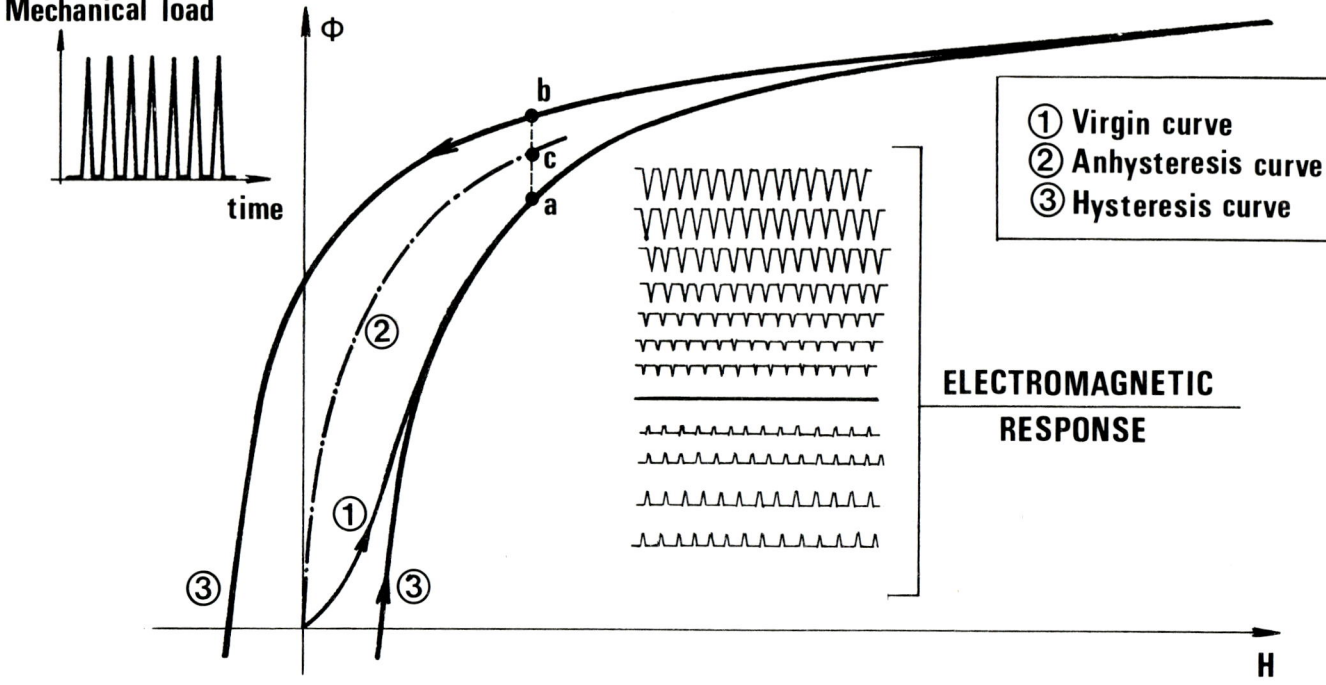

Fig 3 Example of registered electromagnetic curves and electro-magnetic response for a given mechanical load

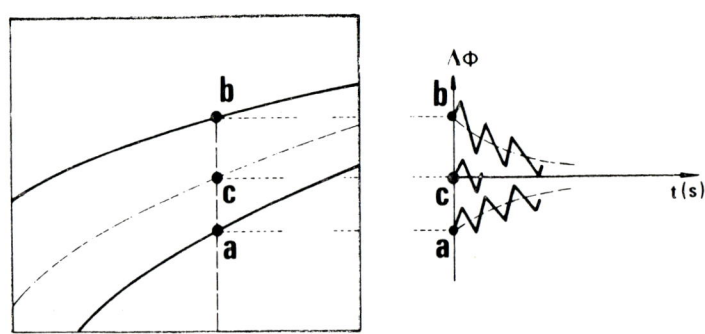

Fig 4 Variation of magnetic flux with time for an applied mechanical load of triangular shape. a and b represent points on the hysteresis curve and c is situated on the anhysteresis curve (see Fig 3).

A significant feature of the flux curve in Fig 5 is that it does not drop to zero as the load is removed. The residual change in flux reflects the fact that the *irreversible phenomena* associated with yielding in the material are registered by the detecting unit, or rather by the "magnetic vectors" of the atoms. The nonvanishing of $-\Delta\phi$ as the load is reduced to zero can thus be used as a measure of whether *yielding* has occurred in the body or not.

Some low-cycle fatigue tests were run at different load amplitudes, which resulted in different lifetimes as expressed by the number of cycles N to fracture. The electromagnetic detecting unit determines

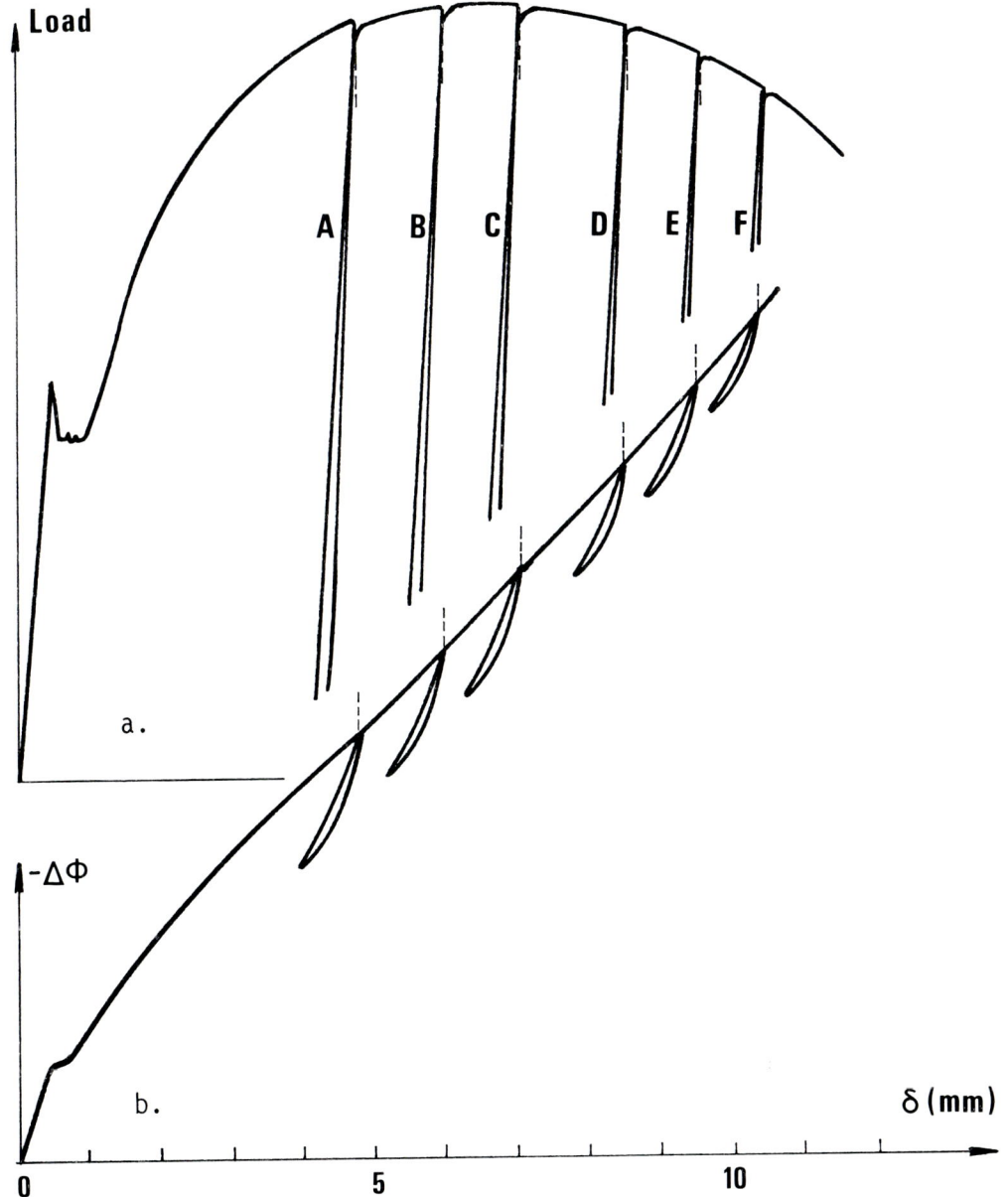

Fig 5   Tensile test with repeated unloading
(a) load P vs elongation δ (All lines A, B, C etc drop to zero load)
(b) change $-\Delta\phi$ in magnetic flux vs elongation δ

the *accumulated damage* to fracture in terms of the negative change $-\Delta\phi_F$ in the magnetic flux. This quantity, i e the total change $-\Delta\phi_F$ in the magnetic flux from zero load to fracture, was remarkably constant at these tests ranging from one quarter of a cycle (static tension) to 1500 cycles, see Fig 6.

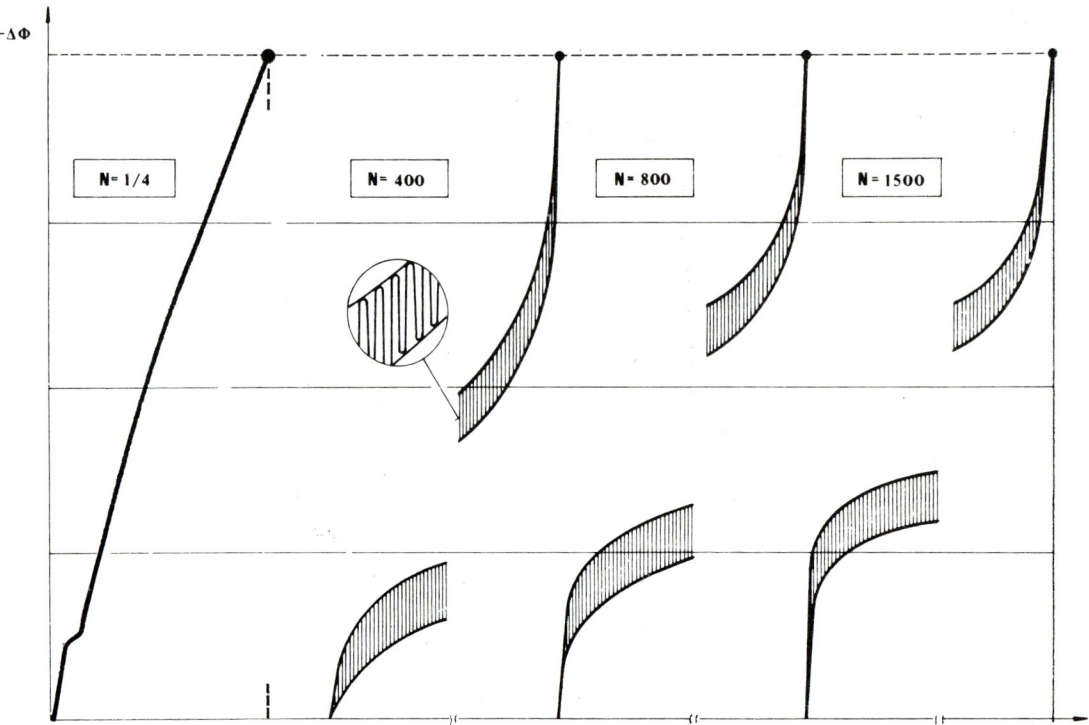

Fig 6 Electromagnetic response at low-cycle fatigue. Four different $-\Delta\phi$ vs time curves are shown corresponding to number of cycles to fracture ranging from 1/4 to 1500 with successively reduced mechanical load amplitudes

4. CHANGE IN MAGNETIC FLUX DUE TO MICRO-ROTATION

In standard continuum mechanics the macro-displacement u_i (i = 1, 2,3) defines the three translational kinematic degrees of freedom of a point, Fig 7a. If we instead look at an atom and treat it as a small elastic body, we need six degrees of freedom to describe its displacements assuming no constraints. These are the macro-displacements u_i as above and the *micro-rotations* ω_i, Fig 7b [5]. Now, we postulate that with each atom is associated a fictitious magnetic vector representing the spins and orbital movements of the electrons, Fig 8a.

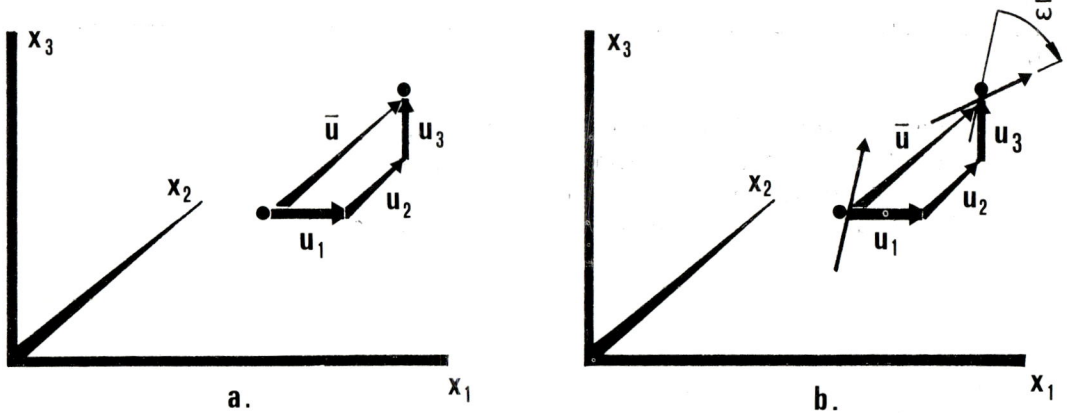

Fig 7 Macro-displacement and micro-rotation

By use of the equipment a magnetic field $\bar{H} = (0, 0, H_z)$ is driven through the specimen. When a mechanical stress field is applied to the specimen, Fig 8b, the direction of each magnetic vector is sligthly changed. This results in a change in the scalar product $\bar{H} \cdot \bar{B}$ (dimension Vs $\cdot$ A/m^3 or N/m^2), where $\bar{B} = \mu_0 (\bar{H} + \Sigma\bar{m})$ is the magnetic flux density.

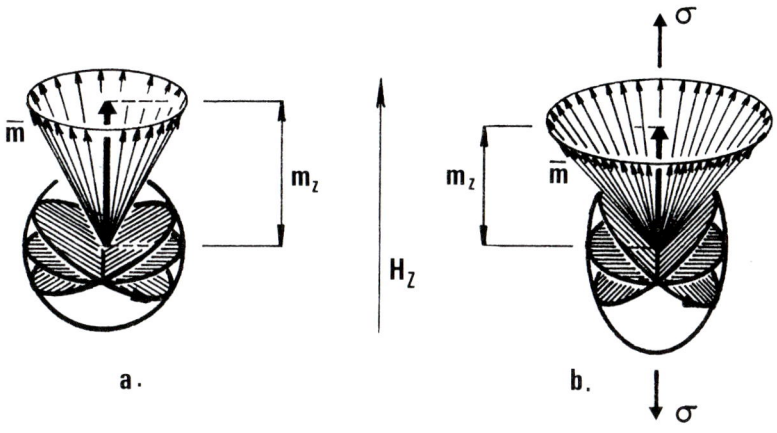

Fig 8 Bohr's model of the atom with magnetic vector $\bar{m}$
(a) inital state
(b) external mechanical stress applied

REFERENCES

[1] Sandor, B., Fundamentals of cyclic stress and strain, The University of Wisconsin Press, Madison, Wisconsin 1972

[2] Jilkén, L., Electromagnetic detection of mechanical phenomena, Linköping Institute of Technology, Department of Mechanical Engineering, Linköping 1978

[3] Jilkén, L. and Bäcklund, J., Electromagnetic measurement in solid mechanics, Proceedings, VIII Symposium Experimental Research in Mechanics of Solids, Warsaw, 4-6 Sept 1968

[4] Jilkén, L. and Bäcklund, J., Fracture toughness of the base material, the heat affected zone and the weldment of two ship hull plate steels, The Swedish Ship Research Foundation, Report No 5610:9, Göteborg 1977

[5] Berglund, K., Three-dimensional models of granular composites as polar continua, Royal Institute of Technology, Department of Mechanics, Stockholm 1977

LOW CYCLE BEHAVIOUR OF CASE HARDENED STEEL

L. Magnusson

Department of Mechanical Engineering, Linköping University,
S-581 83 Linköping, Sweden

ABSTRACT

Case-hardening is used to improve the fatigue resistance of steel components. The beneficial effects arise from that case and core have different chemical compositions which influence transformation kinetics leading to advantageous microstructures and residual stress pattern.

This work is an attempt to approach the case-hardened component in a systematic way. The case-hardened component can be regarded as a composite of high carbon and low carbon steels and these parts are studied separately in homogenous specimens. Besides the high strength, low ductility case material and the low strength, high ductility core material, two materials with intermediate carbon contents were chosen, representing the case-core transition region. All steels were made on SIS 2511 which were realloyed to three high carbon variants. Thus from four steels with carbon contents 0.2, 0.4, 0.6 and 0.8 w/o C specimens were made and heat treated to microstructures representative of carburized steel.

The specimens were subjected to axial strain controlled fatigue. Changes in the volume of retained austenite were measured on some specimens. The results are compared with an earlier work on stress controlled fatigue of case hardened specimens.

INTRODUCTION

Carburized steel can be regarded as a composite material with a hard case and a ductile core. The behaviour of this composite under static bending loads has been analysed very carefully from a mechanical point of view by Ebert et al [1]. Their analytical procedure is capable of determining the stress distribution in bars of composite materials in the elastic-plastic deformation regions. They found that the transformation of retained austenite to martensite in the case plays a major role in determining the stress distribution and that this transformation improve the overall performance. Landgraf [2] used strain-life curves for case and core material separately to predict that in fatigue of smooth specimens the cracks should start at the surface for short lives, while for long lives the cracks should start in the core material. This was in good agreement with observations.

In this investigation Ebert's and Landgraf's works are taken as starting point for a more close look to the fatigue properties of carburized steel in strain cycling. This paper present strain fatigue results on four model materials with a special interest on transformation of retained austenite. A comparison is made with earlier fatigue data.

EXPERIMENTAL DETAILS

The base steel in this investigation is the case hardening steel SIS 2511, with the analysis 0.16 % C, 0.70 % Cr, 0.91 % Ni, 0.89 % Mn, 0.04 % Mo and 0.25 % Si. Specimens from this steel carburized to a surface carbon content about 0.9 % and with an effective case depth of 1.0 mm were tested in axial load controlled fatigue.

Then four steels were made in a laboratory furnace, with the same alloy composition as SIS 2511 but with 0.18, 0.39, 0.59 and 0.74 % C. From these steels smooth specimens were prepared with a reduced cylindrical gage section 8.0 mm in diameter and 14.0 mm long. The specimens were heat treated to structures representative of carburized steel, i.e. for 0.74 % C martensite and retained austenite, 0.59 % C martensite, 0.39 % C martensite and bainite and 0.18 % C ferrite plus pearlite. The specimens were before fatigue testing grinded and electropolished in the gage section.

Strain controlled fatigue was carried out in a closed-loop, servo controlled system capable for accurate measurements of loads and strains. Volumes of retained austenite were measured by x-ray diffraction techniques.

RESULT AND DISCUSSION

Figure 1 summarizes the strain fatigue curves for the carbon contents that were examined. As expected the low carbon steels exhibit good low-cyclic fatigue resistance on basis of their greater ductility, while the high carbon steels are superior at long lives.

It can be noted that the 0.74 % C steel has slightly better strain resistance than the 0.59 % C steel at about 100 cycles. This is said to be due to strain-induced transformation of retained austenite into martensite in the 0.74 % C material which creates a compressive mean stress in cycling between fixed strain limits [3]. But at shorter lives, i.e. strain amplitudes > 1 % where the baised stress is even greater, the 0.74 % C - steel has very low strain resistance. Our measurements on retained austenite show that at $\Delta\varepsilon/2 = 1$ % the change in RA-content is small, from 19 v/o (as quenched and tempered) to ~15 v/o after 200 reversals. On the other hand at $\Delta\varepsilon/2 = 1.5$ % the volume of RA diminish from 19 v/o to between 5 and 10 v/o in one or very few cycles. That means that the retained austenite at the onset of plastic deformation and transformation provides a deformation mechanism for the otherwise brittle high carbon material. However, if plastic deformation cause most of the austenite to transform into martensite the ductility decreases dramatically leading to direct fracture.

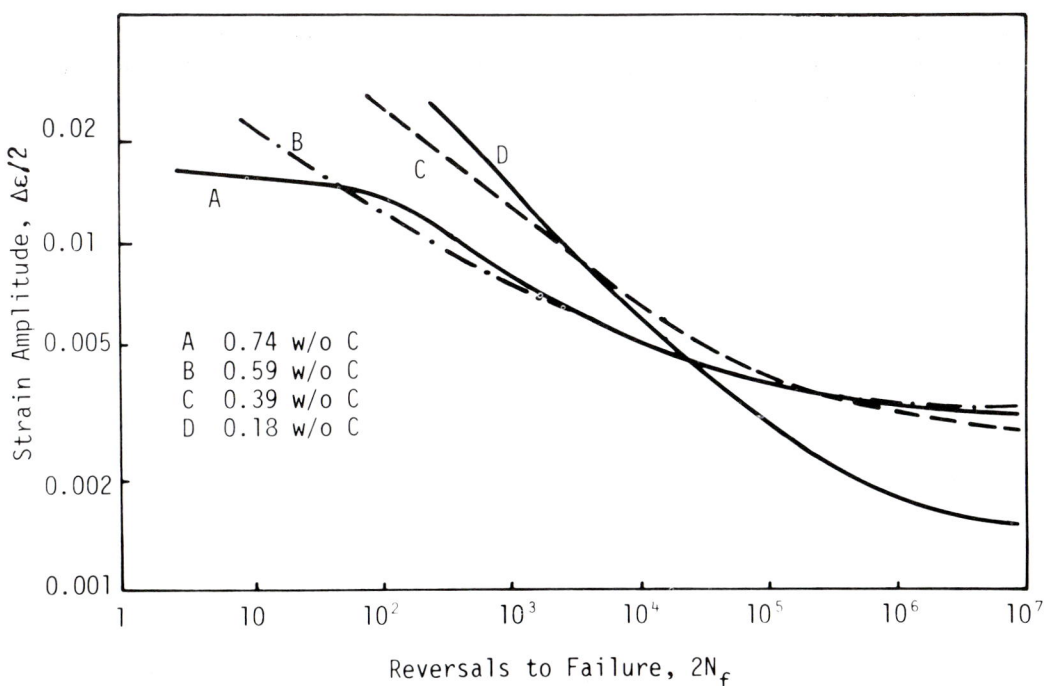

Fig 1. Strain-life curves for four steels (out of SIS 2511) with carbon contents according to figure. Each curve is based on 15 specimens tested at different strain amplitudes, to maximal 0.026.

Figure 2 shows typical stress-strain loops for the 0.74 % C-steel. The large plastic deformation in the first quarter cycle results in the compressive mean stress. After changes in the stress-strain response in the beginning the majority of life is spent under stable conditions. The material cyclically harden which is the case also for the two intermediate carbon content steels while the 0.18 % C-steel undergoes cyclically softening. In figure 3 monotonic and cyclic stress-strain curves for two steels are shown.

Tendencies on beneficial effects from retained austenite in carburized layers have been observed before in a work on stress controlled axial fatigue, figure 4. Smooth specimens from the base steel were given different amounts of retained austenite in case by slightly different carbon contents and quenching intensities. Still the surface hardnesses (and the core hardnesses, too) were the same. For all three series the fatigue cracks start in the core material for low loads, i.e. low strain amplitudes where the stronger case is more resistant to fatigue, figure 1. But at short lives, i.e. high strain amplitudes the more brittle case cracks first and the failure is of the usual surface-initiated sort.

The content of retained austenite in the case seems to influence on the transition point, see figure 4. The case with highest austenite content resists the highest stress amplitude before that surface initiation occurs. This fact must be closely related to the positive effects from the developed compressive mean stress in the strain controlled fatigue on the 0.74 % C-steel. Also in the surface of a carburized specimen, the transformation of austenite which is caused by cyclic deformation builds up compressive stresses and reduces the relaxation effect on the original residual compressive stress in the surface.

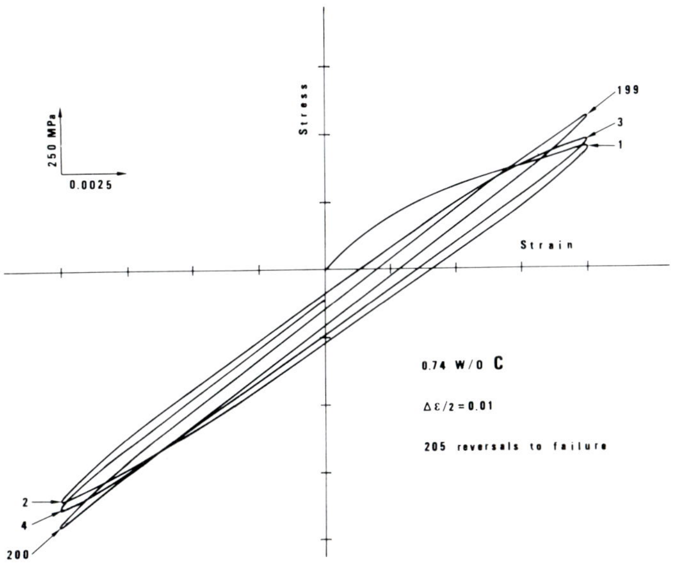

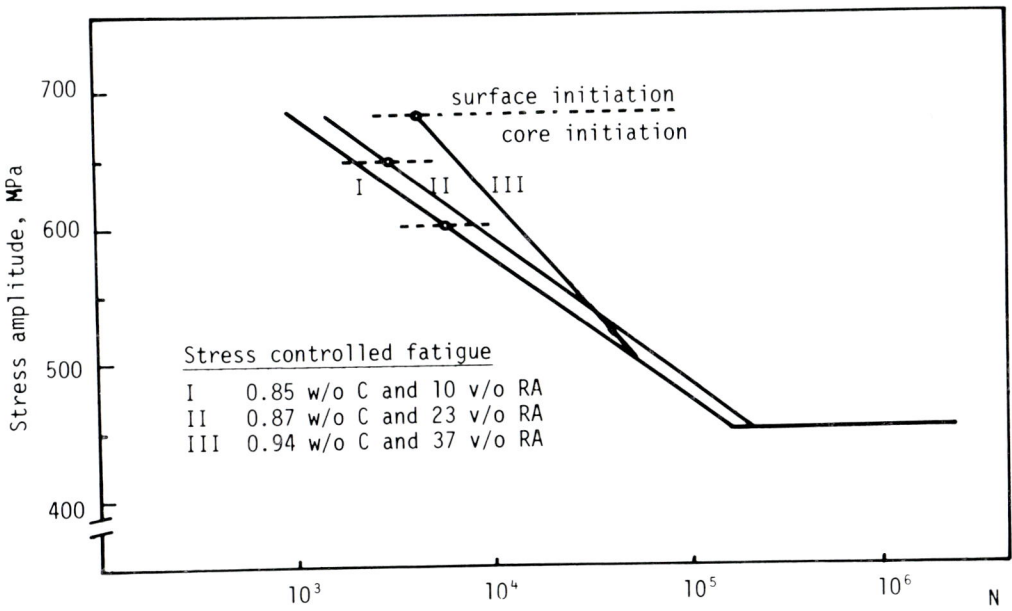

Fig 2. Stress-strain response of quenched and tempered 0.74 % C - steel. Loop tips are labelled with the reversal number.

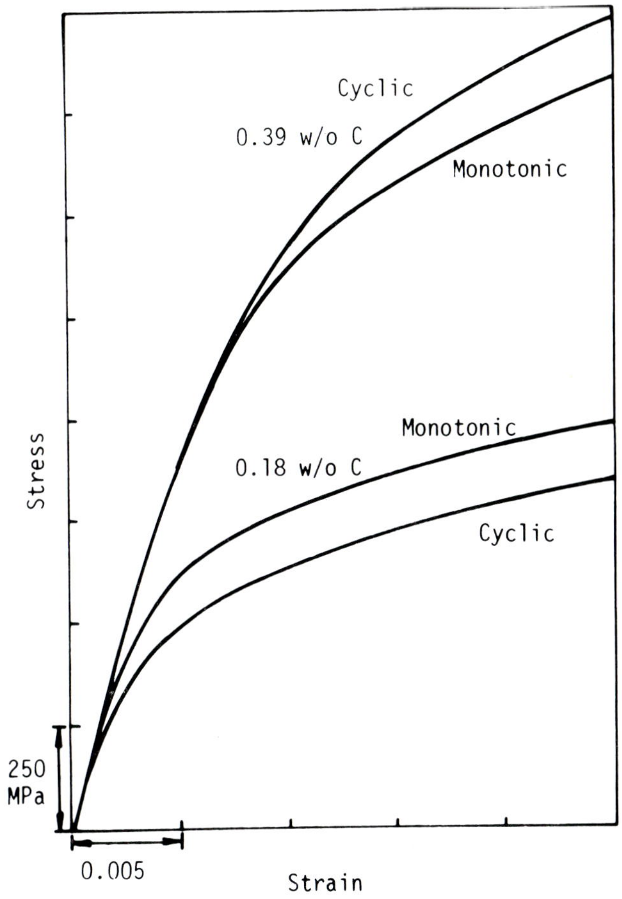

Fig 3. Monotonic and cyclic stress-strain curves for two carbon contents (composition SIS 2511).

Fig 4. S-N curves for case-hardened specimens of SIS 2511[4]. ∅15.0 mm. Surface carbon contents and volumes of retained austenite are shown in the figure. Point on each line marks the transition amplitude from core to surface crack initiation.

REFERENCES

1. McGuire, M.F., Troiano, A.R. and Ebert, L.J., Transactions, Series D, American Society of Mechanical Engineers, Vol. 93, 1971, p. 699.

2. Landgraf, R.W. and Richman, R.M., in "Fatigue of Composite Materials", ASTM STP 569, Am. Soc. Testing an Materials, Philadelphia, 1975.

3. Richman, R.M. and Landgraf, R.W., Metallurgical Transactions, Vol. 6A, No. 5, May 1975, pp. 955-964.

4. Magnusson, L., Utmattning av cythärdat material, LiTH-IKP-R-80, University of Linköping, 1976.

AUTOMATIC FATIGUE THRESHOLD VALUE TESTING

L. Jilkén[*], J. Bäcklund[*] and H. Knutsson[**]

[*]Department of Mechanical Engineering
[**]Department of Electrical Engineering,
Linköping Institute of Technology, S-581 83 Linköping, Sweden

ABSTRACT

Fracture mechanics concepts have recently come to an extensive use in the analysis of and design against fatigue. The stress intensity factor K_I plays an important role in the mathematical modelling of crack growth. It can also be used for predicting when an existing crack does not grow under dynamic loading. The threshold value ΔK_{th} of the stress intensity factor range ΔK_I is the limit, under which crack growth ceases. ΔK_{th} is a parameter which for most materials depends on the temperature and the stress ratio. It is of course of great practical significance when structures have to be designed in such a way that no cracks may propagate. However, the experimental procedure for determining ΔK_{th} is time consuming both with respect to manual effort and usage of testing machines. The main problem is to bring a propagating crack to stop propagating by reducing the load in a proper way. If the load is reduced too slowly, the testing will take too long time and will hence be too expensive, and if the load is reduced too quickly the result will be a false value of ΔK_{th}.

This paper describes a method for threshold value testing where the load is reduced automatically and where the load ratio $R = K_{Imin}/K_{Imax}$ is kept constant throughout the testing. The latter feature of constant R is essential since the threshold value is strongly dependent on R.

1. INTRODUCTION

The cause of most failures of structures subjected to variable loading is the initiation and propagation of cracks. It is often interesting, from the designer's point of view, to estimate the load amplitude that can be acting on a given component with existing voids or cracks without propagation of the cracks. Especially for parts subjected to a very large number of cycles or small components, the cracks should not be allowed to grow and hence the design value of the stress intensity factor range ΔK_I should be below the threshold value ΔK_{th}.

Test methods for determining threshold values have been developed and investigated by a large number of researchers during the last ten to fifteen years [1-4]. Of these, the method proposed by Jerram and Priddle [3] seems to be the most attractive one with reference to manual and testing machine effort. In their method the difficult business of following the crack tip during testing is avoided. The ticklish task of manually decreasing the load in steps, having registered crack growth before the load is reduced, is also circumvented, since the method is fully automatic. The only drawback of the method is that the load ratio $R = P_{min}/P_{max}$, where P is the load, is not held constant during the test. In the present work the method if Jerram and Priddle [3] has been improved in this respect, such that the value of R is held constant during the test. A special controlling unit was developed, which automatically reduces the load under constant value of R. This unit has been designed to fit an MTS servo-hydraulic testing machine, but it could be modified to suit other servo-hydraulic systems or electromagnetic vibrator machines such as the Amsler Vibrophore. Furthermore, analytical tools have been developed for estimation of the initial values P_{max0} and $P_{min0} = R \cdot P_{max0}$ of the load. These values should be chosen in such a way that the crack starts to propagate, i e they should produce a stress intensity range ΔK_I above the threshold value ΔK_{th}. On the other hand, the initial load and the accompanying clip gauge displacement CGD must not have such a large magnitude that the crack runs through the entire specimen without stopping.

2. AUTOMATIC LOAD-REDUCING SYSTEM

Fig 1 shows a picture of the experimental set-up with an 50 kN MTS testing machine (1), a box (2) containing the electronics that control the load and automatically reduces it and an oscilloscope (3) showing the load vs time variation at a certain time. The test specimen (4) is provided with a standard clip gauge.

The objective of the load level controller (2) in Fig 1 is to reduce the load continuously in such a way that the ratio $R = P_{min}/P_{max}$ is kept constant and so that the maximum and minimum clip gauge displacements are kept constant, Fig 2. The controlling system is described schematically in Fig 3 and it is directly connected to the MTS system without any modifications at all of this system.

Fig 1 Picture of testing machine (1), load level controller (2), oscilloscope (3) and test specimen (4) with clip gauge

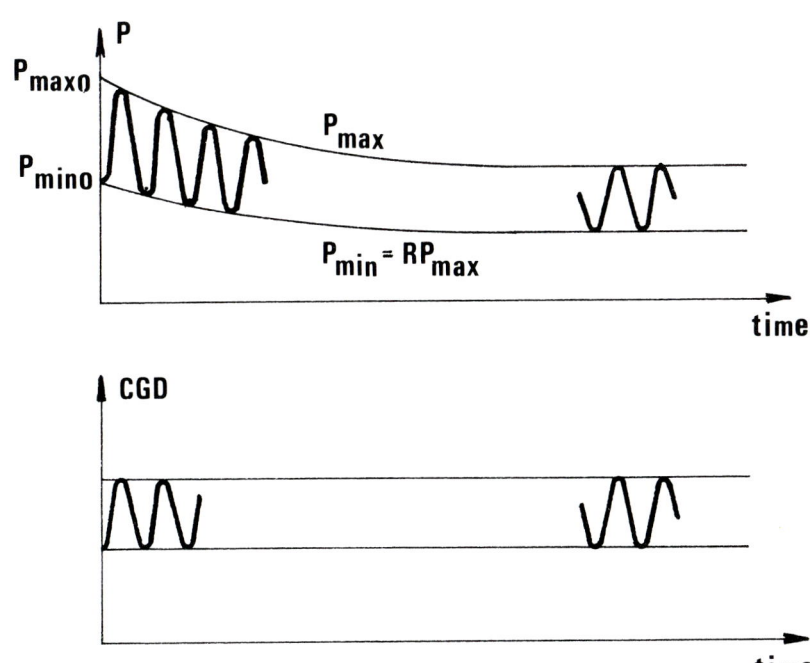

Fig 2 Load P vs time and clip gauge displacement CGD vs time. The curves are drawn for the value 0.5 of the load ratio R, but any value of R in the interval $-\infty < R < 1$ could be chosen

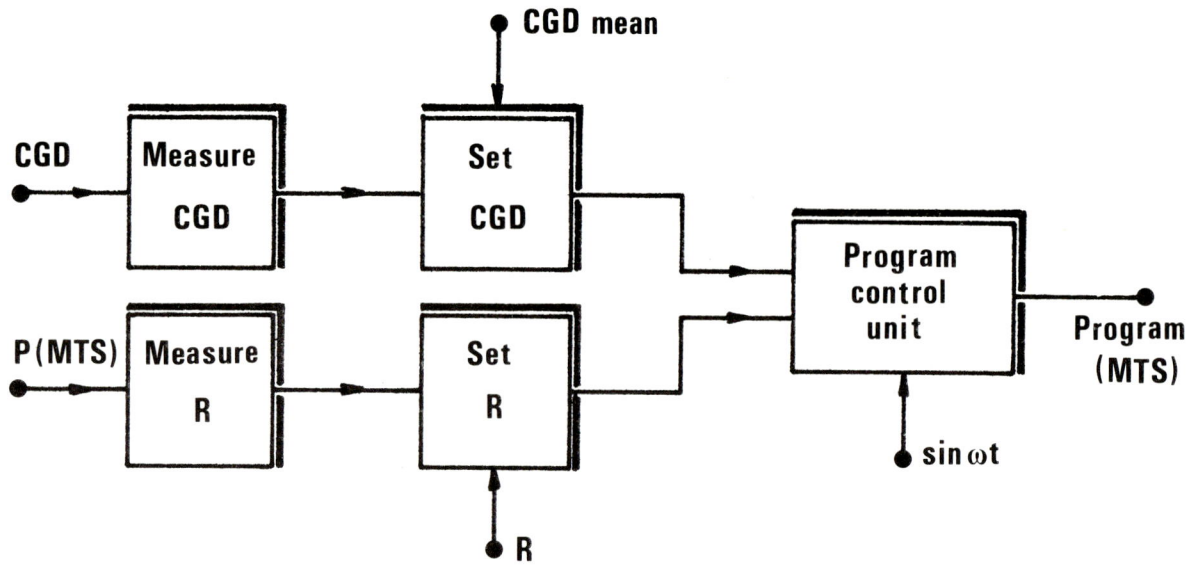

Fig 3   Schematic block diagram of the controlling system

3. COMPACT TENSION SPECIMENS

Compact tension specimens were used at the tests since they are easy to handle and require a minimum of material, Fig 4. Two most central parameters at the threshold value testing are the stress intensity factor K_I and the clip gauge displacement CGD, the latter being the widening of the distance c in Fig 4. Recent curve-fitting [5] to some numerical results gave the following two simple formulae for these parameters,

$$K_I = \frac{P(2+\alpha)}{BW^{1/2}(1-\alpha)^{3/2}} (1.562 - 0.328\alpha + 0.067\alpha^2) \tag{1}$$

$$CGD = \frac{P(1+\alpha)^2}{E'B(1-\alpha)^2} (9.25 - 4.28\alpha - 6.05\alpha^2 + 6.12\alpha^3) \tag{2}$$

where $\alpha = a/W$, $E' = E/(1-n^2)$, E = Young's modulus, n = 0 for plane stress (small thickness B) and $n = \nu$ for plane strain (large thickness B), ν = Poisson's ratio.

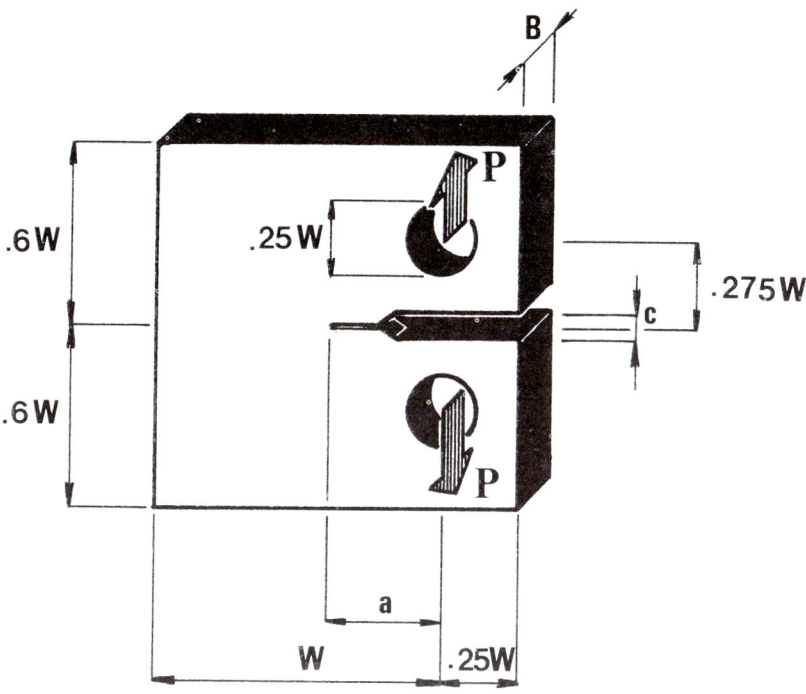

Fig 4   Compact tension specimen

4.   TESTING STRATEGY

The only crucial part of the testing is the determination of the initial values P_{max0} and $P_{min0} = R \cdot P_{max0}$ of the load and hence the values CGD_{max} and CGD_{min} of the clip gauge displacement, which are to be kept constant during the testing, Fig 2. If an approximate value ΔK_{th}^a of the threshold value ΔK_{th} is known or guessed, the corresponding load ratio ΔP_0^a and thus $P_{max0}^a = \Delta P_0^a/(1-R)$ can be calculated from equation (1) for $\alpha = a_0/W$. Here a_0 is the initial crack length of the specimen after fatigue precracking according to ASTM standards for LEFM testing. (It should be mentioned that the use of an eddy current probe greatly simplifies the precracking procedure [6]). The initial maximum load P_{max0} should then be higher than P_{max0}^a.

In order to determine an upper bound P_{max0}^b to the initial maximum load, we can proceed in the following way. Suppose that we do not want the crack to propagate more than to the length $a = 0.7 W$ in order to avoid fracture of the specimen, and that the approximate value ΔK_{th}^a should have been reached at this crack length. Inserting $\alpha = 0.7$ into equation (1) and setting ΔK_I equal to ΔK_{th}^a gives the maximum load range ΔP at $\alpha = 0.7$ as

$$\Delta P = 0.0242 \, B\sqrt{W} \, \Delta K_{th}^a \tag{3}$$

The corresponding maximum range ΔCGD of the clip gauge displacement is then calculated from equation (2)

$$\Delta CGD = \frac{173.1\ \Delta P}{E'B} = 4.2\sqrt{W}\ \Delta K_{th}^a\ \frac{1}{E'} \tag{4}$$

Since ΔCGD is constant throughout the testing, equation (4) yields the highest value of ΔCGD, and thus of $CGD_{max} = \Delta CGD/(1-R)$, which should be allowed if the crack is supposed to stop at $\alpha = 0.7$ or before. The corresponding maximum load ΔP_{max0}^b at the start of the test is then easily calculated from equation (2) using the CGD_{max} of above and setting $\alpha = a_0/W$, where a_0 is the initial crack length.

Having determined lower and upper bounds to the initial loads, it is a matter of experience to select the right values giving as short testing times as possible.

5. DISCUSSION

The present technique for stress intensity threshold value testing has some specific advantages over conventional methods

1. The crack length is measured only at the end of the test and not continuously during the test.

2. The load ratio R is kept constant during the test.

3. The load ratio R is preset to the value desired before the test starts.

4. Due to the automatization the testing needs less manual and testing machine effort.

5. The method is reliable. If the initial loads are chosen properly, the testing machine can be left unattended for any time.

6. Threshold value data for a number of load ratios R can be extracted from one specimen.

REFERENCES

[1] Frost, N. E. and Greenan, A. F., Cyclic stress required to propagate edge cracks in eight materials, J Mech Engng Sci 6 (1964) 203-210

[2] Paris, P. C, Bucci, R. J., Wessel, E. T., Clark, W. G. and Mager, T. R., Extensive study of low fatigue crack growth rates in A533 and A508 steels, ASTM STP 513 (1972) 141-176

[3] Jerram, K. and Priddle, E. K., System for determining the critical range of stress-intensity factor necessary for fatigue-crack propagation, J Mech Engng Sci 15 (1973) 271--273

[4] Ohta, A. and Sasaki, E., A method for determining the stress intensity threshold level for fatigue crack propagation, Engng Frac Mech 9 (1977) 655-662

[5] Bäcklund, J. and Mackerle, J, Stress intensity and crack opening of compact tension specimens, submitted to Engng Frac Mech for publication

[6] Jilkén, L. and Bäcklund, J., Fracture toughness of the base material, the heat affected zone and the weldment of two ship hull plate steels, The Swedish Ship Research Foundation, Report No. 5610:9, Göteborg 1976

PROCESSES OF QUARTZ FRACTURE IN NATURE AND THE FORMATION OF CLASTIC SEDIMENTS

I. J. Smalley[*], D. H. Krinsley[**], C. F. Moon[***] and S. P. Bentley[****]

[*]Department of Civil Engineering, University of Leeds,
Leeds LS2 9JT, England
[**]Department of Geology, Arizona State University, Tempe, Arizona 85281, U.S.A.
[***]Department of Civil Engineering, Sheffield Polytechnic,
Sheffield S1 1WB, England
[****]Department of Civil Engineering, University of Wales Institute of Science and Technology, Cardiff CF1 3NU, Wales

ABSTRACT

Quartz particles are released into the sedimentary system by the weathering and fracture of granitic rocks. These quartz particles contain micorfractures which influence their subsequent breakage. Few fractures means stronger particles which become sand - and then tend to resist fracture. Highly fractured particles break down to become silt and clay sized particles. Silt formation may be locally enhanced by glacial grinding and cold weathering during climatic cold periods - and increased amounts of clay sized material are also produced. Loess deposits tend to form as a result of Quaternary cold periods in which large amounts of source rock were fractured by cold weathering; loess particles are not formed by desert processes. It appears that very sensitive soils may owe their properties to a predominance of clay sized primary mineral particles (quartz making a considerable contribution) and that available processes of fracture are actually the critical controls which allow these quickclays and similar sediments to form. Fracture processes at very small sizes can produce plate shaped particles with virtually undeformed edges; some interesting cleavage processes appear to operate.

1. INTRODUCTION

The purpose of this paper is to discuss in general terms some fracture processes which are responsible for the introduction of clastic particles into the sedimentary system. This is a relatively new area of sedimentary geology; for example, the conversion of sand grains into sandstone has been fairly well studied but the processes by which the sand grains themselves are derived from their granitic source rock has not been thoroughly explored. This particular aspect of fracture studies had a purely academic aspect until recently when the problem of the possible breakup of buried nuclear waste came to be studied under the same heading - so that there is now an urgent, contemporary aspect to the problem. Not only the formation of sand sized particles is of interest; there is the continuing problem of how silt-sized particles are formed (what proportion directly from rock weathering and what proportion by sand grain breakage?); how the very fine primary mineral particles in sediments are formed and if unusual deformation mechanisms account for some of the observed shapes.

2. QUARTZ SAND

Sorby [1] made some very pertinent observations (in 1880) on quartz grains derived from granitic rocks but it is probably safe to say that the systematic study of sand began with Kuenen [2, 3]. He was the first to give serious consideration to the way in which quartz sand particles are formed, and he carried out a classic series of experiments on interparticle abrasion which establised the basis for sand particle studies [4, 5]. The detailed examination of actual quartz sand particles had to await the development of the scanning electron microscope, although Krinsley and Takahashi [6] actually initiated the new era with some transmission microscope studies.

Moss [7] stated that "beyond the fact that sand-sized quartz grains are derived somehow from the quartz of crystalline rocks the mode of origin of these particles, as individuals, seems to be very improperly understood". The initial detachment process by which quartz grains are separated from the igneous rocks produces sand sized particles but during their time as sedimentary particles the sand grains appear to become considerably reduced in size. Blatt [8] has proposed that "During the time between sedimentary input and equilibrium outgo, sedimentary processes have reduced mean quartz grain size from approximately 670 μm (0.67 mm; 0.6 ϕ) to 60 μm (0.06 mm; 4.0 ϕ), a reduction of 90%". The sedimentary processes which reduce the mean size from the sand range to the silt range as the 'Blatt interval' is crossed are discussed in the next section; here we consider the problem of the initial breakout of the sand sized particle.

The most comprehensive study of quartz sand grains was that carried out by Moss [7] and he found that certain observations of Sorby [1] were verified. The quartz grains of weathering granite, once freed from surrounding feldspar, usually fell apart into numbers of smaller particles ranging downwards in size to at least the limit of unaided vision, and a number had remarkably rounded surfaces.

Moss and Green [9] were later able to explain much of the fracture behaviour of recently formed quartz particles on the basis of microfractures developed by the quartz in the igneous rocks. They proposed a form of fracture due to "deformation sheeting" which allowed a range of particle sizes to form. Moss [7] had hinted at the possibility of the high-low transformation in quartz having some effect on quartz particle breakout and this idea was developed further in a speculative paper by Smalley [10]. He proposed that geological conditions allowed the quartz in granites to form as high quartz and that the transition to low quartz (requiring a decrease in volume) imposed local tensile stresses in the rock system which influenced the mechanism of break-out and the shape of the product (see also Krinsley and Smalley [11] 1972).

The critical discovery relating to the controls on quartz sand formation appears at the moment to be the observation of microfractures and defects in the quartz crystals. Moss and Green [9] and Riezebos and Van der Waals [12] proposed similar reasons for quartz breakage, and related their observations to the problem of silt formation. This means that the initial product from a granitic parent rock might be considered to consist of two distinct particle populations, (1) sand sized particles which contain no effective internal fractures or weaknesses and which are destined to remain as quartz sand grains, and (2) sand sized particles which contain varied concentrations of internal defects which cause them to quickly breakdown in the sedimentary system into smaller particles; in those with a high concentration of defects the breakdown occurring virtually simultaneously with release from the granitic matrix. The strong particles stay as sand unless some very powerful natural agency is encountered, and glacial grinding appears to be the only way in which the strong particles may be fractured.

3. COARSE SILT

On the MIT size scale coarse silt is defined as the range between 20 and 60 μm. This is the size range which predominates in loess - the most striking and surely the most discussed of the sedimentary silt deposits. In the major loess deposits most of the clastic particles are quartz and the problem of how these massive amounts of silt sized quartz particles were formed is still being discussed (see Smalley [13] for historical background). At the base of the current discussion is the suggestion by Smalley [14] that quartz silt particles were only formed in large amounts by glacial grinding and therefore major loess deposits should be associated with large scale glaciation.

It is now widely accepted (see Selby [15]) that for certain well defined deposits of glacial loess the 1966 [14] criteria do apply; but this leaves unexplained the formation processes for the silt particles comprising the deposits of "desert" loess. It is largely due to Obruchev [16, 17] that loess is divided between cold (glacial) loess and hot (desert) loess, and the nature of desert loess is still rather ill defined. Smalley and Vita-Finzi [18] proposed that since no particle forming mechanism operated in deserts which could produce particles in the required coarse silt size range then the existence of a true desert loess was doubtful. It is however true that loess deposits associated with deserts do exist, but is also apparent that there are large deserts (e.g. the Sahara and Australian deserts) which lack associated loess deposits. It appears now that the fracture processes necessary to produce the abundant silt for desert loess deposits do not occur actually in the deserts but usually in adjacent cold mountains.

Two distinct fracture processes were envisaged in the Smalley 1966 [14] model of loess particle formation: (1) the granitic source was fractured to release sand sized quartz particles, (2) and then the sand sized particles were themselves fractured to give the silt sized product necessary for loess formation. This model can now be seen to be too restricted and in need of modification. It describes quite well the early events in the formation of a straightforward deposit of glacial loess such as the great band of north European loess but it is totally inadequate for the central Asian loess. The north European glaciers were deforming old shield rocks and releasing quartz particles from them, some of these particles escaped further fracture and remained as sand; however when one was trapped in a crushing situation it was fractured completely and a silt sized mode produced. Postglacial events caused the effective separation of sand and silt fractions and the establishment of the north European loess band.

It is now appreciated that two distinct fracture processes are not required to produce silt sized quartz particles from the parent rock and this allows a much more general view of silt and loess formation to be formulated. If the parent rock can be caused to fracture then the product will be a range of variously sized quartz particles; because of the widespread microfracturing of the quartz in the igneous rock there appears to be no need to invoke a separate action to crush the sand size quartz. If the circumstances are right this sand crushing will occur, as in the case of the formation of the north European loess, but it is not a necessary requirement for silt formation.

4. CLAY-SIZE PARTICLES

Two fracture processes produce primary mineral particles in the 1-5 μm size range; glacial grinding certainly produces abundant material in this range and deposits are found in associated areas, and sand grain impact chipping in deserts produces chips of these dimensions. This latter aspect was discussed in the Saharan Dust workshop in Gothenburg in 1977 and it was pointed out that the contribution of Saharan dust to the total burden of tropospheric aerosols is about 60-200 million tons per year (see Morales [19]). One of the interesting things about particle production in deserts is that it is essentially restricted to the very fine sizes. Kuenen's studies of aeolian abrasion produced no quartz particles in the loess size range (Kuenen [5]).

Smalley [20] proposed that highly sensitive soils (the so-called quickclays) owed their peculiar properties to a compositional predominance of clay sized primary mineral particles - and a corresponding lack of actual clay mineral particles. This would account for the limited distribution of such materials since some powerful natural agency would be required to produce fractured particles of the required dimensions in sufficient amounts. It would mean that a mineral fracture process controlled the formation of quickclays rather than any post-depositional process. Attempts at quantitative clay mineral analysis have been made on a series of sensitive soils from Quebec (Smalley et al [21], Bentley & Smalley [22]) and it has been shown that in most cases the clay mineral content is relatively low, of the order of 10% or less. Determination of absolute clay mineral values is difficult but some comparative studies have been made of clay soils from St. Jean Vianney (the site of the great 1971 landslide) and Gloucester (an experimental site of the Division of Building Research). The Gloucester material with roughly twice as much clay mineral content as the St. Jean Vianney had a much lower sensitivity.

If fairly rapid cementation occurred after deposition then the primary
mineral model could exhibit the necessary open structure which
facilitates quickclay structural collapse. If the original short
range bond concept is extended to include the formation of relatively
brittle interparticle bonds, an apparently realistic picture of a
Canadian quickclay is produced (Bentley & Smalley [23]). Another
attempt (rather more speculative) has been made to produce a "picture"
of a quickclay which involves assuming that the structure is composed
predominantly of single clay-sized primary-mineral particles (Smalley
[24]). A random structure can be produced by a very simple Monte
Carlo placement method and this shows varieties of particle environments and, with some more assumptions, allows the compaction process
to be examined. The assumption is made that the primary mineral
particles tend to be blade shaped, i.e. flattish particles like the
classic clay mineral shape of kaolinite; this was proposed by Krinsley
and Smalley [25] but needs some further consideration.

McKee and Krinsley at Arizona State University are still actively
investigating the production of clay sized quartz by crushing, and
Bull at University of Wales in Swansea is currently also investigating
the nature and shape distribution of small fractured particles. The
Krinsley Smalley hypothesis was that fine quartz particles tended to
be flat because certain cleavage planes became available when the
particle size was sufficiently reduced which did not operate in
massive quartz. It is in effect a double hypothesis; it suggests that
flat particles are formed, and it proposes a cleavage process; both
aspects need further discussion. McKee and Krinsley have carried
out a series of experiments on Brazilian quartz and on carefully
selected aeolian quartz grains which were crushed by various methods
to produce fine grained quartz material. A general scanning electron
microscope (SEM) survey of the coarse crushed material showed that
the particle surfaces were covered by numerous smaller clay sized
quartz particles, as had been previously noted for natural samples.
The smaller particles exhibited flat smooth surfaces suggestive of
cleavage.

Coarse crushed particles were manually agitated in acetone to
disperse the clay sized material, which was collected on carbon
support films and examined by high resolution transmission electron
microscopy (HRTEM) and selected area electron diffraction (SAED).
The clay sized particles in some cases could be isolated by SAED in
order to demonstrate their crystallinity. The < 0.5 μm particles
showed little evidence of severe lattice disruption. Closer observation of the particle edges revealed small protruding platelets. It
was impossible to isolate such small areas by SAED; however it was
possible under proper conditions to image the crystal lattice repeats
("lattice images") of the thin platelets by HRTEM. The "lattice
images" obtained in several platelets showed fringes corresponding
to the 4.26A (100) and 3.43A (101) spacings.

In summary, these recent observations have shown that breaking by
crushing produces flat crystalline particles as small as 0.2-0.3 μm
which do not appear to have severely disrupted structures. Lattice
images were observed to within 10A of the particle edges prior to
beam damage. Cleavage probably extends to the finest dimensions
observed. These results appear to agree essentially with the High
voltage electron microscope (HVEM) observations of Smalley and Moon
[25] and Hammond et al [27]. It may be that the neat edges observed

in small broken quartz particles are affected by the Moss defect structure of the quartz crystals rather than a special cleavage process but the observations on Brazil quartz (which was NOT derived from granitic rocks) suggest that there is more yet to be learned about cleavage in quartz particles.

5. DISCUSSION

A sand grain which exists across the "Blatt Interval" is likely to continue existing. The sand sized particles form one mode of the overall distribution curve for quartz particle sediments. Glacial crushing will convert a sand grain to fragments but no other natural process is anywhere near as efficient. There is another mode in the silt range; Blatt placed the major mode for the whole distribution in the region of 60 μm, i.e. in the coarse silt range. The cracked quartz particles from the source rocks contribute much material for this mode, and glacial grinding adds a substantial amount. Silt forming processes were certainly very effective during Quaternary times when a lot of loess and alluvial deposits were formed. In quartz particle sediments this was largely due to direct glacial grinding and to the production of cracked quartz from granitic source rocks by other cold weathering processes. A third, but less noticeable mode develops in the distribution curve - at the very small end where quartz particles near the comminution limit congregate. One product of continental glaciation is a plentiful supply of very fine (< 5 μm) particles as a result of prolonged and efficient grinding processes (the actual processes have been discussed in some detail by Moon [28]). Where natural processes concentrate them they can make a major contribution to the properties of observed sediments. The clay-sized postglacial sediments tend to contain also a significant proportion of fine feldspar particles produced by the same grinding processes and relatively unweathered because of the short amount of time which has elapsed.

Figure 1 is an attempt to demonstrate and summarise some of the relationships discussed in outline in this paper. There are major geological controls, which relate essentially to a series of fracture processes, which control the production of an important range of clastic particles. If we understand how clastic particles were fractured and formed we have taken the first critical step to understanding how the observed deposits were formed, and how their sometimes extraordinary properties may be accounted for.

The final word should perhaps concern the problem of radioactive waste disposal which was touched upon in passing in the introduction. Studies of quartz particle sediments in the last ten years suggest that a moderately optimistic view of waste disposal in glass might be taken. If a major future problem is likely to the the weathering and breakup of the vitrified waste units then the discovery of just how much quartz particle production relies on inherent defects might suggest that if a defect free glass could be produced it would be likely to resist weathering. The results of Morris et al [29] from Harwell suggest that their glasses are relatively defect free. Expected shattering during trepanning operations did not occur and almost perfect cylindrical cores were obtained. However optimism should be restrained because the next paper in that particular issue of Nature showed that hydrothermal alteration of the glass could present a different and unacceptable set of problems [30].

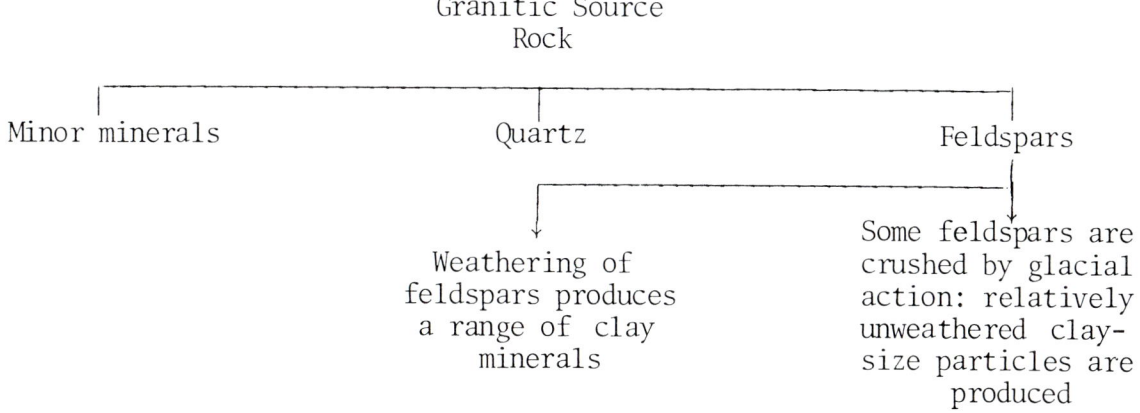

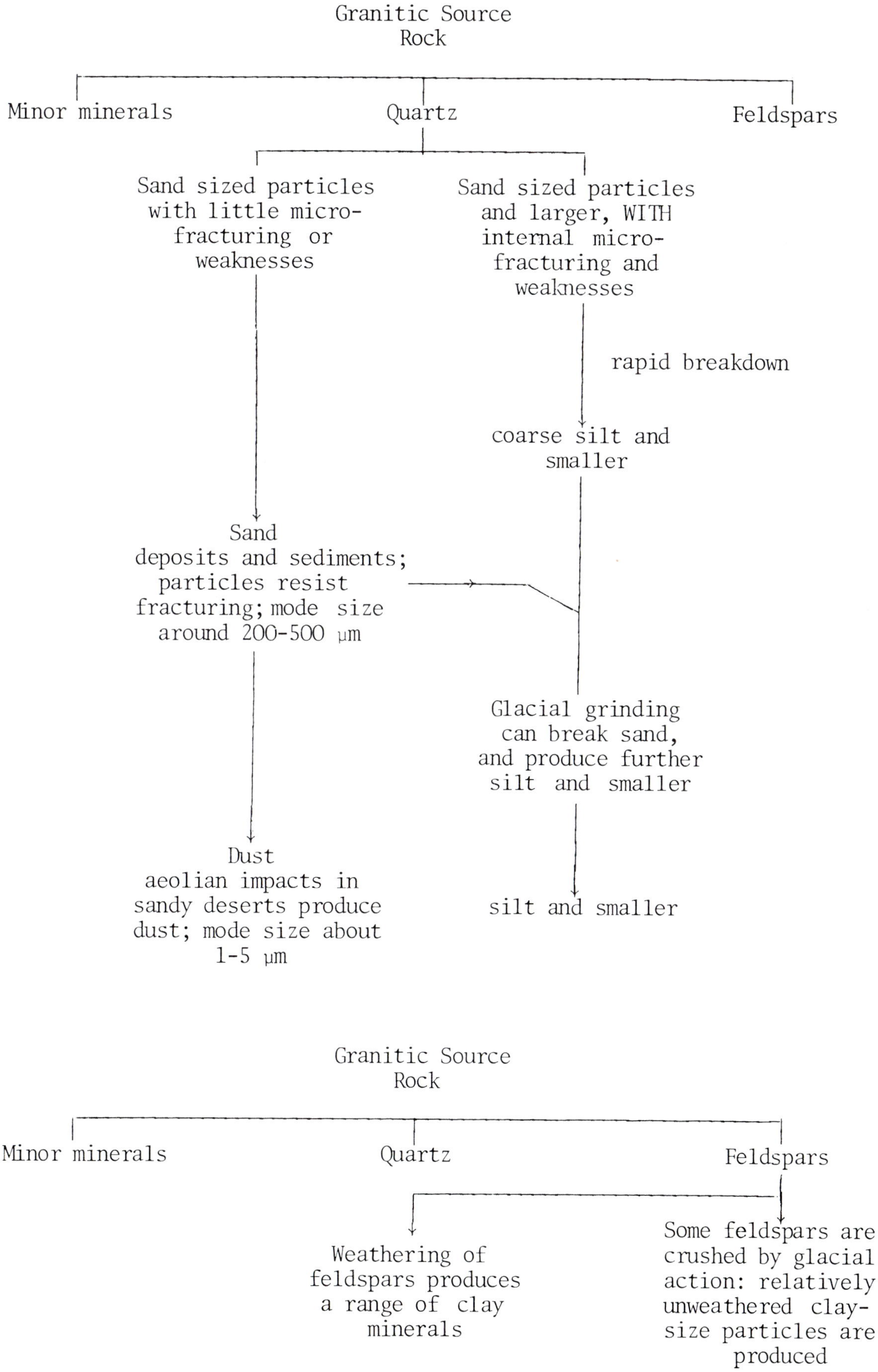

Figure 1: Some Simple Relationships for Clastic Particles Derived from Granitic Source Rocks

REFERENCES

[1] Sorby, H.C., On the structure and origin of non-calcareous stratified rocks. Qu. J. Geol. Soc. London 36(1880) 46-92.

[2] Kuenen, Ph.H., Sand - its origin, transportation, abrasion and accumulation. Trans. Proc. Geol. Soc. Sth. Africa 62 (annexure) (Alex. du Toit Memorial Lecture No. 6), 33 pp., 1959.

[3] Kuenen, Ph.H., Sand. Scientific American 202(4), 21-34, 1960.

[4] Kuenen, Ph.H., Experimental abrasion 3; Fluviatile action on sand. Amer. J. Science 257, 172-90, 1959.

[5] Kuenen, Ph.H., Experimental abrasion 4; Eolian action. J. Geology 68, 427-449, 1960.

[6] Krinsley, D.H. & Takahashi, T., The surface textures of sand grains, an application of electron microscopy. Science 135, 923-925, 1962.

[7] Moss, A.J., Origin, shaping and significance of quartz sand grains. J. Geol. Soc. Australia 13, 97-136, 1966.

[8] Blatt, H., Determination of mean sediment thickness in the crust: a sedimentologic method. Geol. Soc. Amer. Bull. 81, 255-262, 1970.

[9] Moss, A.J. & Green, P., Sand and silt grains: predetermination of their formation and properties by microfractures in quartz. J. Geol. Soc. Australia 22, 485-495, 1975.

[10] Smalley, I.J., Formation of quartz sand. Nature 211, 476-479, 1966.

[11] Krinsley, D.H. & Smalley, I.J., Sand. American Scientist 60, 286-291, 1972.

[12] Riezebos, P.A. & Van der Waals, L., Silt-sized quartz particles: a proposed source. Sediment. Geol. 12, 279-285, 1974.

[13] Smalley, I.J., Loess: lithology and genesis (Benchmark Papers in Geology 26). Dowden, Hutchinson & Ross, Stroudsburg Pa., 1975.

[14] Smalley, I.J., The properties of glacial loess and the formation of loess deposits. J. Sediment. Petrol. 36, 669-676, 1966.

[15] Selby, M.J., Loess. New Zealand J. Geography No. 61, 1-18, 1976.

[16] Obruchev, V.A., Loess types and their origin. Amer. J. Science 243, 256-262, 1945.

[17] Obruchev, V.A., Loess as a particular kind of soil, its genesis and problems to study (in Russian). Byull. Komiss. Chetvert. Perioda No. 12, 5-17, 1948.

[18] Smalley, I.J. & Vita-Finzi, C., The formation of fine particles in sandy deserts and the nature of 'desert' loess. J. Sediment. Petrol. 38, 766-774, 1968.

[19] Morales, C., Saharan dust; mobilisation, transport, deposition. Scope/Ecol. Res. Cttee. NFR Sweden, 24 pp., 1977.

[20] Smalley, I.J., Nature of quickclays. Nature 231, 310, 1971.

[21] Smalley, I.J., Bentley, S.P. & Moon, C.F., The St. Jean Vianney quickclay. Canadian Mineralogist 13, 364-369, 1975.

[22] Bentley, S.P. & Smalley, I.J., Mineralogy of sensitive clays from Quebec. Canadian Mineralogist 16, 103-112, 1978.

[23] Bentley, S.P. & Smalley, I.J., Interparticle cementation in Canadian post-glacial clays and the problem of high sensitivity (St > 50). Sedimentology 25, 297-302, 1978.

[24] Smalley, I.J., Mineralogy, interparticle forces and soil structure of the Leda/Champlain clays of eastern Canada. In Modification of soil structure, ed. W. W. Emerson, R. D. Bond & A. R. Dexter, John Wiley, Chichester 1978, pp. 59-67.

[25] Krinsley, D.H. & Smalley, I.J., Shape and nature of small sedimentary quartz particles. Science 180, 1277-1279, 1973.

[26] Smalley, I.J. & Moon, C.F., High voltage electron microscopy of fine quartz particles from Norwegian marine clay. Sedimentology 20, 317-322, 1973.

[27] Hammond, C., Moon, C.F. & Smalley, I.J., High voltage electron microscopy of quartz particles from post-glacial clay soils. J. Materials Sci. 8, 509-513, 1973.

[28] Moon, C.F., Particle nature in very sensitive soils and its relation to soil structure and geotechnical properties. Ph.D. thesis, University of Leeds, 1977.

[29] Morris, J.B., Boult, K.A., Dalton, J.T., Delve, M.H., Gayler, R., Herring, L., Hough, A. & Marples, J.A.C., Durability of vitrified highly active waste from nuclear reprocessing. Nature 273, 215-216, 1978.

[30] McCarthy, G.J., White, W.B., Roy, R., Scheetz, B.E., Komarneni, S., Smith, D.K. & Roy, D.M., Interactions between nuclear waste and surrounding rock. Nature 273, 216-217, 1978.

DEFORMATION AND SHEAR OF NORMALLY CONSOLIDATED FLOCCULATED KAOLIN

P. Smart and J. W. Dickson

Glasgow University, G12 8QQ, U.K. and SIAE, Bush Estate, Penicuik, EH26 0PH, U.K.

ABSTRACT

The material consisted of domains, i.e., groups of sub-parallel clay plates, which in general retained their identity. Pre-peak deformation involved negligible or slight changes of particle orientation, a small increase in the size of domains, a decrease in inter-domain void ratio, and negligible change in intra-domain void ratio. Post-peak deformation involved breakdown of domains in the failure 'plane'.

1. INTRODUCTION

The structural changes during shear of a normally-consolidated flocculated kaolin which are reported here form the fourth of a series of loosely inter-related experiments studying the structural changes wrought by mechanical disturbance to simple artificial soils. These experiments were undertaken with a view to enabling reasonable hypotheses to be made concerning the behaviour of real soils under various geological and geochemical conditions. Simple artificial soils were studied, because they facilitate control of the conditions of the experiment and eliminate natural variability. Smart (19) reviews work in this field; and it will be sufficient here to recapitulate the three experiments which directly preceeded the present one.

Smart (16) studied a dispersed kaolin. During consolidation the particles slipped to form an anisotropic structure with large domains and no inter-domain voids. During shear box tests at 0.02 mm/sec. under an overconsolidation ratio of 2, the domains increased in size. In slickensided failure zones, complete preferred orientation was seen. Foster and De (7) succeeded in photographing the whole of the width of a failure zone in an electron microscope. McConnachie (10) studied the consolidation of a flocculated kaolin. Ab initio, the particles were in small domains with large voids between them. Under low pressures the domains slipped into new positions forming a more anisotropic structure, with comparatively large voids between the domains. Under higher pressures, the domains moved closer together, reducing the size of the inter-domain voids, but themselves suffering a decrease of intra-domain void ratio as the domains were themselves deformed. The experiment reported here studied the same kaolin as McConnachie at stress levels corresponding to his high pressure range.

A second objective was to study the hypotheses about micro-mechanical theories of soil behaviour. Griffith theory has been used to predict the anisotropy of failure of slate (2). Calladine (3) postulated a model of soil in which crumbs came into contact through asperities on their surfaces. Deformation of the soil was accommodated by plastic deformation of the asperities. Skempton (14) used a model soil in which the particles, e.g., lead shot, were themselves plastic; this model extends at once to the concept of a soil composed of wholly plastic crumbs. Smart (17, 20) postulated a model based essentially on the statistical distribution of a stress transmitting elements within the soil; the material studied here had a different shape of stress-strain curve, and this model will not be discussed further here. Rate-process theory (e.g. 6) also seems inappropriate here.

2. MATERIALS AND METHODS

The kaolin used was similar to that of McConnachie. It was SPS-kaolin, which was supplied as dry powder. The properties are: 93% Kaolinite; 80% finer than 2 μm e.s.d.: c.e.c. 3.5 me/100g; Liquid Limit 67%, Plastic Limit 35%. Samples were remoulded at a Liquidity Index of 5.16 under vacuum in a Z-bladed mixer, consolidated uniaxially to 10 kN/sq m, trimmed, placed in triaxial shear test apparatus, consolidated triaxially to 827 kN/sq m, and sheared at 0.003048 mm/min under drained conditions to the requisite strains. In order to keep the lateral effective stress as high as possible, the samples were tested without backpressure. After shearing, the samples were cut axially into halves using a cheese wire; if a failure plane or other asymmetrical deformation was present the halves were cut to bisect this deformation symmetrically. For optical microscopy, half-samples were impregnated with Vestopal W or Carbowax and thin sections were ground parallel to and about 2 mm from the plane of bisection. For electron microscopy, from each sample examined, four Vestopal W impregnated sub-samples were prepared for ultra-microtomy such that ultra-thin sections were cut from vertical faces located within a few millimetres of the centre of the sample. Two of these faces were parallel to the plane of bisection, and two were perpendicular to this plane. Two batches of sections were cut from each face using knife strokes in the original horizontal and vertical directions. Other half-samples were used for moisture content determination.

3. TRIAXIAL TEST RESULTS

Fig. 1 shows the triaxial test results for two samples. There is good agreement between the tests up to peak stress. Although the observations for pre-peak behaviour in Fig. 1 were best fitted by an hyperbolic curve, the following:-

$$\sigma_1 - \sigma_3 = 262 \sqrt{\epsilon_1} - 212 \tag{1}$$

is shown because some of the other results have significant regressions against $\sqrt{\epsilon_1}$. Post-peak there was a large decrease of stress; this is contrary to the view which is sometimes held that peak and ultimate strength of normally-consolidated soils are equal.

4. OPTICAL MICROPHOTOMETRIC RESULTS

Figs. 2 and 3 show the optical microphotometric results. Each point represents the resultant of up to one hundred observations on a thin section from a triaxial sample. Each observation consisted of measurements of Anistropy Index and direction of preferred orientation (5,15). In order to assess the overall orientation with the samples the vector sum was found. The values of the resultant inclination, θ, and the Overall Anisotropy Index, A, are plotted in Fig. 2. In order to assess the strength of the orientation on a small scale, the arithmetical average $\bar{A}$ was found and is plotted in Fig. 3.

Andrawes et al (1), who made similar measurements on kaolin and illite samples for stress-controlled plane-strain tests, reported a decrease of anisotropy up to peak stress, which seems to have occurred at approximately 8% strain in their samples. This is an attractive idea, as it would suggest that particles rotate before the peak stress is reached, until, at peak stress, sufficient particles have already rotated to permit a failure plane to form. The results presented here may perhaps show a slight decrease of anistropy at about the same strain, 8%; but if so there is a subsequent increase of anisotropy. This, and the large and rapid drop of stress between peak at 24% strain and "final" at 26% strain (Sample 28, Fig. 1), are not in accordance with the idea that rotation of particles pre-peak prepares the samples for failure at peak; and it is more likely that there is no substantial change of anisotropy of this nature in the samples studied here throughout the whole shearing process. Direct examination of the thin sections suggested that the failure planes cut obliquely through relatively strong and uniform horizontal preferred orientation; which suggests that there was no significant rotation of particles pre-peak. If this be so, then pre-peak deformation and post-peak failure would appear to be controlled by different mechanisms.

5. ELECTRON MICROSCOPY

Examination of ultra-thin sections in the electronmicroscope showed that the structure of the soil consisted of domains, with small intra-domain voids within them, and larger inter-domain voids between them (5, 10). As indicated above, one sample was examined per strain, four subsamples per sample, and two sections per subsample. The measurements are based on one micrograph per section. Following McConnachie (9,10), the length, breadth, and orientation of the domains were mapped on the micrographs and measured; however, orientation was defined by the orientation of the particles, and length and breadth were measured parallel and perpendicular to this direction. Shape = breadth/length was calculated for each domain. These results are shown in Fig. 4. Each point is the average for one micrograph; log normal distributions were assumed for length and breadth, and normal for shape.

The following remarks apply to pre-peak conditions, i.e., to the first four samples. During deformation, the average length of the domains increased from 1.1 to 1.6 μm, the average breadth of the domains increased from 0.5 to 0.75 μm, and the average shape increased from 0.47 to 0.53. The curves shown are regressions against the square root of axial strain; these regressions are significant at the 1% level for length and breadth and at 5% for shape. These results seem to suggest that the domains retain their identity during pre-peak deformation of the sample except that some domains amalgamate by being pressed together.

Fig. 4 also shows the results for sample 41 strained to 24% axial strain, i.e., beyond peak stress. The average values for this sample have all fallen to levels comparable to the unstrained samples. However, these results were reobservations of results from sample 18, which had been strained to 26% strain, but which had suffered a small reduction of σ_3 towards the end of the test. The average results for sample 18 were length = 1.35 μm, breadth = 0.63 μm, shape = 0.51. Sample 41 was examined some time after the other samples used in preparing Fig. 4, so some caution must be used in making a direct comparison. However, considering samples 18 and 41 together, it seems that the post-peak reduction of stress is associated with a reduction in the length, breadth, and shape of the domains, which presumably result from minute local deformations similar to those occuring within the failure 'plane'.

The interdomain porosity was also measured from the micrographs which had been mapped. Interdomain porosity = volume of inter-domain voids divided by total volume. The results are in Fig. 5. Fig. 5 also shows the total void ratio, which was obtained from the moisture content measured at the end of the test from those portions of the samples which were not required for impregnation after corrections for the variability of moisture content measured throughout duplicate samples. Inter- and intra-domain void ratios were calculated from inter-domain porosity and total void ratio. The results from sample 41 have been included in the regressions. The curve for total void ratio is significant at the 5% level and that for inter-domain void ratio is significant at the 1% level. Since intra-domain void ratio is the difference between the other two curves, it seems that there is no significant change in intra-domain void ratio during shearing. The curve for inter-domain porosity was calculated from the other curves; inter-domain porosity decreased during shearing.

Comparison of Eqn 1, with the regressions shown in Figs. 4 and 5 suggests that all the quantities involved in these Figures would increase approximately linearly with deviator stress. Fig. 8 shows domain fraction plotted against deviator stress; domain fraction = volume of domains divided by total volume. The regression shown is significant at the 1% level.

The electron micrographs were examined to see whether the crumbs, i.e., domains, came into contact through asperities on their surfaces. The domains, the particles, and the inter-domain voids, were all comparable in size, and contiguous domains abutted over most of their length. There were no asperities of a sort which could be described as small in relation to the domains and large in relation to the particles. It was impossible to see whether there were asperities which were small in relation to the particles, i.e., which were small bumps on the faces of the particles; however, if such asperities existed between domains, they would presumably exist also between particles within domains. Therefore, some slight modification of Calladine's microstructural model seems to be required for this particular soil.

The increase in domain fraction and the decrease in intra-domain void ratio suggest that the domains are rearranged and pressed closer together by shearing. The observations of length and breadth, suggest some amalgamation of domains pre-peak and some breakdown of domains post-peak; however, the constancy of intra-domain void ratio and the smallness of the change of shape suggest either that the domains themselves are not much deformed plastically or that some compress whilst others dilate.

6. DISCUSSION

The structural changes during pre-peak deformation of vertical normally-consolidated flocculated kaolin during slow drained triaxial tests have been examined by both optical and electron microscopy. The observations are as follows.

1. Microphotometer results suggest that apart from within the failure 'plane' itself changes of particle orientation were negligible or slight.

2. Electron microscopy showed that the particles were arranged face-face in small domains, with small voids between the domains, and smaller voids within them. The lengths of the contacts between domains were comparable to the sizes of the domains themselves; no asperities could be identified.

3. Measurements of domains showed a definite small increase in the size of domains as deformation proceeded.

4. Further measurements from the micrographs showed a decrease in inter-domain void ratio.

5. Intra-domain void ratio remained unchanged.

6. After the peak stress had been reached, the deviator stress fell sharply.

The conclusions are:

1. The negligible changes in particle orientation pre-peak and the sharp drop in deviator stress post-peak, suggest that pre-peak deformation and post-peak failure are controlled by different mechanisms.

2. The electron-microscope observations suggest that

    (a) most of the domains retain their identity;
    (b) pre-peak, some domains amalgamate by being pressed together; post-peak, some domains are broken;
    (c) there is no plastic deformation of asperities, although there might have been some plastc deformation of whole domains, and
    (d) some domains may slip without significant rotation.

The observations cannot possibly apply to either the undrained normally consolidated case (void ratio constant) or to the drained over-consolidated case (void ratio increases); and further observations are planned for these two cases.

7. REFERENCES

1. K.Z. Andrawes & D.N. Krishnamurthy & L. Barden; Fabric changes during deformation of orientated clays; Preprint, First Baltic Conf. Soil Mech. Found. Eng.; 1975.
2. P.B. Attewell & M.R. Sandford; Intrinsic shear strength of a brittle anisotropic rock - Parts I, II, III; Int. J. Rock Mech. Min. Sci. & Geomech. Abstr. 11. 1974.
3. C.R. Calladine; A Microstructural view of the mechanical properties of saturated clay; Geotechnique 21, 391; 1971.
4. P.K. De; Kaolin microstructure after consolidation and direct shear; City Univ. Ph.D. Thesis 1970.
5. J.W. Dickson & P. Smart; Some interactions between stress and microstructure of kaolin; 53-57 in W.W. Emerson et al. eds.; Modification of soil structure; Wiley; 1978.
6. R.H. Foster; Behaviour of kaolin fabric under shear loading at low stress; p.81 in P.H.G. Parry ed.; stress-strain behaviour of soils; Henley-on-Thames (Foulis); 1972 for 1971.
7. R.H. Foster & P.K. De; Optical and electron microscopic investigations of shear induced structures in lightly consolidated (soft) and heavily consolidated (hard) kaolinite; Clays and Clay Minerals 19.31; 1971.
8. Gothenburg; Proc. Int. Symp. Soil Structure, Gothenburg; Stockholm (Swedish Geotechnical Society); 1973.
9. I. McConnachie; Electron microscopy of the consolidation of a kaolin; Glasgow Ph.D. Thesis; 1971.
10. I. McConnachie; Fabric changes in consolidated kaolin; Geotechnique 24.207; 1974.
11. R. Pusch; Quick clay microstructure; Eng. Geol. 3.433; 1966.
12. R. Pusch; Microstructural changes in soft quick clay at failure; Canadian Geotechnical J. 7.1; 1970.
13. G.R. Sides; Pore pressure and volume change characteristics of compacted clay; Salford Ph.D. Thesis, 1968.
14. A.W. Skempton; Effective stress in soils, concrete and rocks; p.4. in Pore pressure and suction in soils; London (Butterworths); 1960.
15. P. Smart; Optical microscopy and soil structure; Nature 210.1400; 1966.
16. P. Smart; Particle arrangements in kaolin; Clays and Clay Minerals 15.241; 1966.
17. P. Smart; Discussion; A microstructural view of the mechanical properties of clay; Geotechnique 22.368;1972.
18. P. Smart; Statistics of soil structure in electron microscopy; p. 69 in Proc. Int. Sym. Soil Structure, Gothenburg; Stockholm (Swedish Geotechnical Soc.); 1973.
19. P. Smart; Soil microstructure; Soil Science 119.385; 1975.
20. P. Smart; Exponential stress-strain curves; Civil Eng. & Pub. Works Rev. p. 39, Aug. 1975.

3. CLAY MICROSTRUCTURE

The microstructure of undisturbed illitic clays is characterized by aggregation, which is especially strong in marine sediments. Thus, the majority of the particles are collected in fairly dense aggregates connected by links or bridges of small particles. In principle, the aggregates represent strong structural elements while the particle links represent the weakest elements.

High voltage electron microscopy of dilute as well as concentrated clay/water gels [6,7] has proved that clay particle aggregates exist as stable units, surrounded by free pore water (Fig. 3). This means that external mechanical forces, such as the overburden pressure in nature, are not required to resist disintegration. Cohesion must therefore be responsible for this stability.

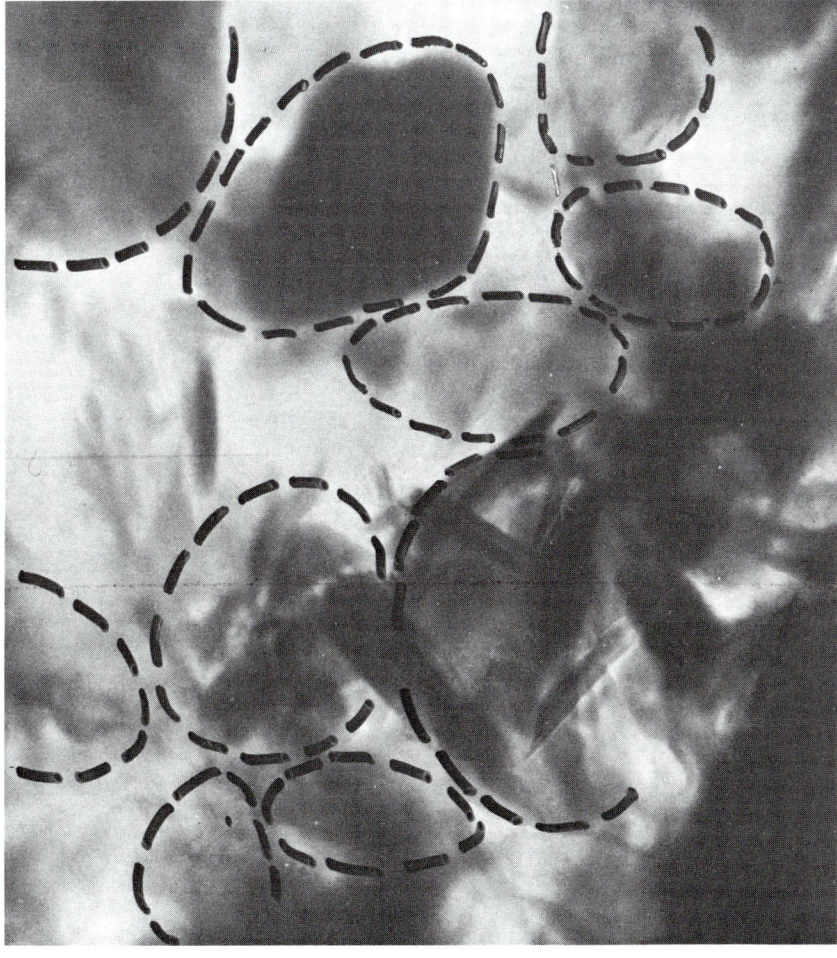

Fig. 3 Electron micrograph of a 1-10 μm microtome-cut section of moist illitic clay with aggregates marked by broken lines. Specimen placed in closed cell in the radiation path of a 1.5 MV microscope [7]. Notice particle links connecting the aggregates.

When a sufficiently high deviator stress is applied, a number of low energy barriers, represented by the links, are overcome and plastic bulk strain is produced. The process involves a break-down of a certain number of particle links in which the members are displaced and oriented to form local groups of parallel particles - "domains". The forced orientation induces repulsive forces between equally charged basal planes of the particles and this produces an increased particle distance in connection with a redistribution of water from adjacent voids into inter-particle space.

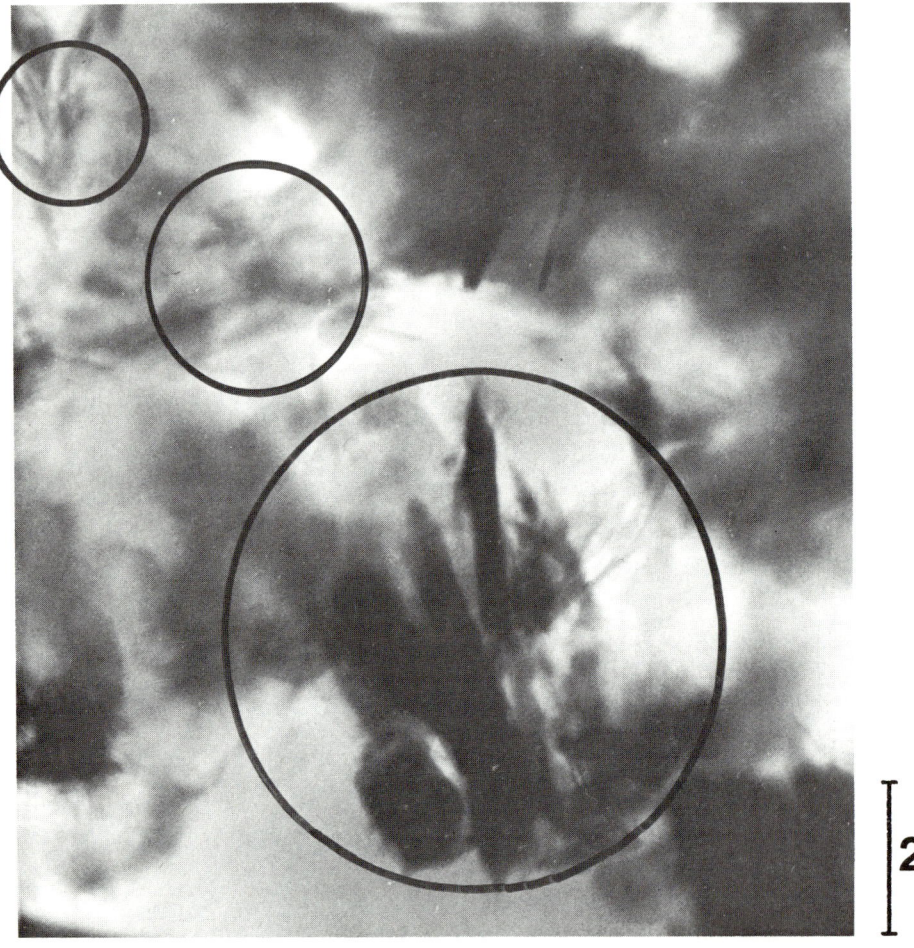

Fig. 4 Electron micrograph of moist illitic clay showing local groups of oriented particles - domains (encircled) - between intact strong aggregates with edge-to-edge and edge-to-face particle associations [7].

The bulk shear strength then becomes very low, especially at low effective pressures at which microdilatancy hardly contributes to the shear resistance.

The integrated strength of all intact links gives the major contribution to the shear resistance of soft clays at small shear strains. The shear strength may therefore be termed cohesion provided that the physical definition of this property is used.

4. THIXOTROPIC STRENGTH REGAIN

A complete remoulding of a soft clay reduces its shear strength to a fraction of the strength in the undisturbed state. This reduction can be tremendous for quick clays meaning that their strength can be reduced to 1/100 or 1/1000 by the mechanical disturbance. If the remoulded specimen is allowed to rest at a constant water content (constant volume conditions) there is a time-dependent strength recovery. This thixotropic process is intimately connected with a microstructural rearrangement of clay particles and a delayed reordering of the water phase. The structural change is illustrated by Fig. 5.

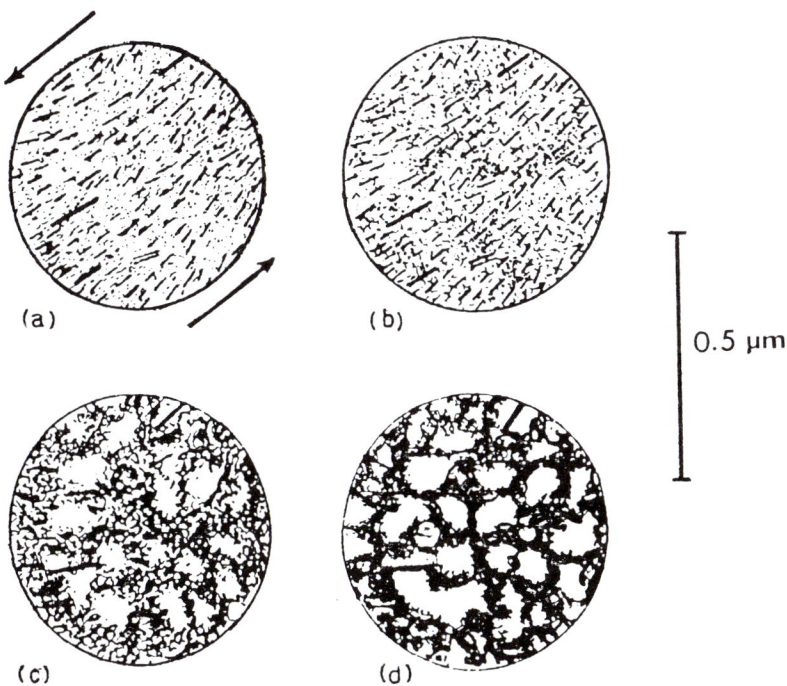

Fig. 5 Hypothetical picture of structural changes in clay resulting from thixotropy [8]. a) State immediately after remoulding, b) and c) stages in course of thixotropic strength regain and d) final state.

A comprehensive study of two Swedish quick clays [9] showed that freshly remoulded specimens contained a large number of dispersed, very small particles which had previously been contained in the aggregates. A large number of particle aggregates and links had been reformed already 30 days after the remoulding. After 4 months the pattern was further changed, as indicated by Fig. 5, by the formation of additional links but the density of the aggregates and links in this state was still lower than in the original, undisturbed state. The main geotechnical properties of the clays are shown in Table 1.

The thixotropic strength regain was determined by means of the Swedish cone penetration test using samples which had been stored in sealed boxes of stain-less steel at +4°C (Fig. 6). The boxes were filled completely with clay which means that the recovery took place under constant volume conditions at constant water content.

Table 1. Geotechnical properties.

Clay	ρ t/m³	τ_{fu}[1] kPa	S_t	ω %	ω_P %	ω_L %	Description
Lilla Edet	1.73	14	81	48	18	31	Grey, clayey silt
Rollsbo	1.71	24	106	58	25	46	Grey clay

[1] Undrained shear strength

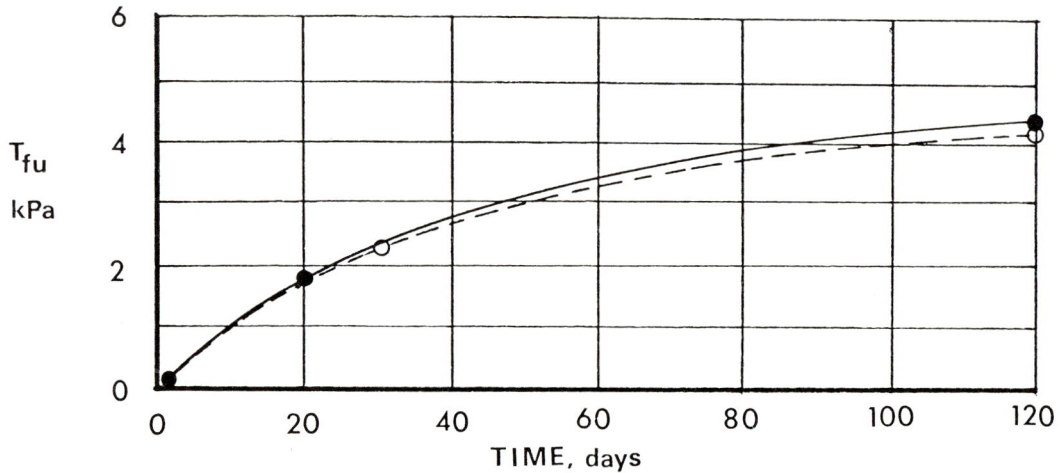

Fig. 6 Undrained shear strength versus time. ● Lilla Edet, ○ Rollsbo.

It is quite obvious that the considerable shear strength increase was primarily caused by the successive formation of a continuous network of aggregates linked together by groups of small particles. The formation of such a particle network through a phase separation process requires the operation of cohesive forces as in the formation of natural clays.

The process involves the approach of particles with adsorbed water envelopes and a final establishment of equilibrium positions with inter-particle distances determined mainly by the mineral lattice properties, and the concentration and kind of electrolytes. This is schematically shown in Fig. 8 where the inter-particle medium, following [8], is assumed to be a water lattice only.

If the particles are not in direct mineral-to-mineral contact the inter-particle medium must exert a disjoining pressure on the minerals which is possible only if it has a strength, which must then be of a cohesive nature. The medium consists of a water lattice with a degree of molecular order which depends on geometry and on the content of electrolytes as well as on the surface properties of the mineral particles. This suggests a low order water lattice character.

Since the displacement of adjacent particles caused by local deviator stress fields takes place in this medium it is therefore reasonable to believe that the shear resistance is of a viscous type.

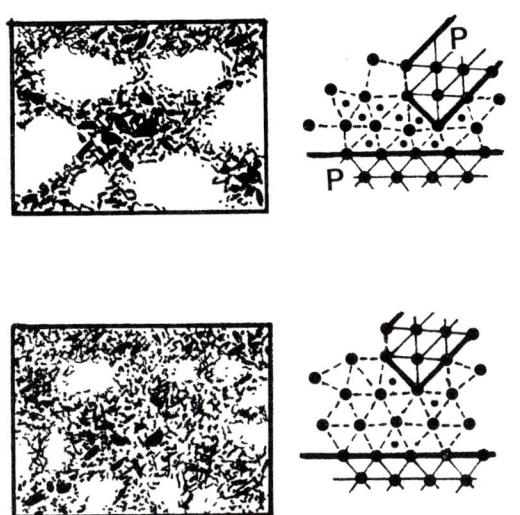

• Oxygen or hydroxyl   • Cation

Fig. 8 Final particle arrangement and generalized particle (P) contact after thixotropic strength regain. Upper line: salt water clay. Lower line: fresh water clay.

5. CONCLUSIONS

Cohesion, which results from the nearness of clay particles, is a main strength-producing factor in clays and clayey soils. It is strongly dependent on the water and electrolyte contents. Bulk shear failure under drained conditions involves local particle displacement associated with a locally increased inter-particle distance and decreased cohesion. Thus, while the majority of the strongest structural elements - the aggregates - are still intact due to cohesive forces, a critical number of the weakest elements - the links - are disintegrated so that cohesive forces are no longer in operation in them. This is compatible with the observation from shear tests at very low effective stresses that the cohesion intercept is almost or completely non-existent for an over-consolidated clay.

REFERENCES

1   Mc Graw-Hill Publ. Co., Encyclopedia of Science and Technology, vol. 3, 3rd edition, p. 287.

2   Larsson, R., Basic behaviour of Scandinavian soft clays. Swed. Geot. Inst. Report no. 4, 1977.

3   van Olphen, H., An introduction to clay colloid chemistry. Interscience Publishers, 1965.

4   Pusch, R., Influence of salinity and organic matter on the formation of clay microstructure. Proc. Int. Symp. Soil Structure, Gothenburg 1973, pp. 162-167.

5   Forslind, E., Water association and hydrogels. Proc. 2. Int. Congr. Rheology, Oxford, Butterworths Sci. Publ. London, 1953, pp. 50-63.

6   Pusch, R., Clay microstructure. Document D8:1970, Nat. Swed. Build. Res. Council, Stockholm 1970.

7   Pusch, R., A technique for investigation of clay microstructure. J. Microscopie. Vol. 6, 1967, pp. 963-986.

8   Arnold. M., A study of thixotropic action in bentonite clay. Thesis, Faculty of Engineering, Univ. of Adelaide, Austr., 1967.

9   Jacobsson, A. & Pusch, R., Thixotropic action in remoulded quick clay. Bull. Engineering Geology, Paris, no. 5, 1972, pp.105-110.

THE DEFORMATION OF GRANULAR MASSES WITH PARTICULAR REFERENCE TO PARTIALLY-MELTED GRANITE

M. S. Paterson and I. van der Molen

Research School of Earth Sciences, Australian National University, Canberra 2600, Australia

ABSTRACT

Experiments on a granite at 300 MPa confining pressure and 800°C with added water indicate that at relatively low melt fractions the specimen becomes effectively disaggregated and behaves as an assemblage of crushable particles with fluid filling the interstices. The behaviour of such specimens is discussed in terms of compaction, dilatancy hardening and cataclastic flow, comparing the observations with the theoretical behaviour of close-packed spheres.

1. INTRODUCTION

A recent study of partially-melted granite (van der Molen and Paterson, submitted for publication) has indicated that its rheological behaviour is basically that of a particle/viscous fluid two-phase system. Such two-phase systems are of considerable interest in various fields, including ceramics (hot glass ceramics), metallurgy (partially-melted alloys), chemical engineering (slurries), and soil mechanics (wet sands, liquefaction), as well as in geology (sedimentology, magmatic processes in volcanoes and igneous bodies, weak zones in the upper mantle). The present paper will aim to bring out some important aspects of the rheology of granular materials with viscous pore fluid and to illustrate these aspects with the results for partially-melted granite. Aspects of particular interest are the definition of a critical fluid fraction for sustainable flow and the nature of the stress-strain characteristics at fluid fractions below the critical value.

A large range of flow behaviour can be expected between the extreme cases of completely solid and completely fluid material. However, within this range it is convenient to distinguish two rheological regimes. In the one regime, the material behaves as a suspension in which the rigid grains suspended in the viscous fluid can pass each other without becoming "locked-up" and in which the flow properties are mainly controlled by the fluid's viscosity and by the volume fraction of suspended grains. In the other regime, the grains interlock after a small deformation and further deformation at constant volume can only be achieved with fracturing or deformation within the grains; the flow properties in this case tend to be controlled mainly by the properties of the grains themselves. With increasing fluid: solid ratio, a transition can be expected from the latter regime into the former at a critical volume ratio which we term the <u>critical fluid fraction</u>. At a given strain rate, it can be conveniently defined as the fluid volume fraction at which the flow stress changes most rapidly with fluid fraction, that is, at the inflexion in the curve of Figure 1.

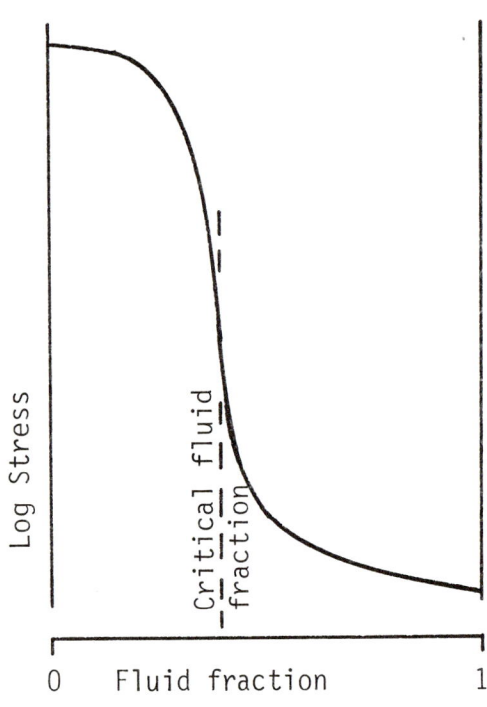

Figure 1.

The trend in flow stress at fluid fractions greater than the critical can be estimated from measurements on the relative viscosity of suspensions, which indicate a concave-upward curve as shown in Figure 1 (see reference 1 for a review). At fluid fractions less than the critical, the situation is more complex and needs more detailed definition and analysis.

We are concerned with a granular material in which all the pore space is filled with a homogeneous fluid usually initially at a substantial pressure. In general, the compressibility of the fluid will be taken to be relatively low, often within an order of magnitude of that of the grains, so that during the deformation it can be assumed that the initial fluid volume fraction is approximately maintained. The closed or undrained case is also assumed unless otherwide stated.

Conceptually, the deformational aspects of such a two-phase material with fluid fraction initially less than the critical fall into three groups:
 (1) Consolidation aspects, concerned with the initial relative movements of the particles until they become locked-up, and with the hardening consequent upon this locking-up
 (2) Yielding aspects, concerned with the manner in which the locking-up is overcome and substantial macroscopic flow achieved
 (3) The stability of the macroscopic flow, that is, whether the material will continue to flow uniformly or whether it will fail by the development of localized shear zones.
In order to clarify these concepts further and to make order-of-magnitude estimates of the stresses and strains involved, we first consider the properties of some spherical particle models and the modifications to these models that might be called for to account for the behaviour of real systems.

2. CONSIDERATIONS ARISING FROM RIGID SPHERE MODELS

After the initial movement of particles into a locked-up configuration, there will be a continuous framework of particles in contact and any further straining will then involve sliding or rolling of the particles over one another. Except for some rearrangements into more efficient packing, this further relative movement of particles will involve dilatancy of the mass accompanied by a reduction in pressure in the fluid phase. The flow stress, or resistance to further deformation, is now determined primarily by the work needed to bring about this reduction in pressure, that is, it is determined by "dilatancy hardening".

Calculations for an assemblage of equal rigid spheres in hexagonal close-packing, shortened normal to the hexagonally-packed layers, show that the increase in porosity, for small strains is approximately equal to 3ε where ε is the shortening strain. If ϕ_0 is the initial porosity and $\overline{K}$ the bulk modulus of the fluid, the decrease in fluid pressure is then $3\varepsilon\overline{K}/\phi_0$, requiring a uniaxial compressive stress σ in the direction of shortening which can be calculated from the balance of forces between the spheres, in the absence of friction, to be

$$\sigma = 12\varepsilon\overline{K}/\phi_0 \text{ for small strains} \tag{I}$$

This formula represents a stress-strain relationship for dilatancy hardening in the case of close-packed equal rigid spheres. In extending it to more general cases, the following considerations can be applied.

(1) If two sizes of spheres are considered for alternative layers, the numerical factor in (I) is reduced, for example, from 12 to 3.1 for a diameter ratio of 0.7. This suggests that the rate of dilatancy hardening will be reduced in case of some variation in particle size but not by more than an order of magnitude.

(2) If sliding friction between the spheres, with coefficient μ, is taken into account, the stress σ is increased by a factor of $(1+\mu)/(1-\mu)$, again less than an order of magnitude modification. Taking into account the viscosity of the fluid will also tend to raise the estimated stress because of resistance to shearing of films between particles; a more detailed model would be needed for a reliable estimate but in cases where the macroscopic stress supported by the assemblage is many orders of magnitude greater than the flow stress in a body of the fluid itself at the same strain rate it can be expected that the role of the fluid's viscosity will also be secondary.

(3) The relationship (I) can also be expected to give a rough indication of the behaviour to be expected in the case of particles of more general shape that still make point contacts. The essential aspects of the sphere models are the configuration of the points of contact between the spheres, and the orientations of the tangential planes at the points of contact. For other particles shapes, these configurations and orientations will not, in general, be radically different, leaving unchanged the general form of (I) and the order of magnitude of its numerical coefficient. In the case of particle shapes closer to space-filling shapes, as in slightly-melted polycrystalline materials, the porosity ϕ_0 may be considerably lower than for spheres and give a higher rate of dilatancy hardening.

It can therefore be concluded that if the fluid has a bulk modulus within an order of magnitude or so of that of the particles, the slope of the stress-strain curve arising from dilatancy hardening will be comparable to that of the elastic deformation of the particles themselves. That is, very substantial macroscopic stresses will tend to be required for even small strains in excess of those associated with bringing the particles into contact. However, the application of substantial macroscopic stresses will lead to very high stresses at contact points due to the Hertzian loading there, and hence tend to lead to flow in the particles or to their fracture. The effective result will be either new particle shapes or a new distribution of particle sizes, thus affecting the possibility of further flow. We now consider the consequences of particle fracturing with specific reference to the behaviour of partially-melted granite.

3. OBSERVATIONS ON PARTIALLY-MELTED GRANITE

Constant strain rate compression experiments have been carried out on sealed specimens of fine-grained granite at 800°C, 300 MPa confining pressure, in which various melt fractions have been generated by adding appropriate amounts of water (details in van der Molen and Paterson, op.cit.). Only about 10 volume percent of melt can be formed under the initial hydrostatic conditions owing to the limited initial porosity, but additional peripheral melt can be drawn into the specimen when it is strained, through a process of "dilatancy pumping" based on the pressure drop associated with the dilatancy hardening discussed above (Figure 2).

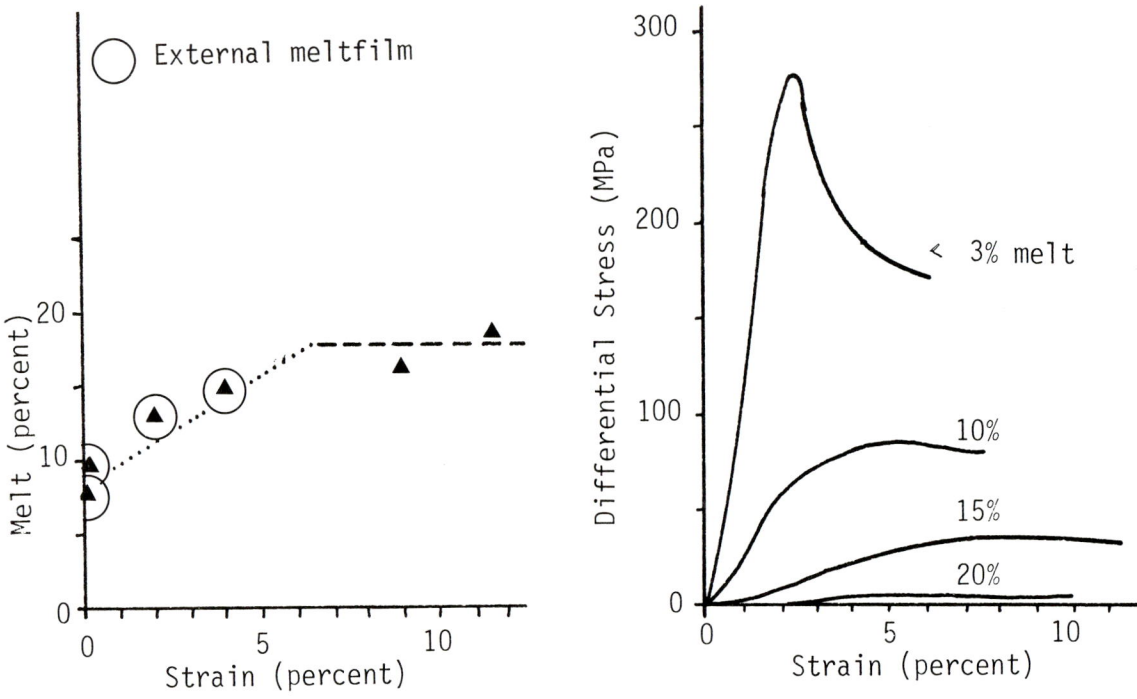

Figure 2. External melt fraction in a set of specimens with the same amounts of water added, deformed at the same rate to different amounts of strain.

Figure 3. Stress-strain results (average) for deformation of partially-melted granite at 800°, 300 MPa at a strain rate of $10^{-5} s^{-1}$ for specimens with different final internal melt fractions.

The stress-strain curves for various melt fractions are shown in Figure 3. In general, a shear fracture or zone of localized shearing forms beyond the peak stress, the zone being broader the higher the melt fraction. Microscopical observations show that in all cases there is melt at almost all grain boundaries, giving effective disaggregation of the specimen so that it can be expected to behave as a two-phase particle-fluid system such as discussed above. The observations further indicate that the deformation involves three concurrent mechanisms: (1) melt film redistribution into a preferred orientation parallel to the shortening direction, (2) axial fracturing of grains thus brought into contact (there is little evidence of plastic flow in the grains), and (3) some relative movement of grains and grain fragments over grain boundaries and cracks.

The sigmoidal form of the stress-strain curves in Figure 3 corresponds to what could be expected on the basis of the discussion of fluid-particle models above. The initial concave-upwards part of the curve will mainly represent, in strain, the initial bringing into contact of the grains and, in stress, the dilatancy hardening associated with locking-up. However, fracturing of grains occurs at relatively small macroscopic stresses and is evidently associated with the yielding, leading into a concave - downwards shape.

It is notable that lower peak stresses are reached at higher melt fractions. This observation suggests that the flow that is finally achieved, whether distributed uniformly or in a shear zone, is made possible by the fracturing introducing a wider range of particle sizes and so reducing the melt fraction corresponding to the critical fluid fraction needed for large-scale flow as defined previously. One of the principal problems posed by the present study is to derive a theoretical relationship between critical fluid fraction and particle size distribution that could provide a justification of this conclusion. Observations such as those on critical void ratio in sands suggest that for non-fracturing equi-sized particles the critical fluid fraction for mobility is close to 40 percent, whereas plotting the granite results on a diagram of the type of Figure 1, together with estimated suspension behaviour, suggests a critical fluid fraction nearer to 30 percent. Does the explanation of this difference lie in the wider range of particle sizes introduced by the grain fracturing?

REFERENCE

[1] Jeffrey, D.J. and Acrivos, A., The rheological properties of suspensions of rigid particles. Am.Inst.Chem.Eng.J. 22 (1976) 417-432.

CHARACTERIZATION OF MICROCRACKS

Gene Simmons, Michael Batzle, Herman Cooper, Robert Siegfried[1]
and Michael Feves[2]

Department of Earth and Planetary Sciences, Massachusetts Institute of Technology, Cambridge, Massachusetts 02139, U.S.A.

ABSTRACT

Microcracks in rocks can be characterized by their effects on various physical properties. One technique which we have developed recently, termed differential strain analysis (DSA), uses high precision strain (2 or 3×10^{-6}) measured as a function of hydrostatic pressure. DSA yields the strain at zero pressure (or any other pressure) due to cracks that close completely as a function of crack-closure pressure. By-products of DSA are compressibility and the open microcrack porosity that remains at any pressure. Strains measured in any three orthogonal directions are sufficient to yield volumetric quantities; strains measured in six independent directions yield tensor quantities.

The data are useful for characterizing cracks in rocks, for testing theoretical expressions for the effect of cracks on physical properties, and for correlating empirically physical properties with various crack parameters. From the data on fifty rocks, obtained during the past three years, we conclude that: (1) the cracks produced by different processes in the laboratory as well as in nature yield cracks with differing characteristics; (2) most natural cracks in terrestrial igneous rocks are closed completely at pressures equal to or less than 500 bars; and (3) shocked lunar rocks contain cracks that do not close completely at pressures of several kilobars.

---

[1] present address: Corning Glass, Corning, NY, USA.
[2] present address: Dowell, Tulsa, OK, USA.

1. INTRODUCTION

Microcracks are present in most rocks in quantities that range from below 0.01% to as much as a few % by volume. Their characterization is important for several reasons. First, the microcracks that occur naturally in rocks have diverse origins and may be determinable from the characteristics of the crack. They can be caused by the linear and volumetric strains that occur in individual grains produced by the change in pressure and temperature when the rock is transported to the surface of the earth. They can be produced by non-hydrostatic stresses, such as the stresses that produce faults. They can be produced by the thermal stresses associated with the cooling of igneous intrusions, such as dikes and batholiths. They can be caused by the passage of shock waves due to the impact of meteorites on the earth, moon, and other planets. In order to determine unambiguously the origin of a particular crack or a set of cracks that was produced naturally, we must have a set of criteria based on observable characteristics. We have begun to develop sets of such criteria for a few microcracks [1,2,3,4]. A second reason for characterizing cracks in rocks is to improve theoretical models for the calculations of physical properties of rocks. Several authors [eg, 5,6,7] in geophysics have attempted to calculate such physical properties as velocity of compressional waves, attenuation, and thermal conductivity of cracked media on the basis of simple geometrical shapes. In our experience of working on many rocks (mainly igneous and metamorphic rocks) with the SEM such simple geometrical shapes are rare. We suggest that significant improvements in the theoretical models might result from developing models based on the observed characteristics of the microcracks. A third reason for the characterization of microcracks is the desire to establish empirical correlations between various physical properties and the characteristics of microcracks (eg, see [8]). Until the theoretical models are sufficient to predict properties as well as they can be measured, then the possibility of using empirical correlations is attractive.

At the present time, we use several different techniques to characterize the microcracks in rocks. Differential strain analysis (DSA) and the scanning electron microscope (SEM) are used for open microcracks. The SEM and the petrographic microscope (PM) are used with crack sections (100 micron thick slices of rock) and ordinary pettrographic thin sections to characterize sealed and healed microcracks. An energy dispersive x-ray (EDX) system is normally used with the SEM to obtain information on the elemental composition of phases that occur within microcracks. In this manuscript, we discuss DSA.

Terminology. The following definitions may be useful in this manuscript to facilitate communication: A microcrack is an opening, or site of a former opening, in rock with one dimension much less than the other two dimensions. The width, typically a few microns, must be less than 100 microns. A sealed microcrack contains mineral phases that differ in composition from the host grains; it may be partially or completely sealed. A healed microcrack contains mineral phases that are identical in composition to the host phase and may be partially or completely healed.

2. DIFFERENTIAL STRAIN ANALYSIS

General. Differential strain analysis (DSA) is based on very precise measurements of strain as a function of hydrostatic pressure. It was described in detail by [9] and [10]. It has been used in our group to characterize approximately 250 specimens.

The principle upon which DSA is based is that rocks containing cracks are more compressible than the same rocks would be without cracks. Schematic compression curves are shown in figure 1. The difference between the measured strain curve and the straight line projection to lower pressure of the linear portion of the strain curve is due to the presence of cracks in the rock, an observation apparently noted first by [11]. This same concept was used by [5] to determine the total crack porosity of a rock at room conditions. [12] extended the concept to show how to use compression data to calculate the distribution of crack aspect ratios on the basis of a specific model for the shape of microcracks, namely, that of ellipsoidal cracks.

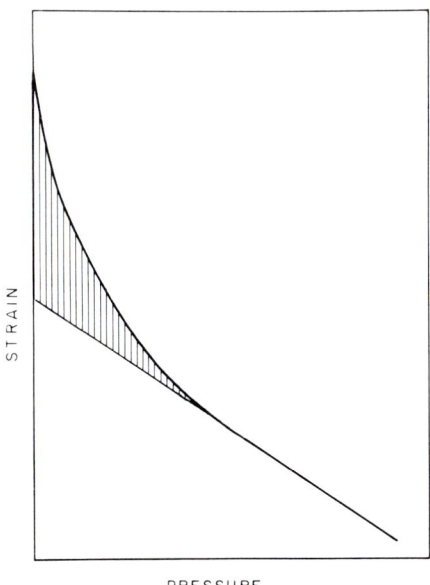

Figure 1. Schematic compression curve for rock that contains microcracks.

Useful experimental parameters can be obtained from the simple compression curves, provided that the strains were measured with very high precision [10]. The necessary precision is obtained in DSA by using differential strain, defined as the sample strain minus the strain of a reference sample exposed to the same experimental conditions. We have used fused silica for the reference samples of all DSA work to date. It has these advantages: (a) linear compressibility is adequately known [13] and approximately the value of many igneous and metamorphic rocks; (b) thermal expansion is very low; (c) it is readily available; (d) a single specimen is sufficiently homogeneous; and (e) the compressibility of a particular specimen can be determined readily with ultrasonic methods.

The strains of the fused silica reference sample and the rock are recorded separately; the differential strain is calculated during data processing. We note that the intercept at zero pressure of the tangent to the differential strain curve is equal to the intercept at zero pressure of the tangent to the strain curve. This observation allows us to use the differential strain curve for the analysis of DSA data.

See figure 2, a set of DSA data typical of granitic rocks.

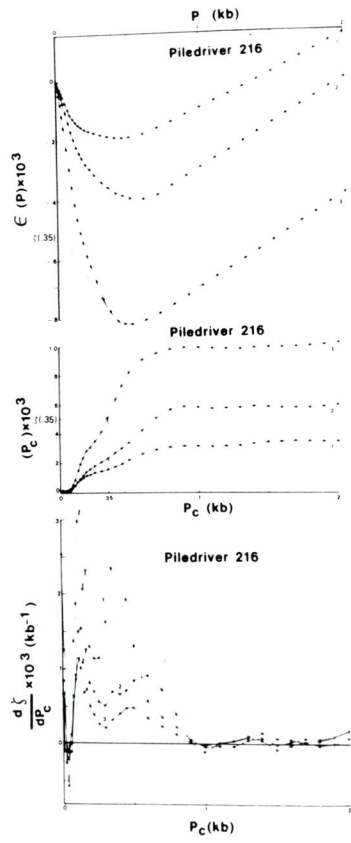

Figure 2. Interpretation of DSA. Upper curve is differential strain. Middle curve is ζ as function of pressure. Lower curve is $d\zeta/dP$ as function of pressure, termed crack spectrum. The graphical construction of ζ, the zero-pressure intercept of the tangent to the differential strain curve, is indicated by the values for P=0.35kb. In this example, the three gages are orthogonal. The orientation of cracks must be anisotropically distributed but the principle values of the tensors are not determinable from these data. After [14].

The parameter ζ, the intercept of the tangent to the differential strain curve, is the strain at zero pressure due to all cracks that close completely by the pressure P_c. It is a second rank tensor. For strains measured in any three orthogonal directions, the sum of the resulting ζ is the volumetric crack porosity of [5]. The parameter ζ is the cumulative distribution function; its derivative with respect to pressure is an ordinary distribution function which we have termed crack spectrum. The meaning of the parameter ν (that is, the crack spectrum) is the strain at zero pressure due to all cracks that close completely between pressure P and pressure P+dP. The plots of ζ and of ν can be converted to distribution functions for crack aspect ratios, if one chooses.

Sample Preparation for DSA. Specimen preparation for DSA is rather simple. Electric resistance foil strain gages (SR-4) mounted on epoxy backing are attached with epoxy. The sample is encapsulated in a material that transmits pressure but excludes the pressure medium, for example, Dow Corning Sylgard 186.

Some DSA Results. Typical crack spectra of virgin igneous rocks are shown in figure 3. Most cracks close

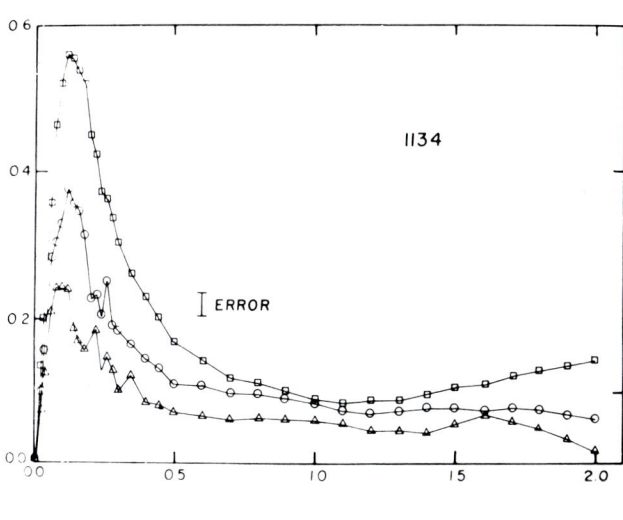

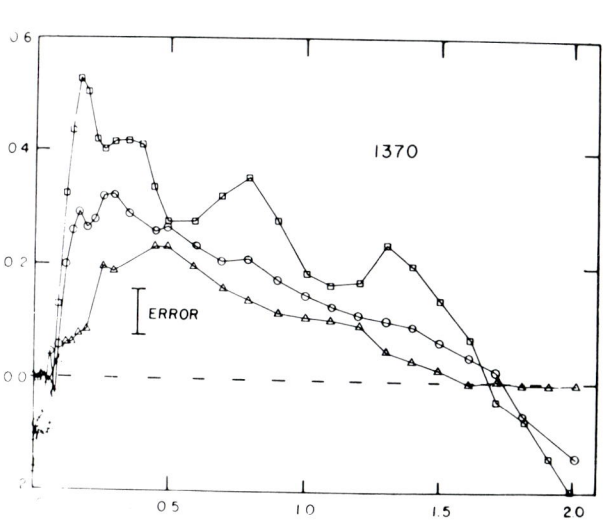

Figure 3. Typical crack spectra obtained with DSA on terrestrial igneous rocks. Sample 1134 is Westerly (RI) granite. Sample 1370 is Red River (WI) quartz monzonite. After [8].

by pressures of 0.5kb in most igneous rocks. Total crack porosity ranges from less than $2-4 \times 10^{-6}$ (the lower limit of our measurement) to about 0.1% [8].

Crack spectra of shocked rocks are shown in figure 4. Comparison of the spectra for lunar samples with the spectra for the samples from Ries crater, the shocked granite block, and the site of the underground nuclear explosion Piledriver shows that the spectra for returned lunar samples are significantly broader than the spectra of all other shocked samples. Such a large difference casts some doubt on the validity of the physical property data measured on the returned lunar samples.

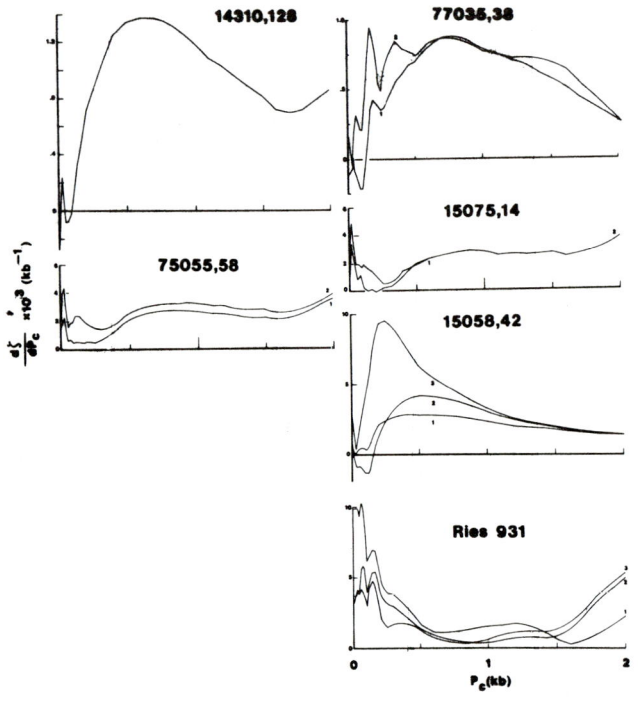

Figure 4. DSA spectra for shocked samples. Those samples with numbers greater than 10,000 are lunar samples. The Ries sample was collected from outcrops in the Ries (Germany) impact crater. The sample with no number was obtained from a granite block that was shocked by the impact of a high speed missle, peak shock pressure was approximately 6kb. After [4, 14].

The crack spectra for thermally cycled samples of Westerly granite are shown in figure 5. The effects of heating rocks are (1) to increase the total crack porosity and (2) to broaden the spectra. These effects can be used in the laboratory to produce suites of rocks for experimental studies in which the volume of cracks is controlled and reproducible.

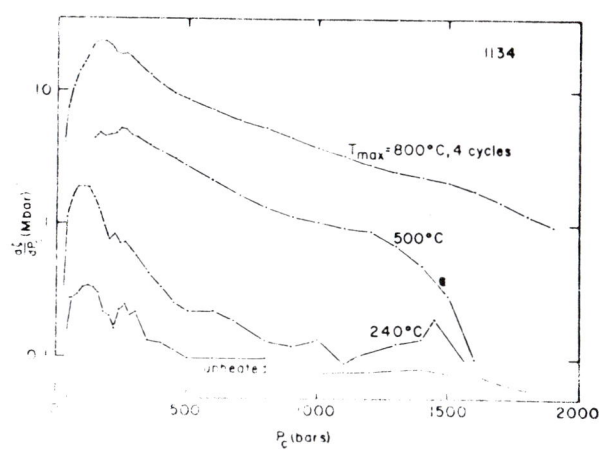

Figure 5. Crack spectra of thermally cycled Westerly granite. Note that a log scale is used for the abscissa. After [3].

REFERENCES

[1] Simmons, G. and D. Richter, Microcracks in rocks, in The Physics and Chemistry of Minerals and Rocks. Wiley-Interscience, New York (1976) 105-137.

[2] Feves, M. and G. Simmons, Effects of stress on cracks in Westerly granite. Bull. Seism. Soc. Am. 66 (1976) 1755-1765.

[3] Simmons, G. and H. Cooper, Thermal cycling cracks in three igneous rocks. Int. J. Rock Mech. Min. Sci. (1978) in press.

[4] Siegfried, R., G. Simmons, D. Richter, and F. Hörz, Microfractures produced by a laboratory scale hypervelocity impact into granite. Proc. Lunar Sci. Conf. 8th 3 (1977) 1249-1270.

[5] Walsh, J.B., The effect of cracks on the compressibility of rock. J. Geophys. Res. 70 (1965) 381-389.

[6] Kuster, G.T. and M.N. Toksöz, Velocity and attenuation of seismic waves in two-phase media: part I-theoretical formulations. Geophysics 39 (1974) 587-606.

[7] O'Connell, R.J. and B. Budiansky, Seismic velocities in dry and saturated cracked solids. J. Geophys. Res. 79 (1974) 5412-5426.

[8] Feves, M., G. Simmons, and R. Siegfried, Microcracks in crustal igneous rocks, in The Earth's Crust: Its Nature and Physical Properties. AGU, Washington (1977) 95-117.

[9] Simmons, G., R. Siegfried, and M. Feves, Differential strain analysis: a new method for examining cracks in rocks. J. Geophys. Res. 79 (1974) 4383-4385.

[10] Siegfried, R. and G. Simmons, Characterization of oriented cracks with differential strain analysis. J. Geophys. Res. 83 (1978) 1269-1278.

[11] Adams, L.H. and E.D. Williamson, The compressibility of minerals and rocks at high pressures. J. Franklin Inst. 195 (1923) 475-529.

[12] Morlier, P., Description de l'état de fissuration d'une roche à partir d'essais non-destructifs simples. Rock Mech. 3 (1971) 125-138.

[13] Peselnick, L., R. Meister, and W.H. Wilson, Pressure derivatives of elastic moduli of fused quartz to 10 kb. J. Phys. Chem. Solids 28 (1967) 635-639.

[14] Simmons, G., R. Siegfried, and D. Richter, Characteristics of microcracks in lunar rocks. Proc. Lunar Sci. Conf. 6th 3 (1975) 3227-3254.

DEFORMATION STRUCTURES IN THE TÄNNÄS AUGEN GNEISS

Kennert Röshoff

Department of Rock Mechanics, University of Luleå, S-951 87 Luleå, Sweden

ABSTRACT

The microstructures observed by optical and transmission electron microscope in the progressivly deformed Tännäs augen gneiss, Swedish Caledonides, are presented. The deformation was mainly by creep within a temperature range of 300-550°C. The augen gneiss was developed by progressive deformation from a K-feldspar megacryst bearing granodiorite of Precambrian age. The augens are porphyroclasts, which were deformed just as much as the matrix. The divergency in response to deformation between matrix and augens is explained by the difference in deformation mechanisms.

1. INTRODUCTION

Augen gneisses of various character and origin appear within the Central Scandinavian Caledonides. Of these the Tännäs augen gneisses (Fig 1) forms a separate, contineous tectonic nappe unit, which can be followed from the eastern Caledonian front to west of the political border between Sweden and Norway [1, 2, 3]. Local zones of augen gneisses also occur within the Veman Nappe. In Norway, augen gneisses are observed at the same tectonic level as the Tännäs augen gneiss in the vicinity of Dombås [4, 5], in the Oppdal area [6] and along the west coast of Norway from Orkanger [7] in the north to Surnadal in the south [8].

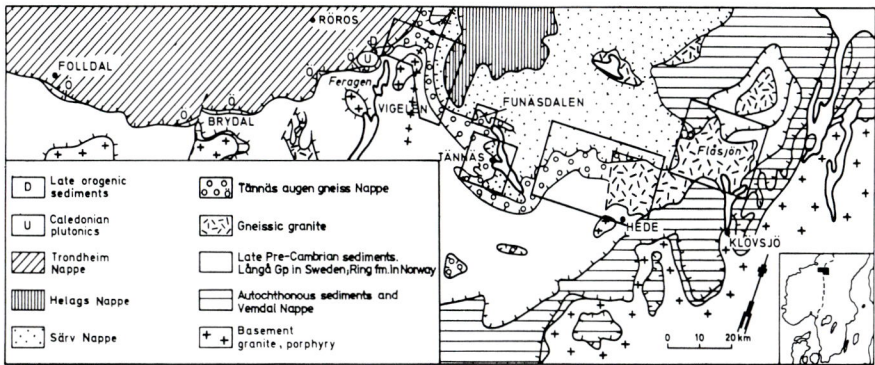

Fig 1 The Tännäs Augen Gneiss Nappe in the Central Scandinavian Caledonides

A major problem of these augen gneisses has been what geological implication do they represent and what is their origin. Törnebohm [1], Högbom [9], Carstens [10], Nystuen [11] and Point [12] favour a development by progressive deformation of Precambrian rock material, usually porphyritic granites. Others [5, 6, 13, 14] interpret the augen gneisses as Postcaledonian metasomatic products formed from late Precambrian sediments. One important proof for the latter interpretation is the reported observation that the augens in the augen gneiss appear to be undeformed, while the surrounding matrix is strongly foliated.

Earlier works on the augen gneisses were mainly based on field observations and interpretations of geochemical data. In this study the presented Tännäs augen gneiss have been studied using optical microstructural analyse on the main mineral assemblage and a pilot study of some minerals in the transmission electron microscope (TEM) in order to analyse the substructures in the minerals. The identification and interpretation of substructures formed both in naturally and experimentally deformed rocks and minerals have considerably contributed to the understanding of the deformation mechanisms in rock materials.

The undulose extinction, deformation bands, polygonization, subgrains and new grains (recrystallization) found in quartz and quarzitic rocks [15, 16, 17, 18, 19, 20, 21, 22, 23] are, when observed in the TEM, comparable to dislocation structures developed in metals. Consequently the deformation features are described as typical microstructures formed during ductile deformation, dynamic recovery and recrystallization.

Microstructures in feldspars, such as deformation twins, bent twins, undulose extinction and kinking, have so far received little attention. White [24] found little evidence for recovery and subgrain formation in a deformed porphyroclastic oligoclase. The recovery process was interpreted to be inhibited by the superlattice of the mineral and the presence of twin lamellaes. The strain energy was instead released by recrystallization.

The observations by Etheridge and Bell [21] and the experimental works published by Etheridge and Hobbs [25] and Etheridge et al [26] on microstructures in mica show that strain shadowing, shredding of the grains, kinking, serrated kink band boundaries and the development of new grains are caused by plastic deformation and recrystallization processes.

2. GEOLOGICAL SETTING

The Tännäs Augen Gneiss Nappe, is overlain by the Särv Nappe, and underlain by the Veman Nappe in the east and the Hede Nappe to the West (Fig 1). All these nappes display a low metamorphic grade in greenschist - lower amphibolite facies. The Tännäs Augen Gneiss Nappe is dominated by augen gneisses which vary in structure and petrology. Deformed Precambrian granodiorite, granite, gabbro and basic dykes out crop in minor parts. The augen gneisses are characterized by K-feldspar megacrysts, which usually are between 6-8 cm in diameter, set in a wavy, fine-grained greenish-grey matrix. The rock has a very heterogeneous strain. Thus, weakly deformed plutonic rocks are often observed in contact with augen gneisses. The contact may either be transitional or sharp. It is a mylonite in the latter case.

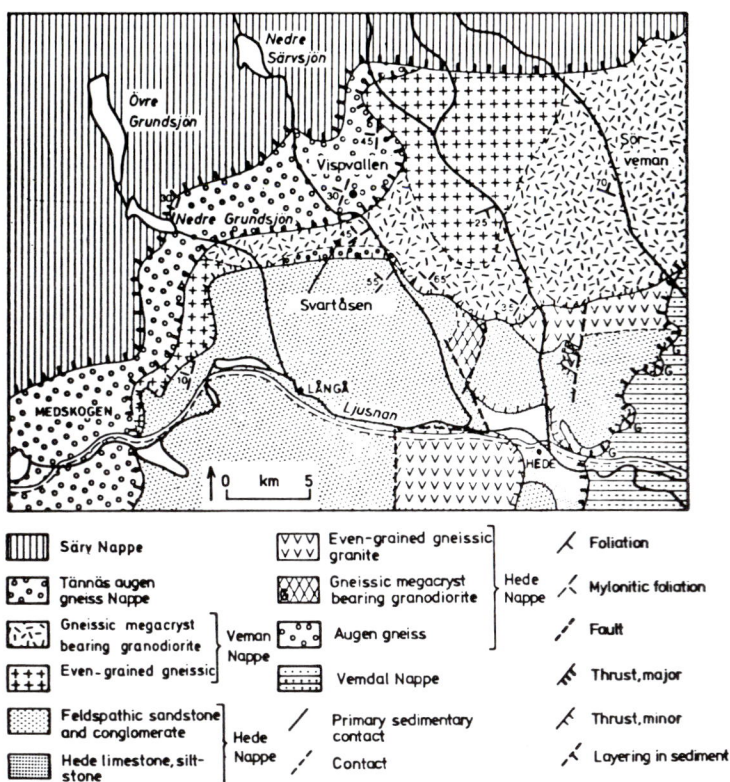

Fig 2 Geological map of the Heda area
       (Large rectangle on Fig 1)

3.   DESCRIPTION OF ANALYSED SAMPLES

      Field observations

The transitional relationship between augen gneisses and weakly deformed plutonic rocks can best be observed at Medskogen and along a 400 m N-S profil east of Lake Nedre Grundsjön at the top of the nappe (Fig 2).

At Medskogen a weakly foliated K-feldspar megacryst bearing granodiorite is progressivly transformed into an augen gneiss within a distance of approximately 75 m. The granodiorite (Fig 3) has a medium-grained matrix of greenish plagioclas, red K-feldspar, quartz and interstitial dark minerals (biotite and hornblend). The red K-feldspar megacrysts, which are 6-8 cm in diameter, are uneven scattered in the matrix. Their shape vary from circular to oval and some have a plagioclase rim. Locally the rock is cataclastically deformed by brecciation, which has caused off sets of the megacrysts.

The least deformed rock at Lake Nedre Grundsjön is a K-feldspar bearing schistose granodiorite which is in contact with an even-grained granite. Some aplitic dykes are observed. The granodiorite is progressively deformed into an augen gneiss and to a banded mylonite at the contact to the overlying Särv Nappe.

The schistose granodiorite (Fig 4) is medium-grained with a matrix of elongated greenish plagioclase, K-feldspar, quartz and dark mineral aggregates. The foliation is bent around the megacrysts. The red K-feldspar megacrysts are microbrecciated and elongated subparallel to the foliation. However, some circular megacrysts are observed.

Fig 3 K-feldspar megacryst bearing granodiorite

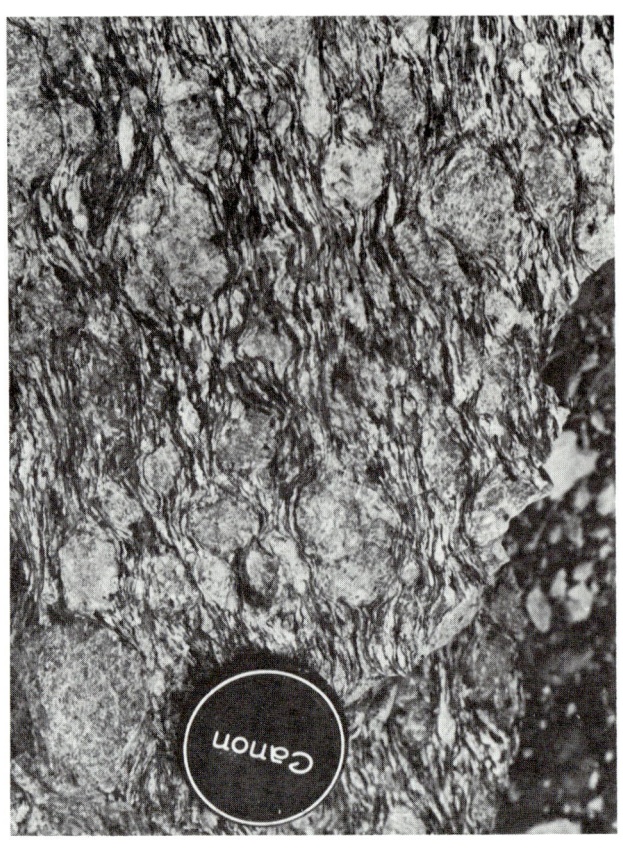

Fig 4 Schistose granodiorite

Fig 5 Augen gneiss

Fig 6 Banded mylonite

The augen gneiss (Fig 5) is characterized by very elongated K-feldspar megacrysts in a strongly foliated greenish-grey fine-medium-grained matrix consisting of lenses of quartz, greenish plagioclase and very thin layers of dark minerals. Small porphyroclasts of plagioclase and K-feldspar occur in the matrix. Strained parts of the megacryst crystals are deformed by shearing and the sheared parts are displaced along the foliation. Besides, some large (approximately 8-10 cm) circular to oval megacrysts are observed.

The amount of matrix will increase with the progressive deformation. Thus, the matrix becomes darker in colour and more fine-grained. The megacrysts will in general be reduced in size and may either, depending on the deformation mechanism, obtain a layer shape or become a round-oval-shaped porphyroclast.

The mylonite, which form the contact towards the Särv Nappe, is banded (Fig 6) with alternating dark greenish and light layers. The rock is dense and even-grained with no visible porphyroclasts or megacrysts.

Optical microstructures

The weakly deformed K-feldspar megacryst bearing granodiorite at Medskogen clearly has been deformed as observed in thin section (Fig 7). Microstructures occur in quartz and mica and the feldspar is bent. Undulose extinction, often banded, deformation bands, small equidimensional subgrains (low angle boundary) and large elongated subgrains are identified in the quartz. Recrystallization is prominent along grain boundaries and in highly strained parts of the mineral. The new grains (high angle boundary) have serrated grain boundaries and display no visible preferred orientation. The mica is a brownish biotite, which is kinked, bent and has strain shadows (Fig 7). The mineral is partly transformed into chlorite. Secondary minerals, mainly sphena and chlorite, appear in the highly strained kink band boundaries. The plagioclase is almost completely altered to sericite. Bent twins are observed in the sericite mass. The K-feldspar consists of both perthite and microcline with albite-pericline twins. The latter mineral is dominating in the matrix, while the megacrysts mainly are composed of perthite. The megacrysts enclose matrix crystals of deformed quartz, sericitized plagioclase and mica, and thus are xenoblastic (Fig 8). The crystals are brecciated and bent. The mismatch between the distorted parts of the K-feldspar is low angled. New grains have developed in highly strained kinks.

The foliated granodiorite have a well developed penetrative schistosity. The plagioclase is strongly altered to sericite (Fig 9). The albite twins are kinked and bent and some new untwined grains, probably of albite, have grown in the kinks. The quartz is recrystallized and consists of elongated host- new grain aggregates. The larger host grains contains deformation bands and polygonal subgrains and are surrounded by an aggregate of strained new grains. The latter are usually elongated and show variable preferred orientations to the foliation. The biotite is kinked and partly recrystallized, mainly along kink band boundaries and 001 planes, to new green biotite, muscovite and sphena. The crystals are sheared off along the 001 planes and the kink band boundaries. The sheared off parts are displaced along the foliation. The kink band boundaries are at a low angle to the foliation. The K-feldspar megacrysts are strongly kinked (Fig 10) with a moderate mis-

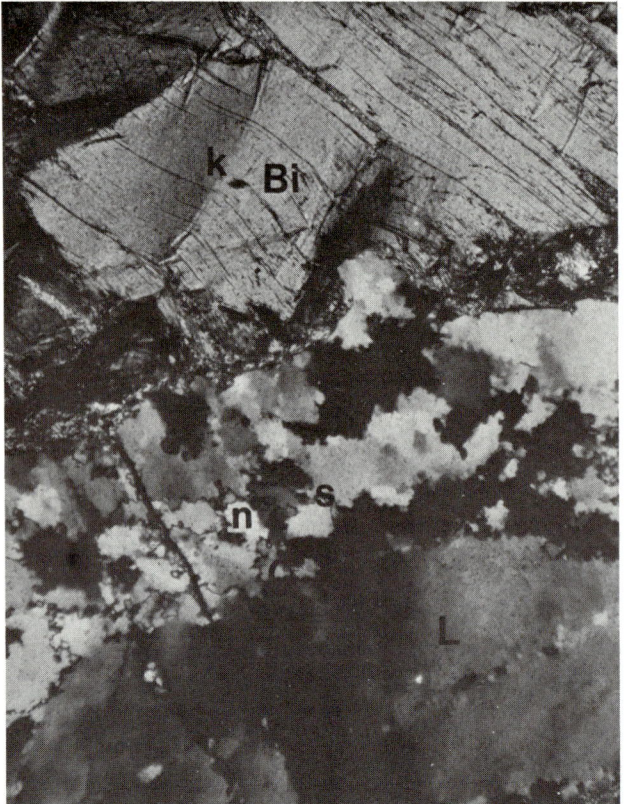

Fig 7 Deformed quartz (Q) and biotite (Bi) in megacryst bearing granodiorite
k = kink band boundary
n = new grain; s = serrated grain boundary; L = undulose extinction in large quartz grain
M = 25, crossed nicols

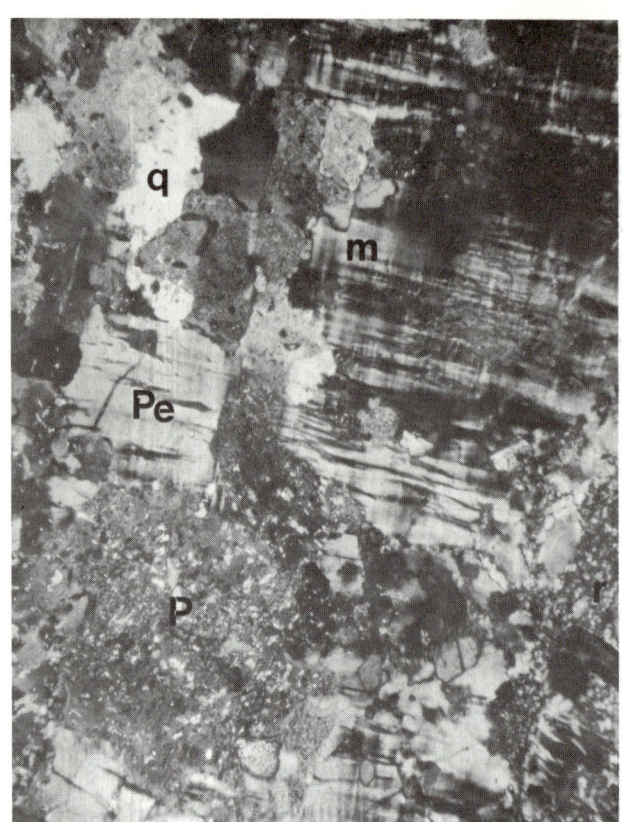

Fig 8 Part of megacryst with distorted microline (m) and perthite (Pe), inclusions of quartz (q) and altered plagioclase (P)
M = 25, crossed nicols

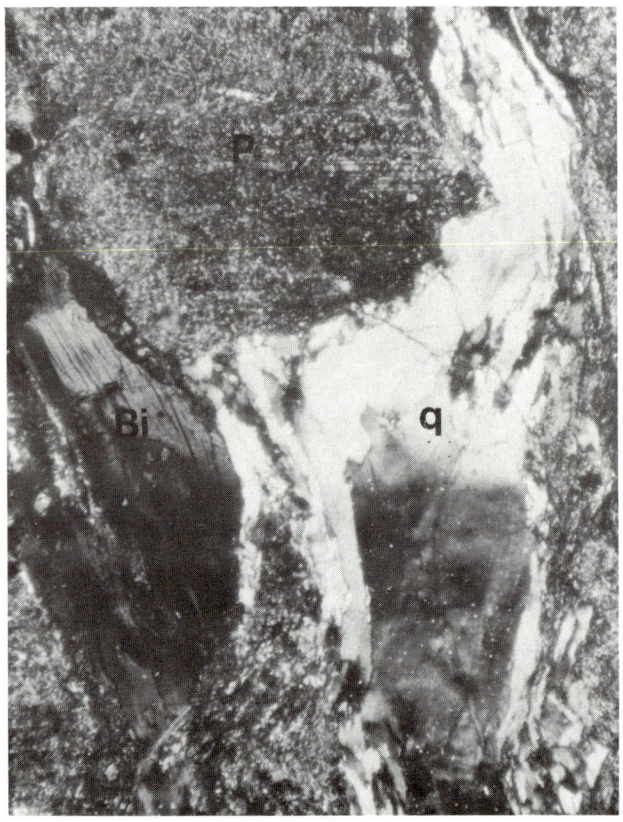

Fig 9 Schistose granodiorite with deformed biotite (Bi) and quartz (q) and bent albite twins in plagioclase (P)
M = 25, crossed nicols

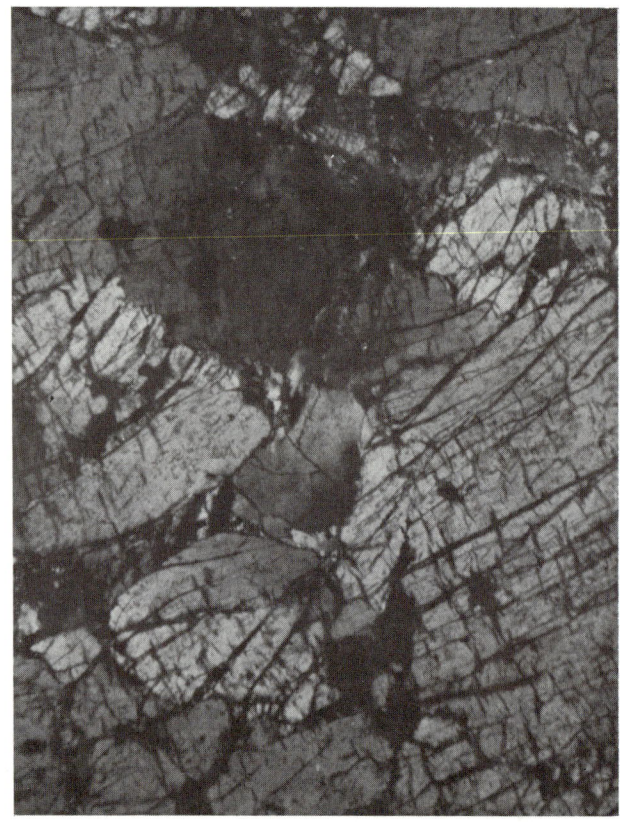

Fig 10 Core of K-feldspar megacryst in schistose granodiorite. The crystal is strongly brecciated
M = 25, crossed nicols

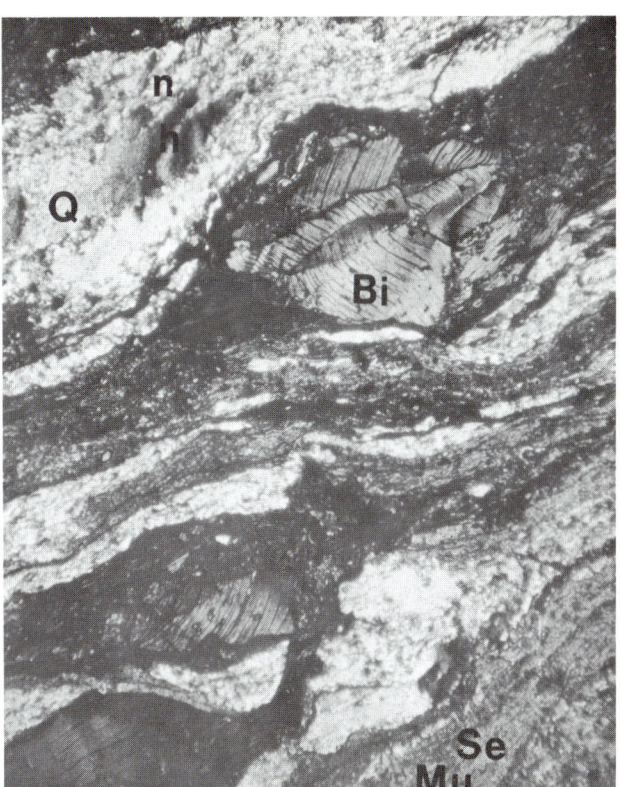

Fig 11 Margin of megacryst in schistose granodiorite.
Pe = brecciated and distorted perthite
r = recrystallized new K-feldspar grains
M = 25, crossed nicols

Fig 12 Matrix in augen gneiss with deformed quartz (Q), kinked biotite (Bi) and aggregates of muscovite (Mu) and sericite (Se). Quartz displays a host (h) - new grain (n) aggregate
M = 25, crossed nicols

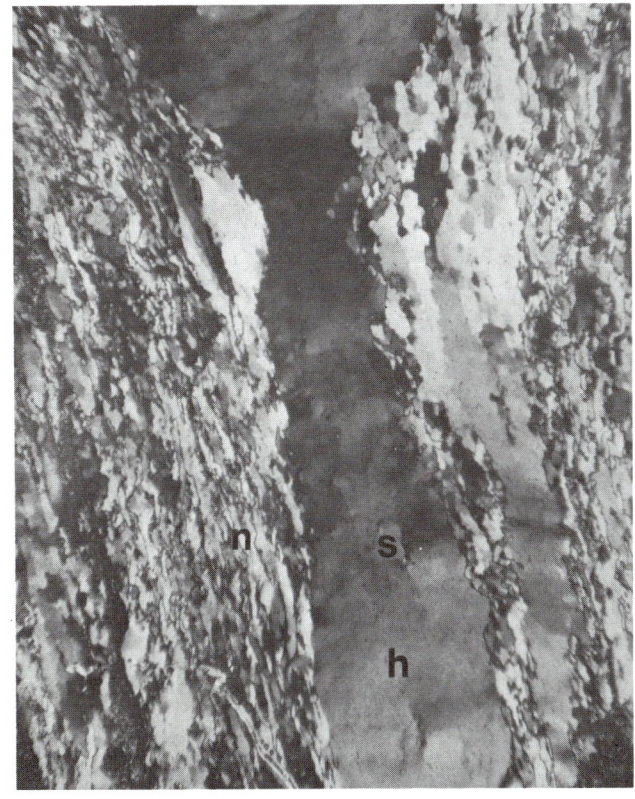
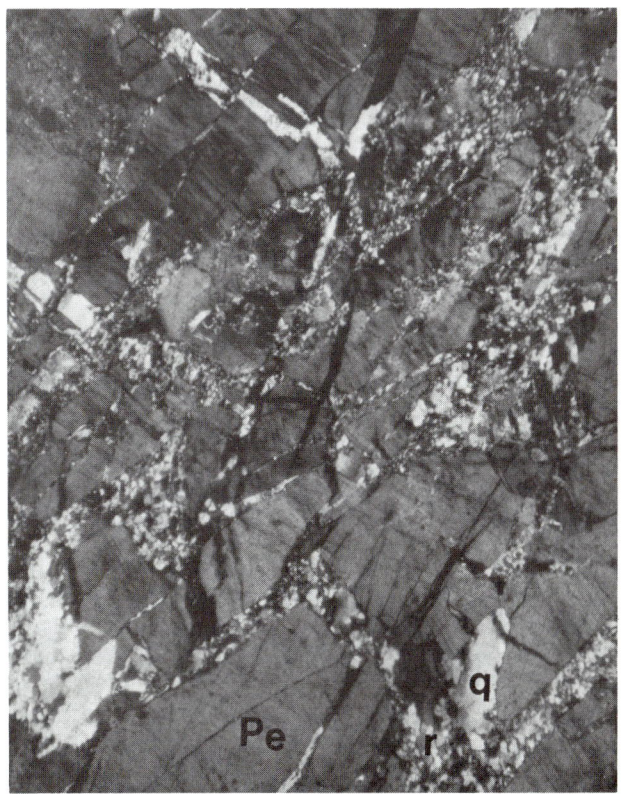

Fig 13 Close up of Fig 12 of quartz grain. The host grain (h) has equidimensional subgrains (s) and deformation bands. The new grains (n) have a preferred orientation
M = 63, crossed nicols

Fig 14 Brecciated core of K-felsdspar megacryst in augen gneiss
Pe = perthite
q = quartz
M = 25, crossed nicols

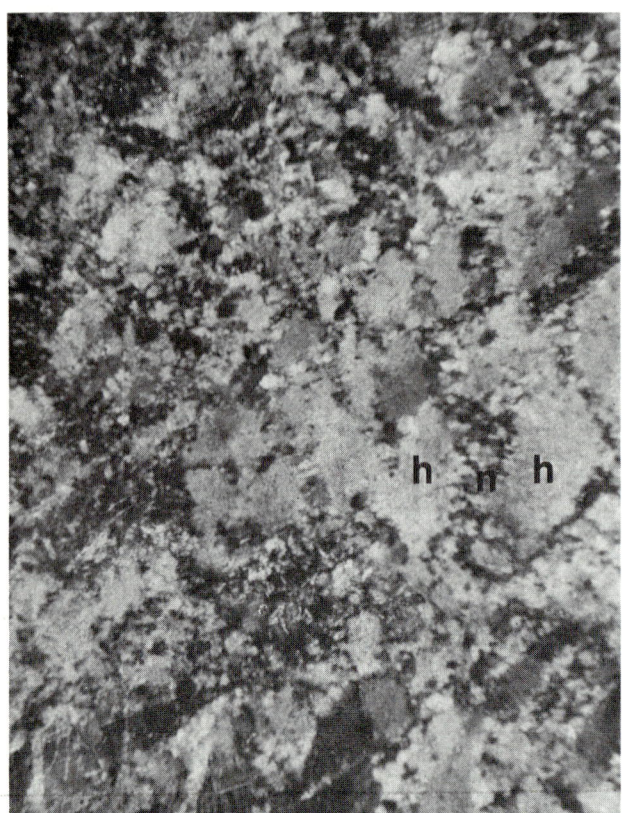

Fig 15 Recrystallized part in a K-feldspar megacryst with host grains (h) and new grains (n). The rock is an augen gneiss
M = 63, crossed nicols

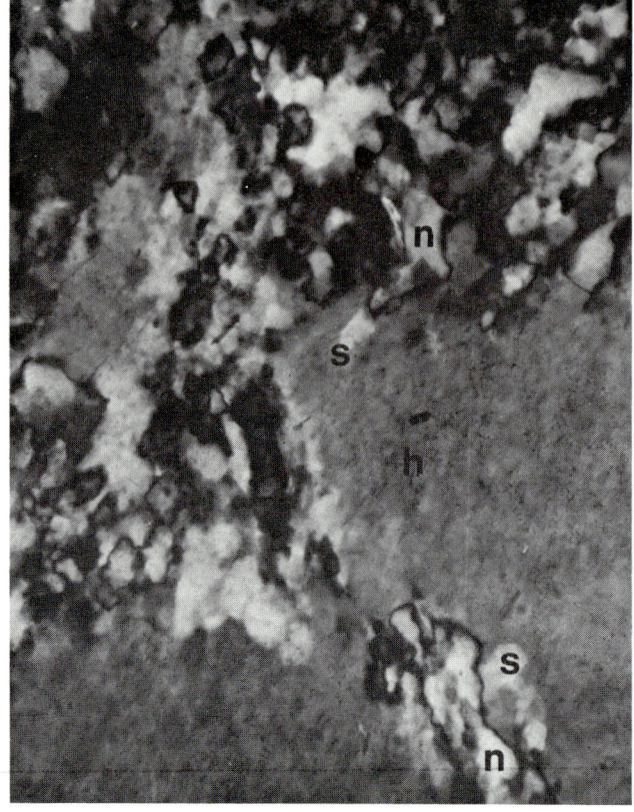

Fig 16 Close up of Fig 15. Subgrains (s) occur at the recrystallized front between the host (h) and new (n) grains
M = 250, crossed nicols

Fig 17 Banded mylonite M = 25, crossed nicols

match angle (10-15°) in the core. The mismatch is greater in the stress shadow regions due to rotation and displacement of sheared crystal fragments (Fig 11).

The augen gneiss is produced in the next stage of the progressive deformation. The matrix shows a mylonitic banding with a porphyroclastic microtexture (Fig 12). The plagioclase is plastically deformed and completely transformed into sericite, which locally is forming a wavy white mica. The quartz is composed of lenses of host - new grain aggregates. The host grains have undulose extinction, deformation bands with equidimensional subgrains (Fig 13). The surrounding new grain aggregates are composed of strained grains with strongly serriated grain boundaries. The new grains have a preferred orientation with low angles to the foliation.

The biotite is strongly recrystallised to green biotite (Fig 12) surrounded by sphena and epidote. Remnants of old biotite grains appear as porphyroclasts. The old grains are kinked or have the 001 planes subparallel to the foliation. The K-feldspar megacrysts are composed of a core of kinked and brecciated perthite (Fig 14) surrounded by elongated aggregates of old and new grains. Recrystallization is intense in strained parts of the crystal (Fig 15 and 16). Subgrains are rare. In addition muscovite, epidote and opague minerals occur as syn-tectonic minerals.

With still higher strains the augen gneiss will become darker and more fine-grained, with scattered megacrysts of varying sizes. The matrix of the rock is layered with light quartz and K-feldspar bands and dark bands of mica, epidote, sphena and sericite. The quartz consists of a very fine-grained new grain aggregates with very few host grains left. The host grains have a polygonal subgrain structure. The new grains usually have a strong preferred orientation subparallel to the foliation, except in protected areas, where the grains usually are strain free. The quartz starts to get ribbon character. The brownish biotite is completely transformed to a greenish biotite, which has a preferred orientation subparallel to the foliation. The K-feldspar megacrysts are strongly brecciated and boudinaged. New grains are formed in the strained areas.

The banded mylonite displays evengrained, alternating light and dark greenish layers, with a few small porphyroclasts remnants (Fig 17). The light layers consist of an aggregate of quartz, albite and K-feldspar and scattered oriented mica flakes. The main part of the quartz is composed of irregular, partly strained, new grains with a preferred orientation. Secondary grain growth is observed. The dark bands are mainly composed of mica, chlorite, epidote and sphena. The small grain size make it difficult to identify the microstructures.

Transmission electron microscopy (TEM)

A pilot study was made in the transmission electron microscope (JEOL 200) in order to analyse some microstructures optically observed in the augen gneisses. The K-feldspar megacrysts were of special interest.

Deformation and growth twins, subgrains, dislocations and new grains were observed microstructures in the K-feldspars. Fig 18 shows a recrystallization front at the margin of a host grain (h). The host grain contains twins, probably deformation twins, and scattered dislocations in the core. The dislocation density (D) increases very rapidly towards the margins of the grain. The new grains (n) develop without any preceded recovery process, as very few subgrains are observed in the high density dislocation area (D). This means that the nucleation of new K-feldspar grains starts without any significant subgrain formation.

The microstructures in quartz were found to be comaprable to the dislocation substructures described by White [16, 17, 18, 19].

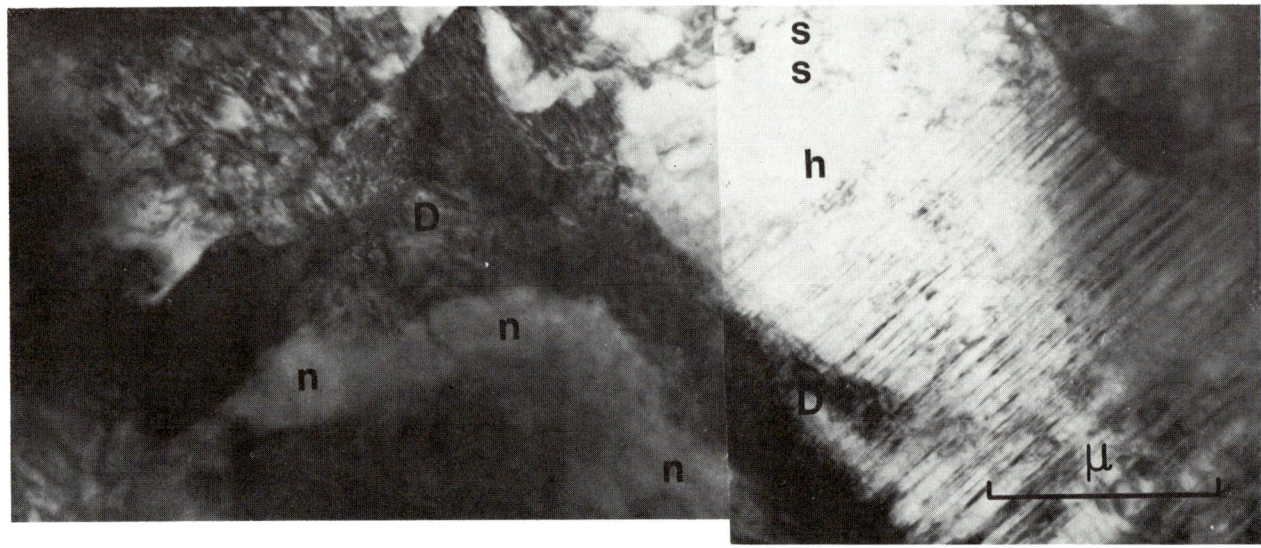

Fig 18   Micrograph of recrystallization front in a K-feldspar
         h = large old grain with twins   n = new grain
         D = dislocation front   s = subgrain

CONCLUSION

The augen gneisses are composed of a polycrystalline mineral assemblage of plagioclase, quartz, mica and K-feldspar. These minerals behave differently under similar physical conditions. Thus, to get a synthesis of the mechanisms and conditions of the deformation in the augen gneisses each mineral has to be treated separately.

The microstructures observed in quartz are comparable to deformation-structures, recovery and recrystallization processes described by Wilson [15], White [16, 17, 18, 19, 24] and Bell and Etheridge [21]. A sequentially development of the microstructures from undulose extinction - deformation bands - subgrains to misoriented subgrains is obvious with increasing strain, see [20]. These characteristics are indicative [20] of a deformation mechanism by primary and secondary dislocation creep.

The recovery process is dynamic as it involves both subgrain formation, recrystallisation and subsequent straining. All these stages of phenomena are observed in varying degrees from the K-feldspar megacryst bearing granodiorite to the most deformed augen gneisses.

The biotite is strain shadowed, kinked and shredded and only partly recrystallized at low strains. Recrystallization seems to be the main strain energy relaxation process at higher strains. That is from the schistose granodiorite and onwards.

The plagioclas behaves brittle in the granodiorite at Medskogen but starts to be plastically deformed in the schistose granodiorite. The microstructures were difficult to identify and analyse because of strong seritization of the mineral.

The K-feldspar, microcline and perthite, displays a brittle behaviour at all stages of the progressive deformation. Recrystallization, however, occurs extensivly in areas of high strain. That is in the megacrysts along brecciated and kinked zones and along the margins of the megacryst, especially in the stress shadow areas. There is only a few observations, both in the optical and electron microscope, of subgrain formation before recrystallization. The recovery process is very probable inhibited by the twins in a similar mannar as described by White [24]. The exclusion of the subgrain formation will cause greater workhardning and brittleness of the K-feldspar megacrysts dompared to the matrix minerals.

Temperature and strain rate have fundamental influences on the character of the deformation. Voll [27] found that recrystallization of quartz and biotite starts at a temperature of 300°C in a variscian granite and plagioclase deformed plastically above 500°C. These temperature estimations correspond well with the metamorphic mineral assemblage and microstructures observed in the augen gneisses. Thus, the plagioclase begins to behave plastic in the schistose granodiorite where also the first syntectonic epidote occurs. The growth of epidote indicates a temperature between 500-550°C.

The microstructures in the minerals, especially quartz, indicate that the augen gneisses were deformed by creep. Thus the strain rate is low. However, as local deformation by cataclasis is observed the strain rate very likely can be characterized as slow to moderate.

In conclusion, the Tännäs augen gneiss was developed by progressive deformation of a K-feldspar megacryst bearing granodiorite of Precambrian age. The deformation was by creep locally by cataclasis at a temperature range of 300-550°C. The strain rate was low to moderate. The augens are porphyroclasts, which have been deformed just as much as the minerals in the matrix. The divergency in response to deformation may be explained by the difference in deformation mechanisms between the augens and matrix minerals.

ACKNOWLEDGEMENT

This work is included in the Swedish Geodynamic Project, and financed by the Swedish Research Council. The TEM analyses were performed on the JEOL 200 microscope at the Department of Materials Technology, University of Luleå under supervision of Dr S White, Imperial College, London, PhD (eng) Bengt Loberg and Mr R Harrysson. Mrs M Lövgren draw the figures and Miss G Törnqvist typed the final manuscript. All are thanked for their contributions.

REFERENCES

[1] Törnebohm, A.E., 1898, Grunddragen af det Centrala Skandinaviens bergbyggnad. Kungl Sv Vet Akad Handl 28.

[2] Strömberg, A., 1961, On the tectonics of the Caledonides in the south-western part of the county of Jämtland, Sweden. Bull Geol Inst Uppsala. XXXIX

[3] Röshoff, K., 1978, Structures of the Tännäs augen gneiss nappe and its relation to under- and overlying units in the central Scandinavian Caledonides. Sver Geol Unders C 739

[4] Wegman, C.E., 1949. La flexure axiale de la Driva et quelques problèmes structuraux des Calédonides scandinaves. Norsk Geol Tidsk 39

[5] Strand, I., 1951. The Sel and Vågå map areas. Norsk Geol Under 178

[6] Rosenqvist, I., 1944. Metamorphism and metasomatism in the Opdal area (Sör-Tröndelag, Norway). Norsk Geol Tidsk 22

[7] Ramberg, H., 1973. Beskrivelse til berggrundsgeologisk kart over ströket Agdenes-Hemmnefjord, Sör-Tröndelag. Norsk Geol Under 299

[8] Hernes, J., 1951. Surnadalssynklinalen, The Surnaldal syncline, Central Norway. Norsk Geol Tidsk 36

[9] Högbom, A.E., 1920. Geologisk beskrivning över Jämtlands län. Sver Geol Under C140

[10] Carstens, C.W., 1920. Rapakiwigesteine und der westlichen Grenze des Trondheimgebiet. Norsk Geol Tidskrift Bd VIII

[11] Nystuen, J.P., 1975. Hovedtrekk av den tektoniske utvecklingen i östre del av sparagmitområdet i Sör-Norge. Rap 2. Institutt for Geologi, Norges Lantbrukshögskole

[12] Point, R., 1975. Mylonites et orogenése tangentielle: nature, géochimie, origine et age de gneiss oellés dans les nappes calédoniennes externes. B.S.G.F. 7, XVII

[13] Holmsen, P. and Holmsen, G., 1950. Tynset. Beskrivelse till det geologiske rektangel kart. Norg Geol Under 185

[14] Rui, I., 1972. Geology of the Röros district, south-eastern Trondheim region, with special study of the Kjöliskarvene-Holtsjöen area. Norg Geol Tidsk 52

[15] Wilson, C.J.L., 1973. The prograde microfabric in a deformed quartzite sequence, Mt.Isa, Australia. Tectonophysics 19

[16] White, S., 1971. Natural creep in quartzites. Nature Phys Sci 234

[17] White, S., 1973a. The dislocation structures responsible for the optical effects in some naturally deformed quartzites. J Mater Sci 9

[18] White, S., 1973b. Dislocations and bubbles in vein quartz. Nature Phys Sci 123

[19] White, S., 1975. The effect of polyphase deformation on the intracrystalline defect structures of quartz. N Jb Miner Abh 123

[20] White, S., 1976. The effects of strain on the microstructures, fabrics and deformation mechanisms in quartzite. Phil Trans R Soc London A

[21] Bell, T.H. and Etheridge, M.A., 1973. Microstructure of mylonites and their descriptive terminolgy. Lithos 6

[22] Carreras, K., Estrada, A. and White, S., 1977. The effects of folding on the c-axis fabrics of a quartz mylonite. Tectonophysics 39 (1-31)

[23] Tullis, J., Christie, J.M. and Griggs, D.T., 1973. Microstructures and preferred orientations of experimentally deformed quartzites. Geol Soc Am Bull 84

[24] White, S., 1975. Tectonic deformation and recrystallization of oligoclase. Contr Mineral and Petrol 50

[25] Etheridge, M.A. and Hobbs, B.E., 1974. Chemical and deformational controls on recrystallization of mica. Contr Mineral and Petrol 13

[26] Etheridge, M.A., Paterson, M.S. and Hobbs, B.E., 1974. Experimentally produced preferred orientation in synthetic mica aggregates. Contr Mineral and Petrol 44

[27] Voll, G., 1977. Recrystallization of quartz from variscan granites along a profile through Aar-and-Gotthard massif (Swiss alps). Abstract. Leiden conference on "Fabric, microtextures and microtectonics

EXPERIMENTAL DEFORMATION OF AUGEN-GNEISSES

Ove Alm and Ove Stephansson

Department of Rock Mechanics, University of Luleå, S-95187 Luleå, Sweden

ABSTRACT

Rock deformation experiments of two samples from the Tännäs augen gneiss nappe, Middle Sweden, are presented. The nappe is characterized by augen gneisses in different stages of natural deformation. The augen gneisses are developed by progressive deformation of Precambrian rocks of granodioritic composition with K-feldspar megacrysts.

The experiments were performed in a Griggs´ hot creep apparatus at maximum temperature of 500°C and pressures up to 700 MPa. The experimental results indicate a plastic flow of the matrix surrounding the megacrysts down to room temperature. The augens of K-feldspar behaved brittle at high pressure and temperature.

1. INTRODUCTION

Most theoretical models and geological models in particular need experimental input data. Knowledge of deformation properties of different rocks at high temperature and high pressure is a necessity in tectonic modelling of large scale structures.

Until recently most of the efforts in experimental rock deformation have been concentrated on rocks containing few minerals, in e.g. quartz [1], [2], [3], olivine [4], [5], diopside [6] and hornblende [7]. Experimental deformation of plagioclases have been reported by Borg and Heard [8] and plastic defects in experimentally deformed potassic feldspar have been studied by Williame and Gandais [9]. A great deal of experimental work has been done on granites in the brittle deformation regime. The influence of pressure and temperature on friction in faulted granite samples have been studied by several authors [10], [11], [12]. Experimental studies of granitic rocks in the ductile field are few and very recent [13], [14], [15]. We know, however, that granitic rocks are studied in several laboratories in different parts of the world and the number of papers will probably increase in the next few years.

We report here the results from a series of high temperature and/or high pressure experiments in augen gneisses from the Tännäs augen gneiss nappe in the province of Härjedalen, Middle Sweden. A comprehensive geological description of the nappe and its relation to under- and overlying units was recently published by Röshoff [16] and only a short summery will therefore be given in this paper. The difference in competence between the augen and the surrounding matrix will affect the deformation properties of the nappe structure. Quantitative values of the equivalent viscosities of the nappe forming rocks at different pressures, temperatures and strain rates and their mechanism of deformation are necessary for the theoretical modelling of the tectonic behaviour of the area. The present study is the first report in this project.

2. A BRIEF DESCRIPTION OF THE GEOLOGY AND THE ROCK TYPES

To study the mechanism of deformation in a nappe structure and to compare the fabric with the one developed in experimentally deformed rock the ideal is to find an area with progressive deformation. Here the rocks can be analysed from the stage of weak deformation to highly strained rocks. From the mapping by Röshoff [16] we were pointed out a suitable area at Hede in the nappe structures of the eastern parts of the Caledonides, Midddle Sweden. Here the rocks form large, low dippint thrust sheets with a direction of movement from the west to the east. The rock samples for this study were collected in the Tännäs augen gneiss Nappe overlain by the Serve Nappe of low grade quartzo-feldspatic rock. The sampling area is indicated in the geological map, Fig 1.

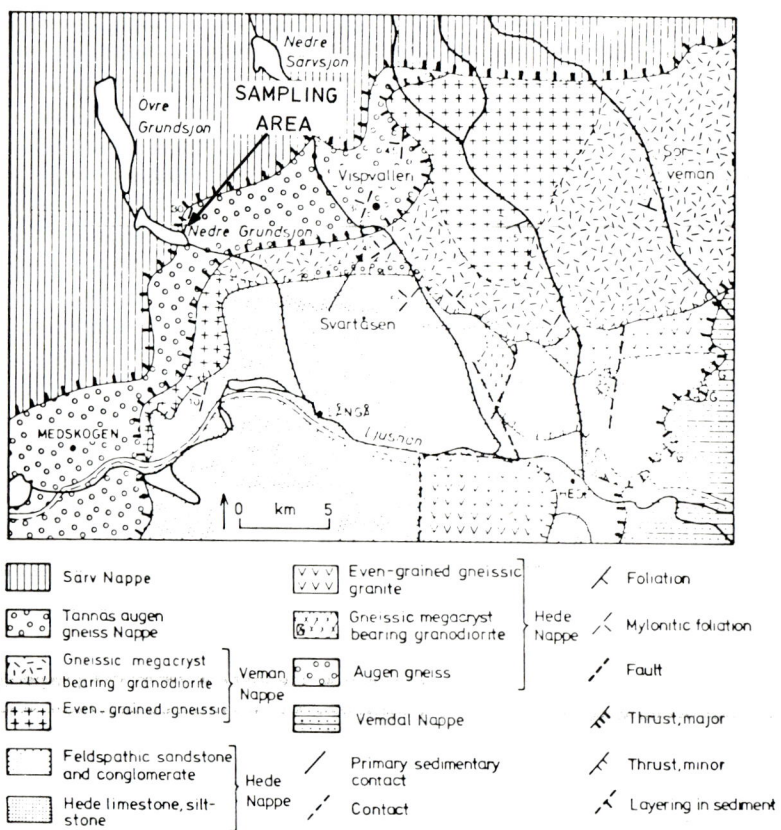

Fig 1 Geological map showing the sampling area
       (After Röshoff [16])

For a length of about one kilometer along a new road the rocks of the Tännäs augen gneiss Nappe are exposed. They varies from a homogeneous, coarse grained granodiorite with granitic texture over to a typical augen gneiss with large porphyroblasts of K-feldspars to the final stage of a complete mylonite. A microphotograph of a thin section of the granodiorite is shown in Fig 2. The rock is mainly composed of sericite, perthite and quartz. The same rock type in a more deformed stage of deformation forms a typical augen gneiss with large porphyroblast of K-feldspar in a matrix of fine-grained sericitesized feldspars and quartz as indicated in Fig 3. A close up of a microcline porphyroblast is shown

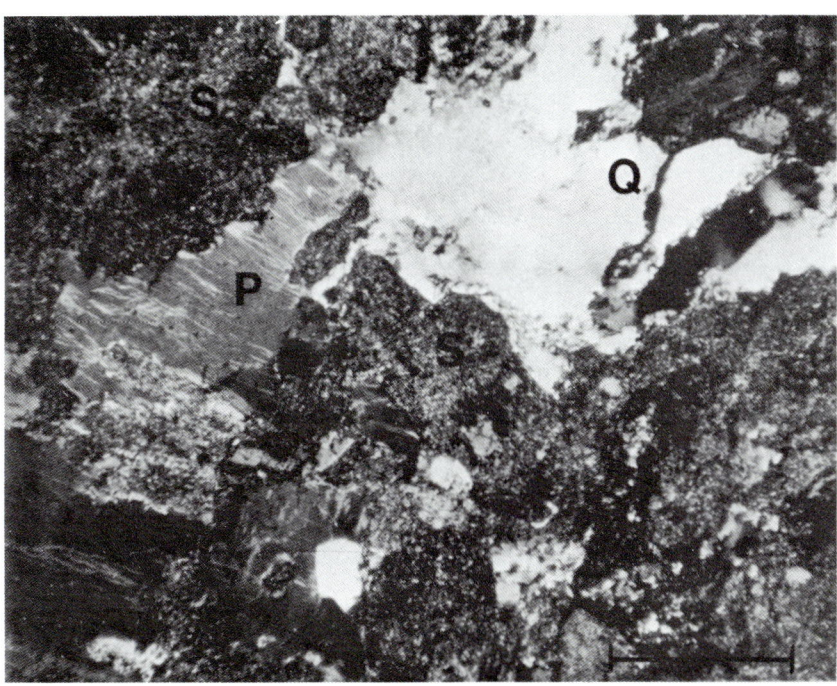

Fig 2   Microphotograph of granodiorite; S = sericite, P = perthite, Q = quartz
         Scale bar 1 mm, crossed nicols

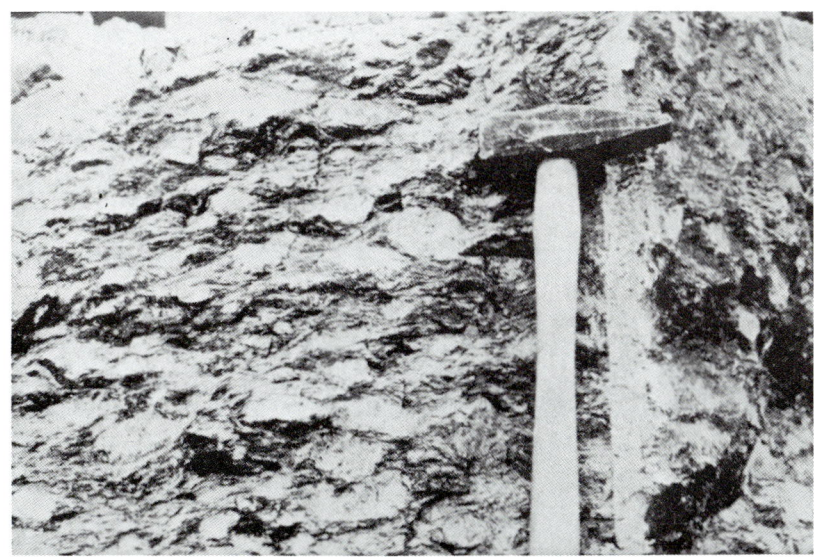

Fig 3   Augen gneiss from the area east of lake Nedre Grundsjön

in Fig 4. The matrix surrounding the porphyroblasts show a strong foliation and the structures very often warp around the more competent inclusion and sometimes brake the edge or outer parts of the inclusions,

A microphotograph of the matrix is shown in Fig 5. For the experimental analysis we have chosen to squeez the rocks with a granitic texture and the matrix and porphyroblast of a typical augen gneiss.

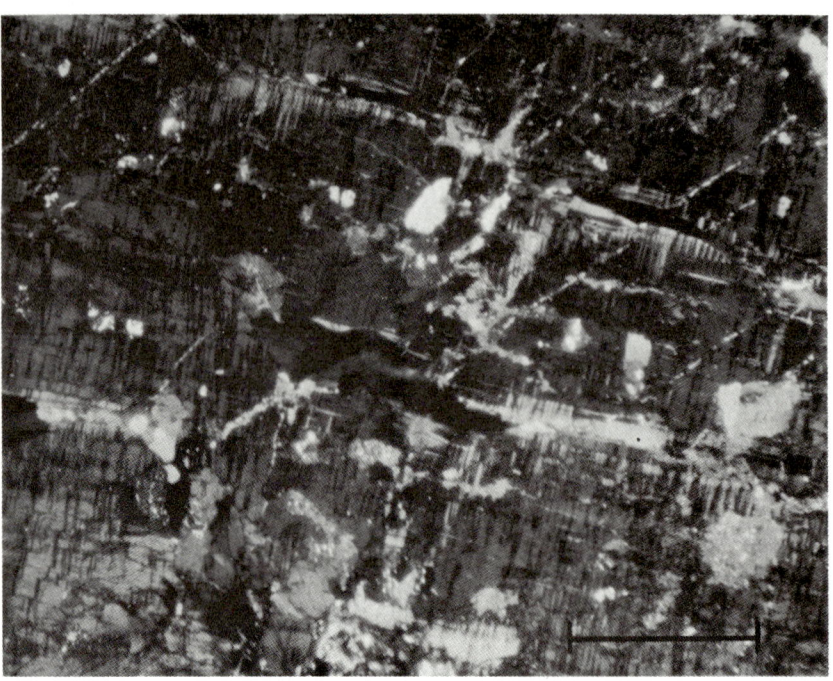

Fig 4   Microphotograph of a microcline perthite auge; scale bar 1 mm, crossed nicols

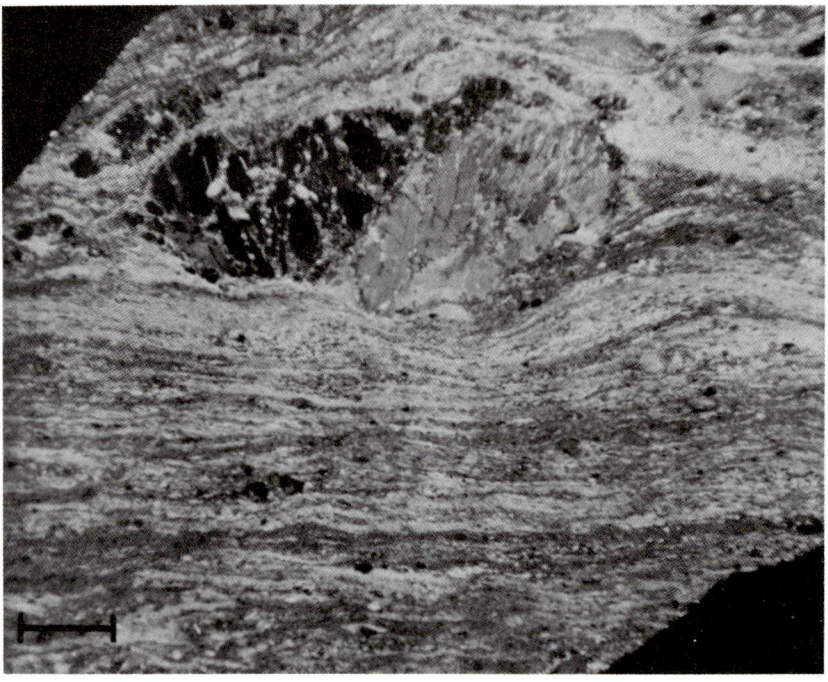

Fig 5   Microphotograph of the dark greyish matrix; scale bar 2 mm, crossed nicols

3. EXPERIMENTAL DEFORMATION OF TÄNNÄS AUGEN GNEISSES

Experimental procedure

The rock specimens used in this study were cored from two samples of naturally deformed augen gneiss - one granodiorite sample with granitic texture and one sample containing one large microcline perthite auge ($\phi \sim 8$ cm) surrounded by a dark, greyish fine-grain matrix. The specimens obtained from the matrix were cored out normal to the foliation. These specimens often contained one or more microcline perthite auge of a diameter of about 0.5 mm. The breccia-like structure, typical of the microcline perthite augen, is caused by the natural deformation of the rock after the formation of the augen. All the specimens cored from the large auge contained visible cracks.

A slide rest grinding machine equipped with a thin diamond blade and mounted on a lathe was used to cut the specimens (ϕ 8 mm) to a length of about 11 mm. No further preparation of the ends was necessary. All the specimens were then left to dry at 100°C for 5 days and thereafter stored in a desiccator until they were used in the experiments.

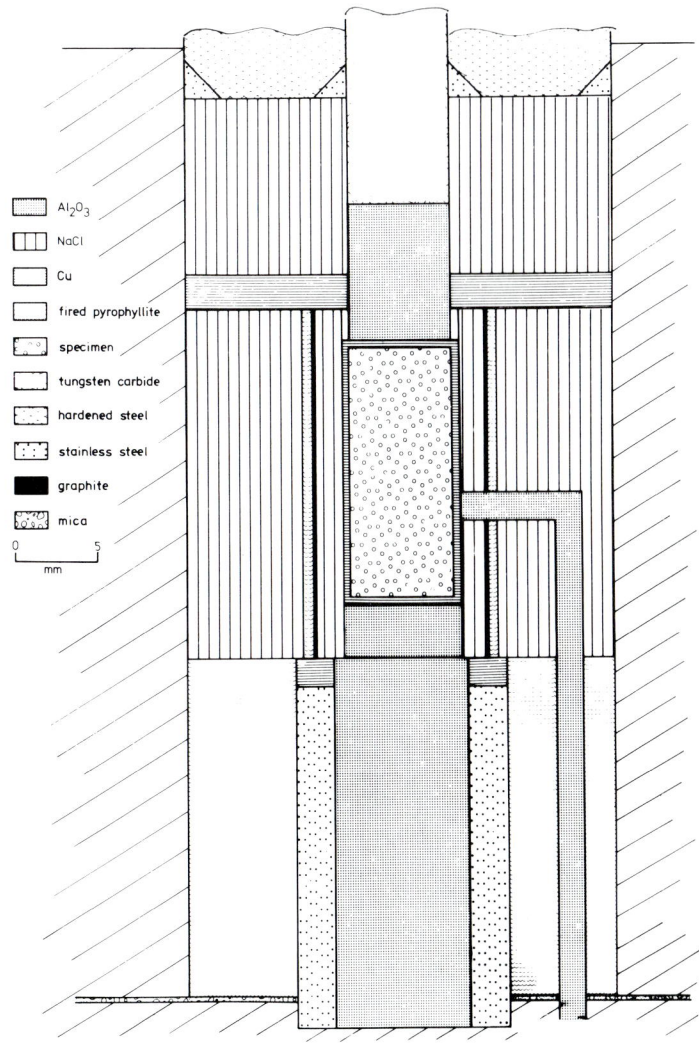

Fig 6   High-pressure cell for rock deformation
        (After Alm [15])

The experiments were performed in a solid medium Griggs' hot creep apparatus operated in the constant strain-rate mode. Lead was used as pressure transmitting medium in the low temperature ($\leq 100°C$) experiments. A slightly modified version of the high-pressure cell described in the paper of Alm [15] was used in most of the high temperature experiments, Fig 6. A new design of the graphite heater was used in some of the experiments. This new heater consisted of a thin-walled, graphite tube short circuited in the central region by a copper ring 4 mm high. The temperature differences along the specimens were reduced to less than 1/4 of the original ones, which could be as large as $100°C$ in unfavourable cases. These temperature differences were determined by measuring the temperature just below the bottom of the specimen simultaneously with a registration close to the central region of the specimen.

The experiments were performed at constant temperature and confining pressure in the range 20 - 600°C and 100 - 700 MPa respectively. All the specimens were deformed at a mean strain-rate of $1 \cdot 10^{-6} s^{-1}$.

The deformed specimens were immersed in epoxy and sawed in halves along the axis of the cylinder and oriented normal to fault planes or to other interesting visible features on the surface. Optical thin sections were then manufactured from one of the halves.

Stress-strain curves

The stress-strain curves shown below are nominal stress-strain curves. These curves do not represent the actual, effective response of the rock to the applied load, but it is as close as it is possible to get because of the non-homogeneous deformation of the specimen. The non-homogeneity is partly due to the experimental conditions - solid pressure medium and piston diameter equal to the diameter of the specimens - and partly to the mineral and textural differences prevailing in each specimen. It is difficult to define the effective area on which the axial load is applied.

In Fig 7 the stress-strain curves for three specimens, are presented, one curve for the granodiorite with granitic texture, one for the microcline perthit auge, and for the matrix surrounding this auge. The strength of the granodiorite seems to be approximately the same as that of the matrix. The microcline specimen failed by a shear fault at a much lower stress and strain than the other two specimens.

The other specimens, both of the granodiorite with granitic texture and the microcline porphyroblast, failed at a much earlier stage of deformation. We believe that the distribution of cracks in the microcline auge and the non-homogeneous distribution of minerals and grain size together with cracks in the granodiorite sepcimen are the limiting factors of strength for these rock specimens. Fig 8 presents the stress-strain data for matrix specimens at three different temperatures. The influence of temperature on the strength of the matrix seems to be small at least at temperatures up to 400°C.

Four experiments have been performed at different perssures at room temperature. The aim is to investigate the pressure dependence of the strength of the matrix. Only the specimen deformed at 100 MPa failed by a throghgoing shear fault. The sterss-strain curves for four specimens are shown in Fig 9. The strength increases with pressure but not as

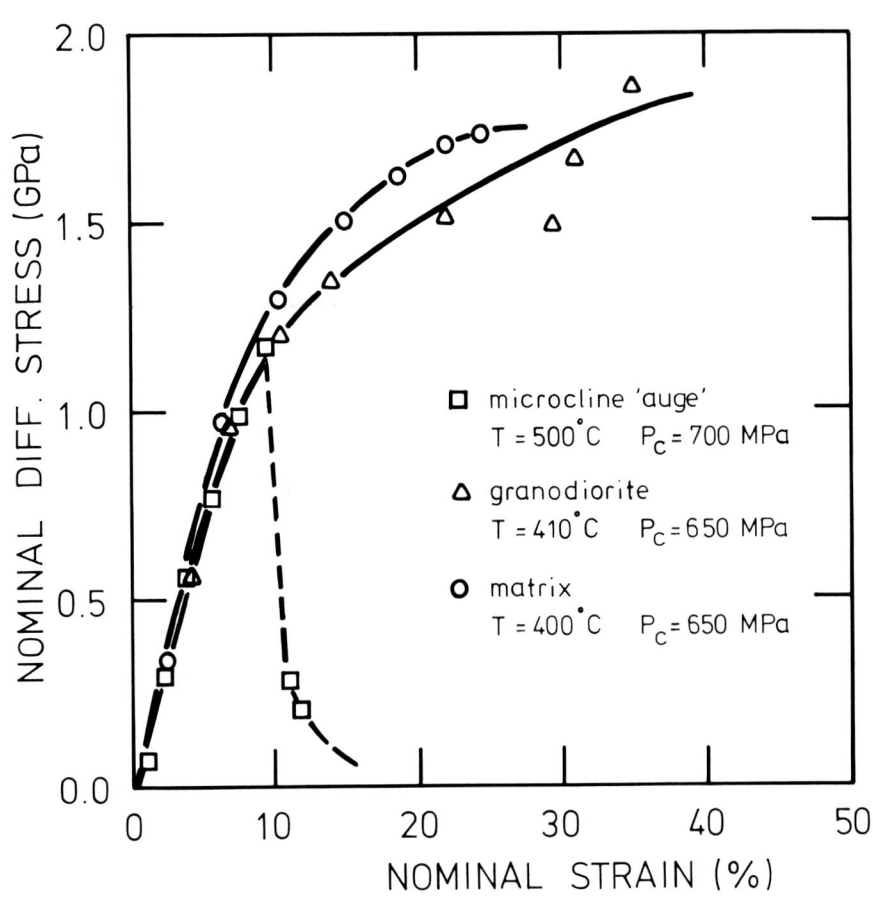

Fig 7 Nominal stress-strain curves for three specimens; $\dot{\varepsilon} = 1 \cdot 10^{-6}$ s^{-1}

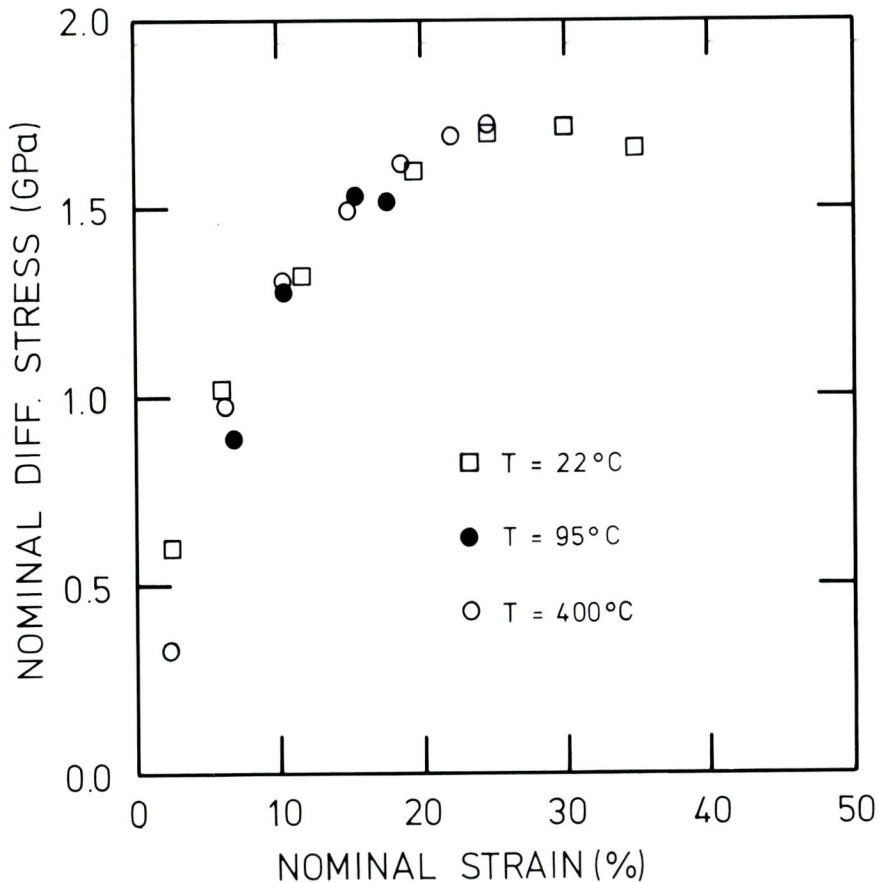

Fig 8 Nominal sterss-strain data for matrix specimens tested at different temperatures; $\dot{\varepsilon} = 10^{-6}$ s^{-1} and confining pressure 700 MPa

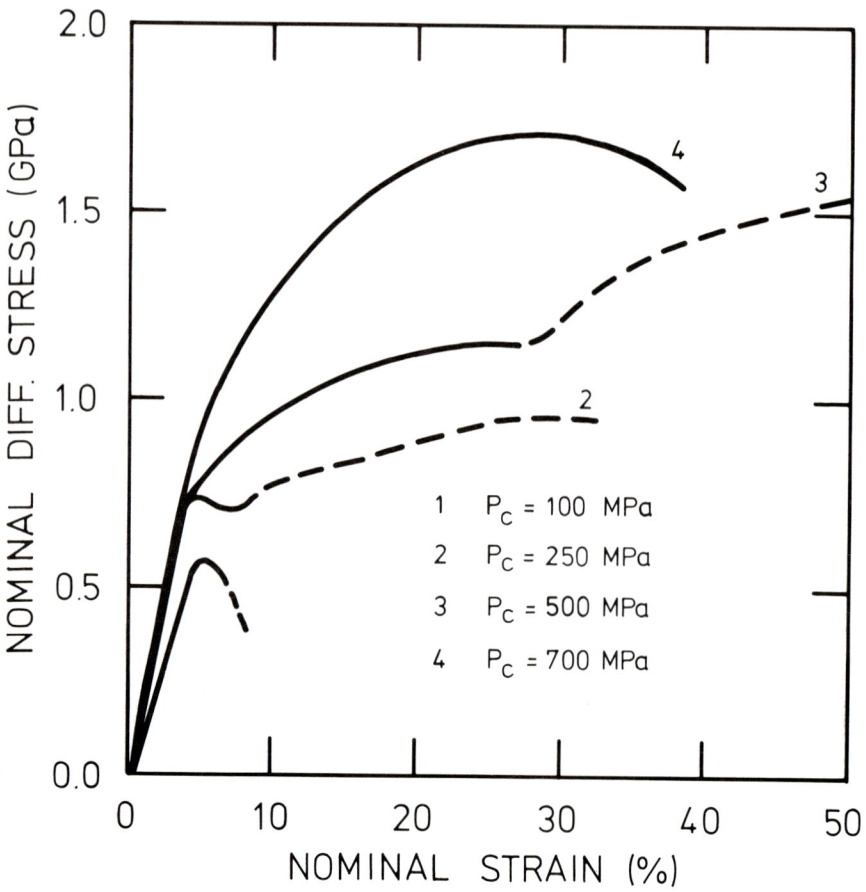

Fig 9 Nominal stress-strain relationship for specimens of the matrix tested at room temperature and various pressures; $\dot{\varepsilon} = 10^{-6}$ s^{-1}

much as for granites [14], [17] and the actual effective strehgth is roughly 30% lower than for these rocks at 700 MPa. The upward trend of curve 2 and 3 is an effect of increase in diameter of the specimens.

Microstructures

In this section we will discuss the difference in microstructures between naturally deformed samples and samples which have an experimental deformation superimposed on the natural deformation. The microstructures of the rocks belonging to the Tännäs augen gneiss Nappe have been studied by Röshoff (this volume).

The specimens of microcline perthite failed by throughgoing fractures along existing open or healed cracks or along grain boundaries. The rock was fragmented in the fault zone and the small fragments were lost in the cutting of the specimen. New cracks were developed at a small distance from the fault zone. One of the specimen was heavily fracture although the deformation was stopped at 7% shortening.

Five experiments were performed on granodiorite specimens with granitic textures. All but one failed by throughgoing fracturing. The fracture pattern was different in the specimens. One of them was fractured in the direction of the largest applied stress σ_1. Another contained large cracks various directions although most of them seemed to form a fault plane approximately 38° to the direction of σ_1. Two specimens failed by shear fractures orientated about 34° to σ_1.

Only one specimen behaved ductile as indicated in the microphotograph of Fig 10. A weak shear zone crossing the specimen is visible. The deformation is mainly cataclastic for the K-feldspar and quartz grains and large grains are fragmented into a number of smaller grains, (Fig 11). The biotite behaved less brittle and the sericite is deformed by intergrain

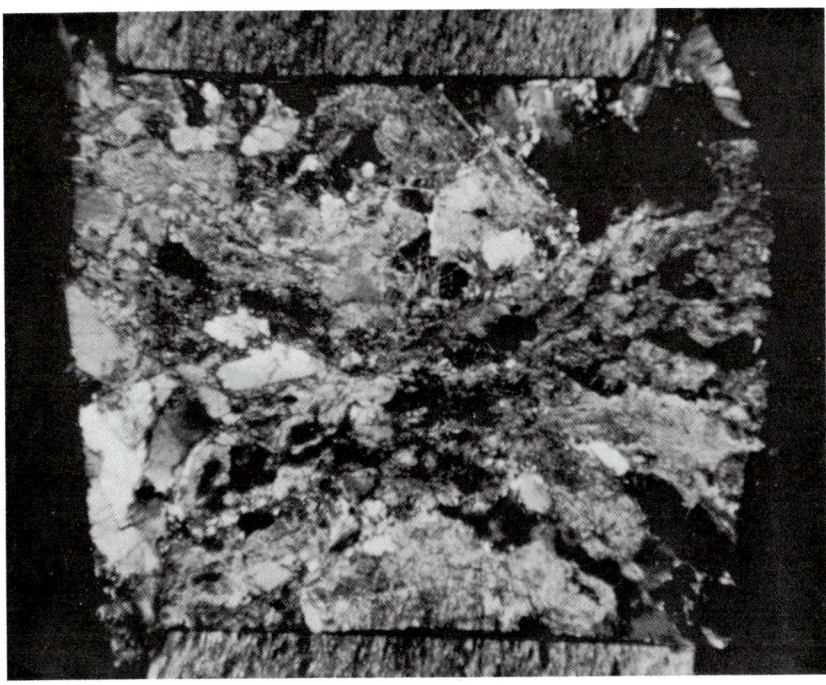

Fig 10   Cross-section of deformed granodiorite; specimen length between the pistons is 7.5 mm after 35% shortening

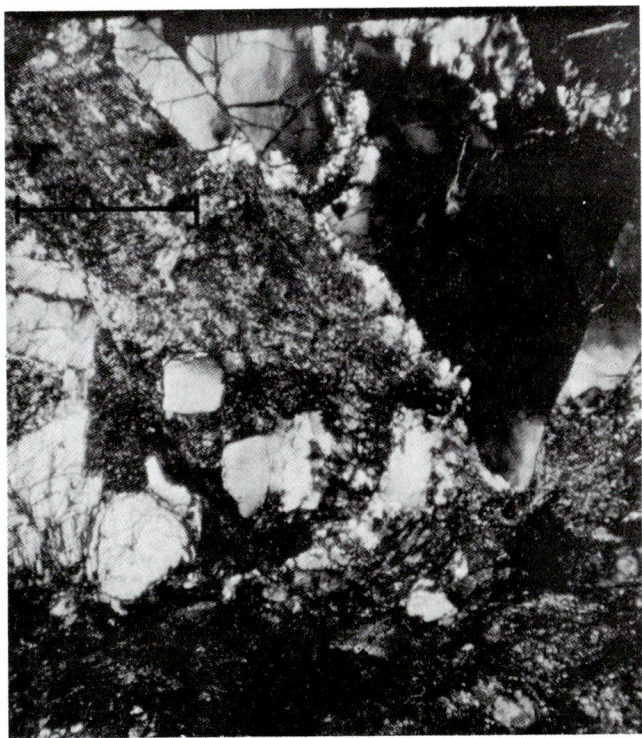

Fig 11   Extensive fracturing in granodiorite; larger magnification of the upper right-hand corner of figure 10, scale bar 1 mm, crossed nicols

gliding and rotation. The grain size of the sericite is too small to optically reveal any fracturing of individual grains.

There may be sereral explanations for the mechanism of deformation of these specimens, but we believe that the non-uniform distribution of the minerals, especially the micas, have a great influence on the development of cracks and the direction of propagation of these cracks. Furthermore, the pressure and the temperature are too low and the strain rate to high to favour any other deformation mechanism than cataclasis for quartz and feldspar. The strain rate is probably the most important factor to account for the difference in microstructure for naturally and experimentally deformed rocks. Notice that the grain size is large compared to the size of the specimens. This usually results in a non-homogeneous stress field around the grains and increases the probability for cracking.

We chose the sample containing one large microcline auge (ϕ 8 cm) surrounded by a dark grey matrix because the matrix itself contained small microcline perthite augen. We can therefore study the interaction between the matrix and the more competent augen. The fine banded structure of the matrix also allows us to follow the shearing of material during the deformation and how the sheared structures varies with temperature and total strain.

Figure 12 shows a microphotograph of a cross-section of a specimen of the matrix deformed at room temperature. The confining pressure was 500 MPa ar specimen was shortened 45%. The figure displays well developed shear zones along which the strain rate has been mych higher than for other parts of the specimen. The white bands in the sample are strongly deformec in these zones. The same features were produced in all experiments at room temperature except the one that failed by a throughgoing fault. Moreover the shear zones and the central area where the zones cross have a different, light green, colour compared to the rest of the specimen. This may be due to cataclasis and changes in orientation of the grains by rotation. One specimen deformed at 400°C do not have such well developed shear zones. A couple of the white bands are much thicker than for the specimens deformed att room temperature, Fig 15. This may, however, depend on the initial microstructure of the specimen. More experiments are needed to be performed at these temperatures before any explatation can be given.

A magnification of the large feldspar grain and of the central region of the specimen in figure 12 are shown in figure 13 and 14 respectively. The already brecciated grain is further broken and the fragments towards right in the figure are rotated away from the host grain and the flaws are filled with fine grained material from the surrounding matrix.

The white quartz and feldspar layers consisting of small recrystalized grains are formed by progressive deformation of larger recrystallized quartz or feldspar grains. These layers are more competent than the extremely fine-grained ground mass on either side and are likely to break up into boudins during the experiment, as is displayed by figure 14 The boudins are separated from each other because of the difference in size and difference in the rate of flow of the surrounding material. Small obstacles may also cause rotation of the smaller boudins. These microscopic structures seen in figure 14 resemble very much the large scale boudinage structures.

Fig 12   Cross-section of a specimen of matrix tested at room temperature and 500 MPa confining pressure; length after 45% shortening 7.6 mm, crossed nicols

Fig 13   Magnification of the large microcline perthite grain in figure 12; direction of maximum stress diagonal as indicated by arrows, scale bar 1 mm, crossed nicols

Fig 14   Boudinage structure developed in the central part of the specimen shown in figure 12; direction of maximum stress is vertical, scale bar 0.5 mm, crossed nicols

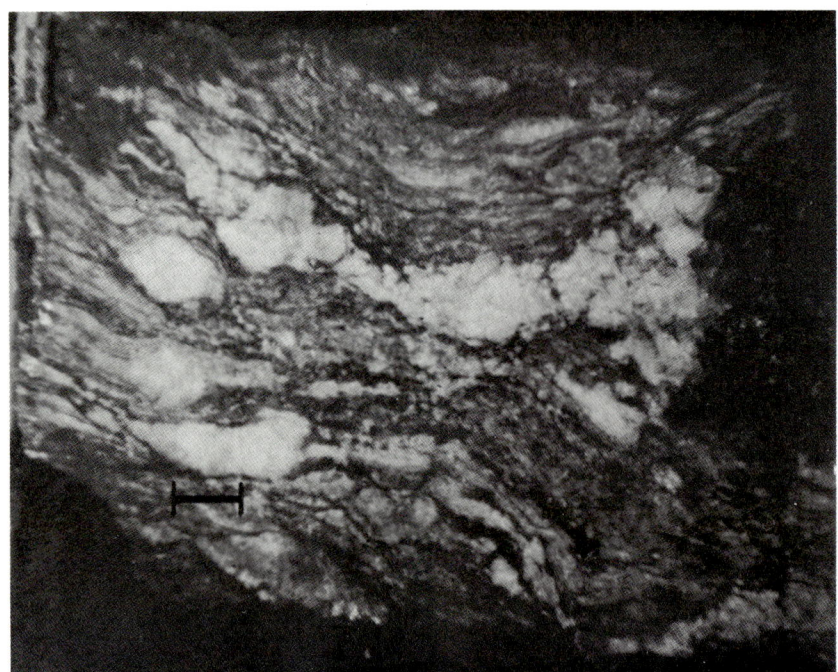

Fig 15   Microphotograph of the polished surface of the cross-section of matrix deformed at $T = 400°C$ and $P_c = 650$ MPa; direction of maximum stress is vertical, scale bar 1 mm

The high temperature specimen shown in figure 15 was shortened 30%, which is enough to allow comparision with the results discussed in the preceding paragraph. The two much wider bande appearing in this specimen indicate a pinch and swell structure typical for an complete body of a boudinage structure.

The different microstructures of the samples shown in figure 12 and 15 indicate that different deformation mechanisms have been active in these two specimens. The material flow shown in figure 12 is mainly caused by intergrain sliding and rotation while twinning, kinking and dislocation flow may contribute substancially in the deformation at high temperatures.

ACKNOWLEDGEMENT

This study is a part of the Swedish Geodynamic Project finanzed by the Swedish Natural Science Research Council, Project nr G 3447-010. We are grateful for valuable discussions with Dr K Röshoff.

REFERENCES

[1] Hobbs, B.E., McLaren, A.C. and Patterson, M.S., 1972. Plasticity of single crystals of synthetic quartz. In: Flow and Fracture of Rocks, Geophysical Monograph 16, Eds Heard, H.C., Borg, I.Y. Carter, N.L. and Raleigh, C.B. American Geophysical Union, Washington, pp 29-53

[2] Tullis, J., Christie, J.M. and Griggs, D.T., 1973. Microstructures and preferred orientations of experimentally deformed quartzites. Geol Soc Am Bull 84, 297-314

[3] Ardell, A.J., Christie, J.M. and Tullis, J., 1973. Dislocation substructures in deformed quartz rocks. Crystal Lattice Defects 4, 275-285

[4] Raleigh, C.B., 1968. Mechanisms of plastic deformation in olivine J Geophys Res 73, 5391-5406.

[5] Blacic, J.D., 1972. Effect of water on the experimental deformation of olivine. In: Flow and Fracture of Rocks, Geophysical Monograph 16, Eds Heard, H.C., Borg, I.Y., Carter, N.L. and Raleigh, C.B., American Geophysical Union, Washington, pp 109-115

[6] Kirby, S.H. and Christie, J.M., 1977. Mechanical twinning in diopside Ca(Mg, Fe)Si$_2$O$_6$: structural mechanism and associated crystal defects. Phys Chem Minerals 1, 137-163

[7] Rooney, T.P., Riecker, R.E. and Gavasci, A.T., 1975. Hornblende deformation features. Geology 3, 364-366

[8] Borg, I.Y. and Heard, H.C., 1969. Mechanical twinning and slip in experimentally deformed plagioclases. Contrib Mineral Petrol 23, 128-135

[9] Williame, C., Gandais, M., 1977. Electron microscope study of plastic defects in experimentslly deformed alkali feldspars. Bull Soc fr Minéral Cristallogr 100, 263-271

[10] Stesky, R.M., Brace, W.F., Riley, D.K. and Robin, P-Y.F., 1974. Friction in faulted rock at high temperature and high pressure. Tectonophysics 23, 177-203

[11] Ohnaka, M., 1975. Frictional characteristics of typical rocks. J Phys Earth 23, 78-112

[12] Stesky, R.M., 1978. Mechanisms of high temperature frictional sliding in Westerly granite. Can J Earth Sci 15, 361-375

[13] Murrell, S.A.F. and Ismail, I.A.H., 1976. The effect of temperature on the strength at high confining pressure of granodiorite containing free and chemically-bound water. Contrib Mineral Petrol 55, 317-330

[14] Tullis, J. and Yound, R.A., 1977. Experimental deformation of dry Weaterly granite. J Geophys Res 82, 5705-5718

[15] Alm, O., 1977. Fine-grained granite rocks experimentally deformed at high temperatures and high pressures. Technical Report 1977:61T, University of Luleå. Paper discussed at the 16th EHPRG Meeting att Paris-Saclay 3-5 Oct 1977

[16] Röshoff, K., 1978. Structures of the Tännäs augen gneiss nappe and its relation to under- and overlying units in the central scandinavian Caledonides. Sver Geol Unders C739, 1-35

[17] Ohnaka, M., 1973. The quantitative effect of hydrostatic confining pressure on the compressive strength of crystalline rocks. J Phys Earth 21, 125-140

ROCK FRACTURE UNDER SUPERIMPOSED STATIC AND DYNAMIC LOADS

G. Swan

Department of Rock Mechanics, University of Luleå, S-951 87 Luleå, Sweden

ABSTRACT

The basic philosophy and reasoning for attempting to superimpose static and dynamic loads in the field of rock crushing, where failure occurs by fracture under compressive loading, is presented. A discussion then follows on the choice of experimental method from which an evaluation of the effects of such load superpositioning may be made. Tentative results from experimental work are given next, followed by an assessment of this data.

1. INTRODUCTION

Typically when concerned with fracture mechanics an engineer is giving attention to the prevention of fracture or failure in a body. From the point of view of rock fragmentation however one seeks in some way to reverse preventive fracture notions and so hopefully effect an increased efficiency of fracture (here defined as the ratio of work done by forces extending crack tips to the work done by the external load(s)) or an improved fragmentation.

Shown in Fig 1(a) and 1(b) are two simplified external static load types which do or could apply when crushing particles (spherical or disc form)

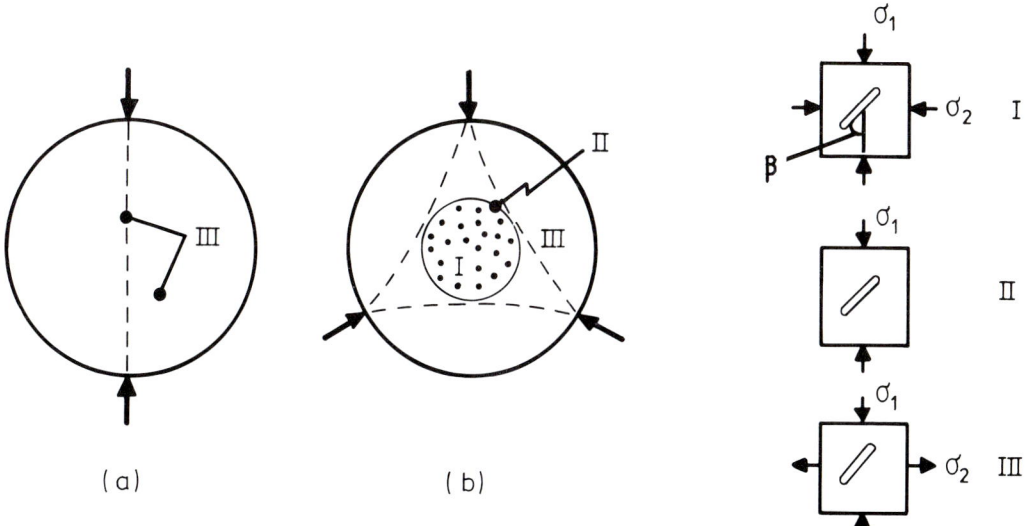

Fig 1 Two simplified load types for crushing particles, I-III showing local characteristic principal stress field

under practical conditions. In the case of Fig 1(a) the external load will, for the majority of rock types, simply cause fracture on a single diametral plane [1]. This result is in contrast to the observation in [2] where glass particles are found to shatter into many pieces. (Such a contrasting result between rock and glass has also been noticed by the author in the case of the "blow-out" test of Gorham and Rickerby [3] and [4]). The observed shattering in glass has been used as evidence for the existence of internal flaws [2]. However, the fact that rock does not shatter cannot, it is clear, be used as evidence against there being flaws in rock. Rather, to account for the comparatively non-brittle behaviour of rock in respect of the above, some brief attention must be given to

(a) the flaw size distribution

and (b) the crystalline structure of rocks

Considering the first of these a recent paper [5] goes on to discuss the physical meaning of the Weibull modulus m in brittle fracture. Among other things it is shown that for materials with smaller m values the variability of flaw size per unit volume is the greater, i e the number of most dangerous flaws in a given volume is an inverse function of m. With this interpretation for m and from the data of Table 1 it can be expected that glass will fragment so much better than rock at failure, there being for equal stressed volumes many more dangerous flaws in glass.

Table 1

	Energy Release Rate, G_c N/m	Weibull m	Source
Glass	8 - 9	2 - 4	[11], [12]
Bohus Granite	80 - 120	20	[8]

This being the case, a method for treating material so as to increase the density of dangerous flaws present prior to its failure in a fragmentation process would be clearly attractive. This essentially provides the motivation in part for attempting to superimpose dynamic loads to an otherwise statically loaded particle in the process of size reduction.

Probably of equal importance in the matter of accounting for the comparatively non-brittle behaviour of rock is the rôle played by the crystalline structure in which flaws are embedded. It being almost certain that cracks propagate in rock by a process of coalescence [6, 7], it is not unreasonable to expect certain crystals to act as obstacles thereby slowing down the rate at which a new fragment may be created. Thus, while recently testing 6 mm x 99 mm ϕ Bohus Granite discs in the "blow-out" test mentioned earlier [8] cracks were observed to extend with apparent velocities the order of (mm/s). In addition to obstacles, frictional forces are no doubt acting on flaw surfaces owing to the existence of compressive principal stresses bringing the surfaces into contact. Such surface tractions will certainly diminish the efficiency of fracture.

2. THE SUPERPOSITION EFFECT

The usefulness of stress waves per se in rock fragmentation has received considerable attention elsewhere [9]. Suffice to say here that a spalling process may occur in brittle materials where tensile pulses are generated at free or reflective boundaries. Where strong wave interaction occurs both grain rotations and separations have been observed [10]. However, the thought behind the present work is not that of a dedicated stress wave spalling process, but of a partial relieving mechanism to friction forces which otherwise (here speculated) lock flaw surfaces. Thus, assuming certain cracks under a static load could not initiate, at the arrival of a stress wave (now partially tensile by inversion) superposition occurs so causing a seperating influence of flaw surfaces. Given such a influence the prospect of initiation will increase as can be seen from Fig 2.

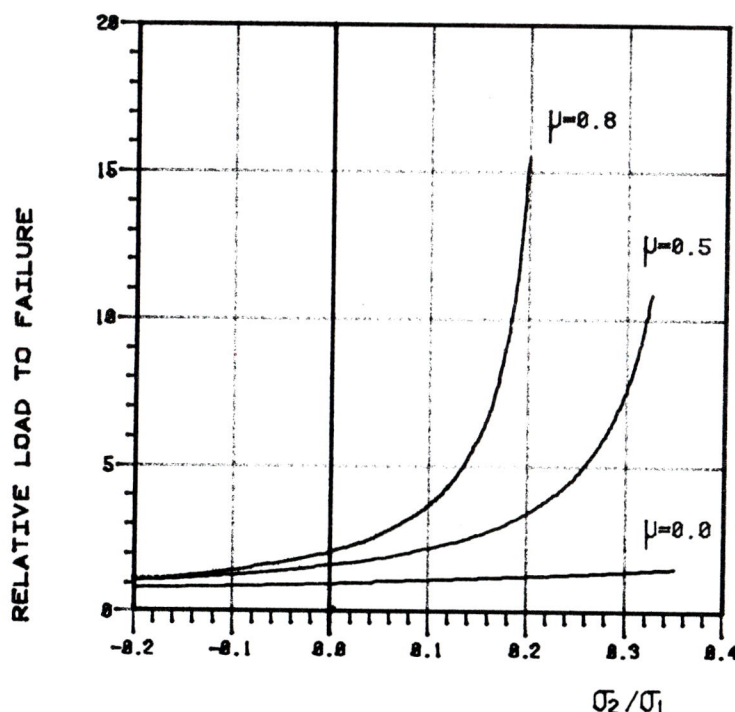

Fig 2 Relative load to failure vs principal stress ratio for an inclined crack (see Fig 1(c)-I) with different surface friction values

In addition to this favourable effect upon crack initiation, a further benefit can be thought to occur given propagation conditions. This argument originates from the known stabilising influence of the stresses which act upon a crack propagating under the conditions of Fig 1(c)-I. The benefit to be looked for is that of deferring this stabilising influence by momentarily disturbing the alignment of the major compressive principal stress trajectory. In relation to the crack axis, this would mean decreasing the angle β as the crack extends, so enabling the crack to move through a greater arc than would otherwise be possible. Although the ability to control the action of the disturbance upon a given crack will be practically impossible, one should expect a certain proportion of cracks to be preferentially "over-extended", analogous to the results in [9].

3. EXPERIMENTAL

Basically the experimental work has as its objective to observe whether or not the statistical distribution of flaws in a rock can be favourably modified by a superposition treatment. Once the rock - here machined to the shape of a thin disc, 51 mm dia x 6mm - has been treated, it remains only to examine residual strengths in order to formulate a result. The philosophy follows closely that reported in [3] where glass discs treated by water jet impaction were finally tested using the apparatus shown in Fig 3. This apparatus, as well as possessing many practical

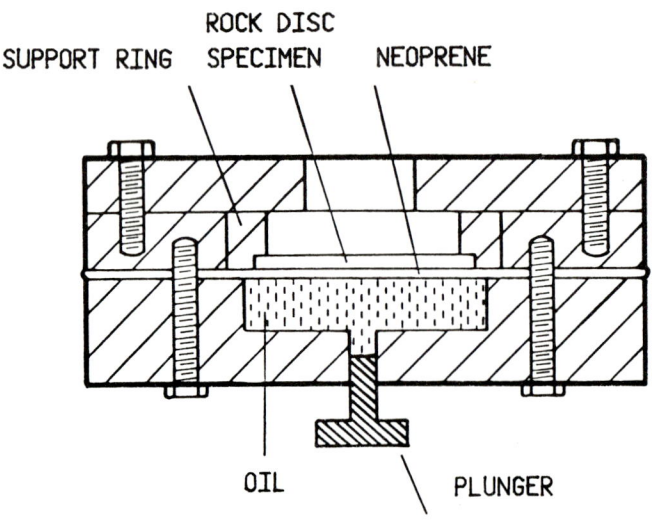

Fig 3 Biaxial tension apparatus used for residual strength testing

advantages over other strength testing methods (see [3]) is featured by:

(a) sensitivity to strength impairing flaws greatest at the specimen (disc) centre

(b) surface condition or superficial distribution of cracks dominates in strength determinations obtained

Where the feature (a) is considered in conjunction with a superposition treatment based on the static load configuration of Fig 1(b), a symmetry of desirable conditions is apparent. The superposition of a dynamic pulse to the central core static stress field so as to obtain a new flaw distribution more favourable to fragmentation (i e containing a greater proportion of dangerous cracks) is, in particular, our objective. The method of applying the dynamic pulse is that of exploding a thin wire encapsulated in a fuse, more details of which are given in [4]. The energy delivered for a single explosion is small (10-30 J) compared to the energy of the static load, the latter at the moment of superposition being just under that necessary for failure. Typical pulse magnitudes and wavelengths obtained in rock are 50-150 MPa and 5-10 µs respectively.

Three experiments were performed, each using a number of discs, so obtaining strength data for the following:

(I) prepared rock discs without any pre-loading treatment

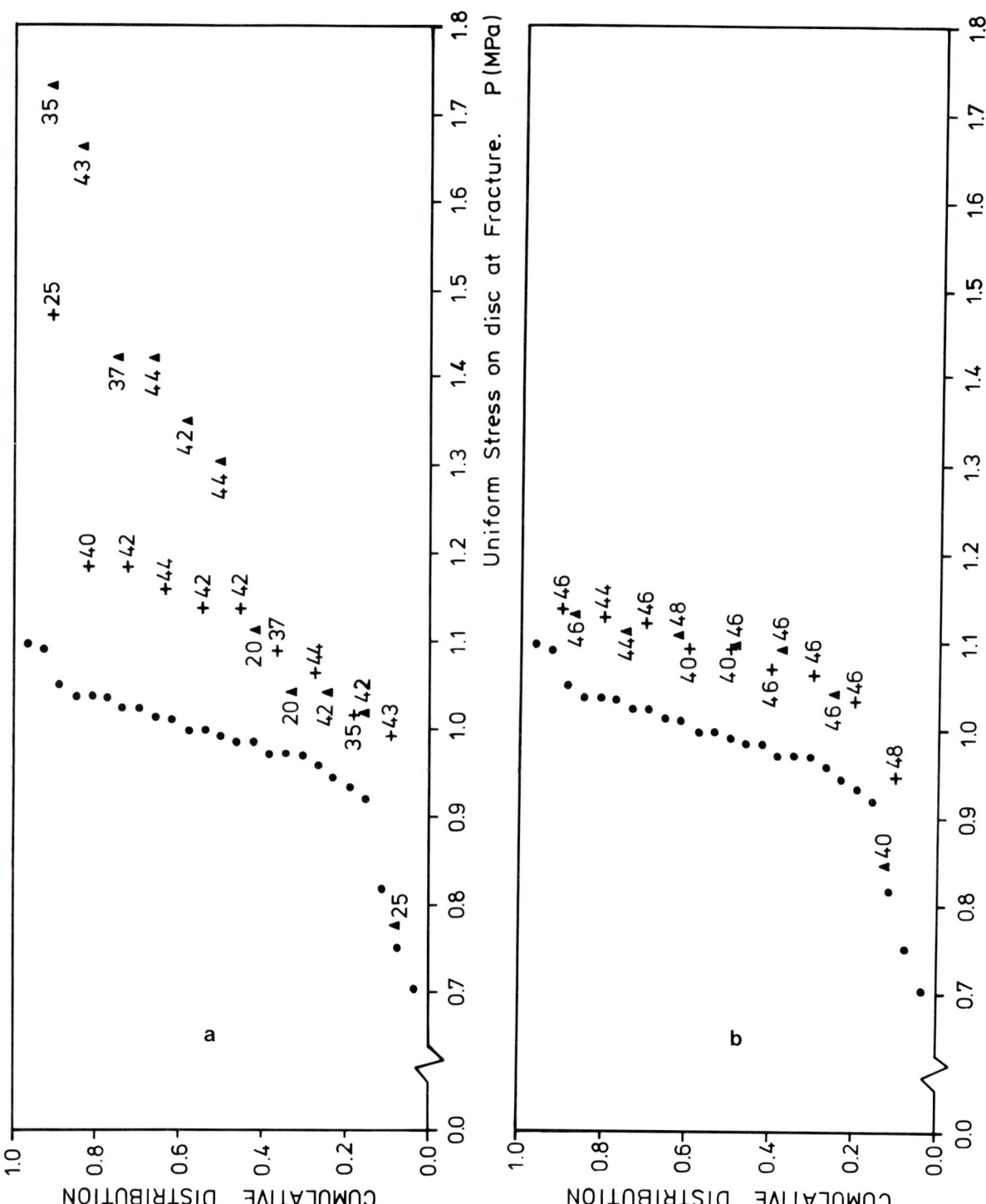

Fig 4 (a) Residual strength data, initial test series

(b) Residual strength data, second series after "improving" test apparatus

Symbols: • discs without pre-treatment

▲ discs pre-loaded statically, figures give load in kN (50 kN is failure load)

+ discs pre-treated with superimposed loads

(II)   prepared rock discs pre-loaded statically to within 95 % failure load

(III)   prepared rock discs pre-loaded with a superimposed static and dynamic treatment, the former adjusted to within 95 % failure load

4. RESULTS

The entire experimental results are shown plotted in Fig 4(a) and (b) as distributions of failure stress. In each case the data (·) are for those discs without any pre-load treatment, from which a Weibull modulus m of approximately 20 may be calculated. The remaining data of Fig 4(a) applies to an initial series of experiments where the conditions (II) (Symbol (Δ)) and (III) (symbol (+)) were fullfilled. After making certain modifications to the equipment (eg replacing support ring in Fig 3 with a hardened steel finish) a further series of experiments were performed, the results of which appear in Fig 4(b). The method of preparing the discs in each series was unchanged.

5. DISCUSSION

It is apparent from the data of Fig 4, considering the two treated disc conditions (II) and (III), that whilst the initial series suggests a strength reduction by superposition on the average of about 12 %, the second series shows no effect of significance. The difference between the two series is thought to be due to the wearing of the mild steel support ring, see Fig 3, which was hardened for the second series. Although it thus appears that superimposed loads under the conditions of the experiment were of none or little significance to improving energy to fracture requirements, in retrospect it should be remembered that

(a)   the pulse amplitude and energy containment were arbitrarily chosen. As such Fig 4 is yet incomplete; increasing the dynamic load contribution may well led to a significant result

(b)   as pointed out above, the strength testing apparatus is biased towards the outer disc surface. This means that although "in volume" effects may be present they would tend to be unnoticed

Finally as an incidental comment to the overall data of Fig 4, it is significant that for both test series the untreated discs appear weaker in the strength test than either of the pre-load cases. This is curious since the pre-loads range from between 50-96 % the 3 line loaded disc (Fig 1(b)) failure load. Thus it seems that the stressed core shown in Fig 1(b), which extends to 0.4 the disc radius, strengths the rock (upon unloading) superficially and probably (though not demonstrated for certain in Fig 4) volumetrically.

REFERENCES

[1] Jaeger, J.C., <u>Failure of Rocks under Tensile Conditions</u>. Int Jnl Rock Mech Min Sci, Vol 4 (1967), pp 219-227

[2] Gilvary, J.J. and Bergström, B.H., <u>Fracture of Brittle Solids</u>. II, Jnl App Physics, Vol 32, 3 (1961), pp 400-410

[3] Gorham, D.A. and Rickerby, D.G., <u>A Hydraulic Strength Test for Brittle Samples</u>, Jnl Physics E, Vol 8 (1975), pp 794-796

[4] Swan, G., <u>Some Observations conserning the Superpositioning of Static and Dynamic Loads in Rock Crushers</u>, Research Report TULEA:13, University of Luleå, Sweden

[5] Jayatilaka, A., De, S. and Trustrum, K., <u>Statistical approach to Brittle Fracture</u>, Jnl Mats Sc, Vol 12 (1977), pp 1426-1430

[6] Swan, G., <u>The Observation of Cracks Propagating in Rock Plates</u>, Int Jnl Rock Mec Min Sci, Vol 12 (1975), pp 329-334

[7] Peng, S. and Ortiz, C., <u>Crack Propagation and Fracture of Specimens loaded in Compression</u>, in Proc. Int Conf Dynamic Crack Propagation, Lehigh Univ, Noordhoff publishing (1972), pp 113-129

[8] Swan, G., Unpublished data from bending tests on Bohus Granite (1978)

[9] Field, J.E. and Ladegaard-Pedersen, A., <u>The Importance of the Refleeted Stress Wave in Rock Blasting</u>, Int Jnl Rock Mech Min Sc, Vol 8, pp 213-226

[10] Steverding, B., <u>Fracture and Dislocation Dynamics</u>, in Proc. Int Conf Dynamic Crack Propagation, Leigh. Univ, Noordhoff Publishing (1972), pp 356-357

[11] Oh, H.L. and Finnie, I., <u>On the Location of Fracture in Brittle Solids - I</u>, Int Jnl Frac Mech, Vol 6 (1970), pp 287-300

[12] Davidge, R.N. and Tappin, G., <u>The Effective Surface Energy of of Brittle Materials</u>, Jnl Mat Sci, Vol 3 (1968), p 165

DEFORMATION OF MUDDY ROCKS AT LOW METAMORPHIC GRADES

A. W. B. Siddans

Department of Earth Sciences, University of Leeds, Leeds LS2 9JT, U.K.

ABSTRACT

Compaction of a muddy sediment by burial is largely achieved by reduction in porosity and expulsion of pore-fluid. Reactions produce denser, less hydrous minerals and new pore-fluid. Decomposition of organic material also produces pore-fluid. If permeability is low new pore-fluid may cause high pore-fluid pressures, so increasing the solubility of quartz and calcite, facilitating rotation and grain boundary sliding, and perhaps inducing hydraulic fractures. Clay minerals recrystallise on a massive scale. Other phases may be produced or consumed during prograde reactions. The dominant deformation mechanisms are rotation of clay floccules, grain boundary sliding, recrystallisation and pressure solution.

The same mechanisms operate during tectonic deformation, possibly with the addition of internal buckling if the rock is mechanically anisotropic. Tectonic deformation is more complex than compaction because the strain-path is more variable and additional heat may be supplied. It seems likely that the various processes do not occur uniformly, or make uniform contributions to the overall deformation. They probably operate unevenly according to thermal events, periods of recrystallisation and production of new pore-fluid, pore-fluid pressure behaviour and permeability.

1. INTRODUCTION

The deformation history of a muddy sediment starts just below the sediment water interface, when further sediment is deposited on it. Temperature and load pressure increase with depth of burial, causing lithification and burial metamorphism. The physical and chemical processes that lead to these events are complex and interactive. They are outlined in the first part of this contribution, together with the attributes of the resulting rocks.

The burial processes may be accompanied or followed by slow tectonic deformation, caused by the superposition of additional non-hydrostatic stresses. The response of a muddy rock to tectonic deformation depends upon the extent to which the rock has been buried, the time relationship between burial and deformation, the orientation of the strain-path with respect to any mechanical anisotropy in the rock, whether or not the deformation is accompanied by additional heating, and what is happening in adjacent rocks. The controls of these factors on the deformation processes are outlined in the second part of this contribution, together with the attributes of the resulting low-grade metamorphic tectonites.

2. PROCESSES ACCOMPANYING BURIAL

Physical Environment

The variations of temperature, T, and load pressure, P_{load}, with depth of burial are shown in Fig. 1. Values for the geothermal and lithostatic gradients are taken as $25°C/km$ and $0.25 kb/km$ (1). Variation in pore-fluid pressure, P_{fluid}, with depth is shown in Fig. 1 as $0.1 kb/km$. This figure would apply to a permeable sediment whose pore-spaces have free access to the surface, so that $\lambda = P_{fluid}/P_{load} = 0.4$. For a muddy sediment this figure is likely to be highly variable and λ is known to range from 0.4 to 1.0 (1).

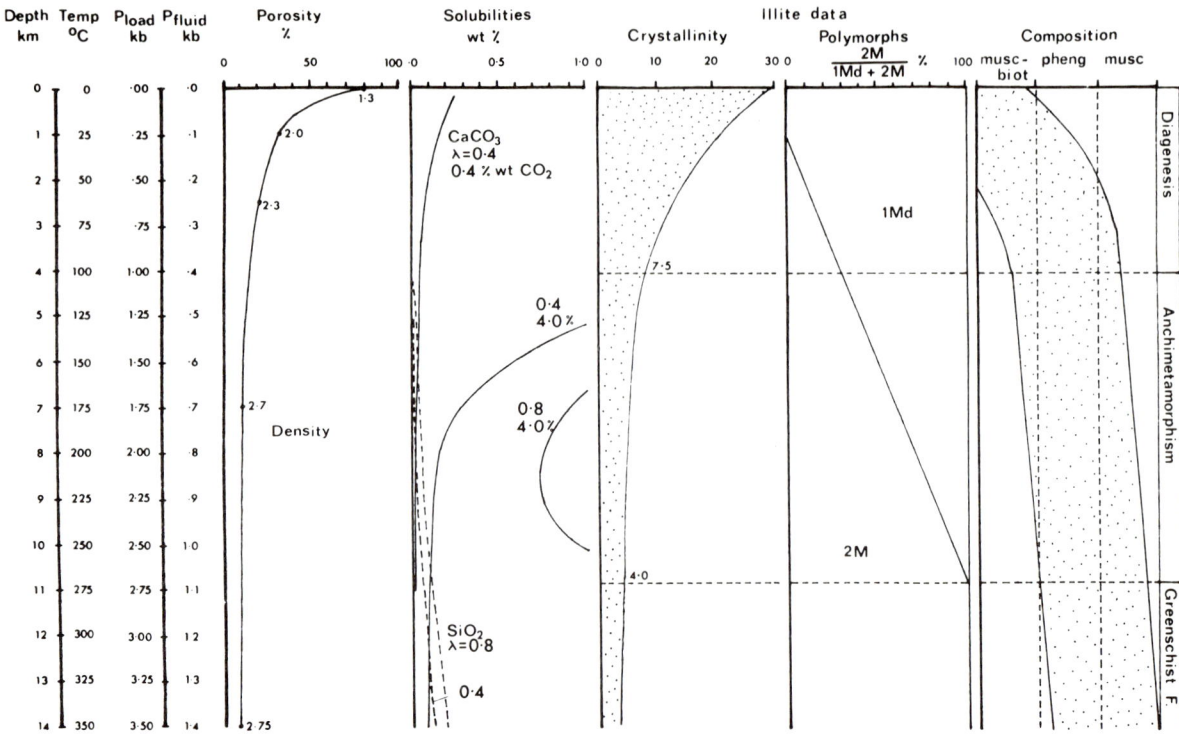

Fig. 1

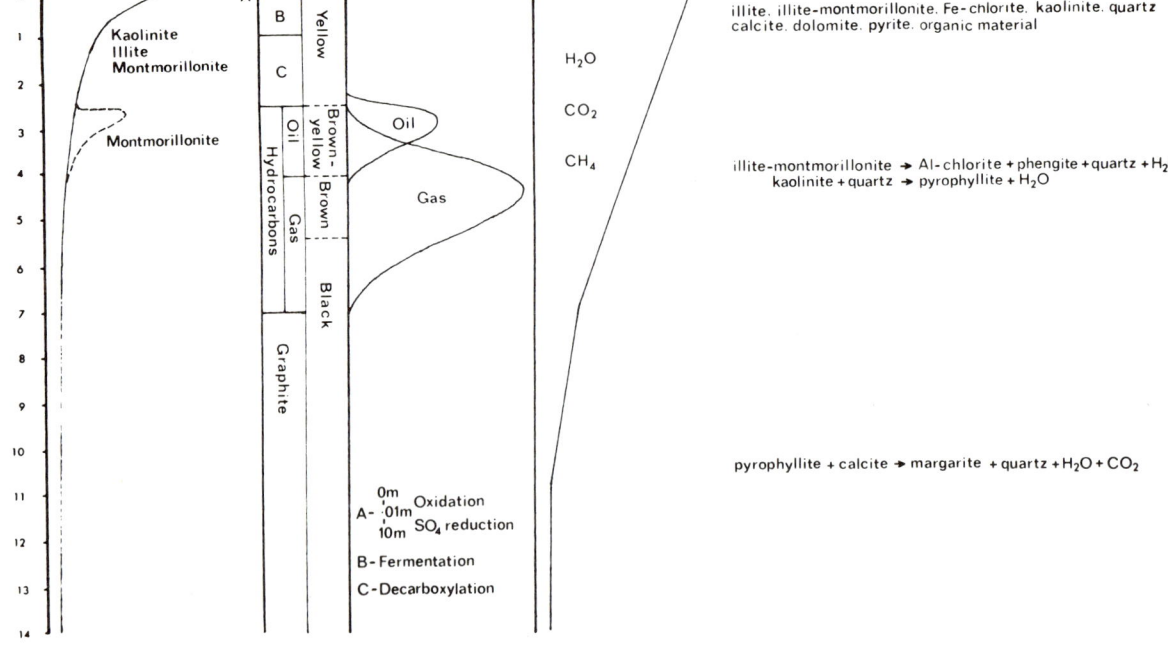

Fig. 2

Compaction and Pore-fluid Expulsion

Compaction of a muddy sediment due to pore-fluid expulsion results in reduction of pososity and increase in density. Representative values (2) are shown in Fig. 1. Pore-fluid may be created during burial by the breakdown of some clay minerals. Montmorillonite, for example, near the surface contains 2 semi-bound and up to 10 free water interlayers in its atomic structure. These represent 14% and 58% respectively of the total volume of the montmorillonite. With burial the water interlayers are expelled to form new pore-fluid in characteristic stages as the reaction

$$\text{montmorillonite} \rightarrow \text{illite} + H_2O \qquad (1)$$

occurs (3). The water escape pattern for montmorillonite-rich muds is compared with that for kaolinite- and illite-rich muds (4) in Fig. 2. Other sources of new pore-fluid are from the breakdown of organic matter and other dehydration reactions.

Illite Data

During progressive burial of illite-rich muds systematic changes occur in the crystallinity, polymorphs and chemical composition of illite (5, 6,7). These properties can be quantified by standard XRD methods and illite is common in muddy sediments, so they are used as good guides to low metamorphic grade, Fig. 1. They also indicate that burial is accompanied by recrystallisation of illite on a massive scale.

Organic Material

During burial a muddy sediment is a dynamic, reacting system. If rich in organic material bacterial action, thermal decomposition and migrating pore-fluids, containing dissolved species, participate in the evolution of the chemically open system (8). Such a sediment may become red clay, black, pyritous shale, grey, iron-rich carbonate mudstone, or a possible source-rock for hydrocarbons, according to the rate of burial, Fig. 2. Vegetable matter becomes progressively coalified during burial. Higher-grade coals are produced as their volatile content (H_2O, CO_2, CH_4) is reduced, Fig. 2 (9).

Changes in Mineralogy and Dehydration Reactions

With increase in burial the clay mineral assemblages stable in muddy sediments change. For example the low-grade assemblage in the marly Lias of the S. Schwarzwald changes southwards in the Glarus Alps (10). Some of the dehydration reactions producing these changes are shown in Fig. 2. The reaction

$$\text{kaolinite} + \text{quartz} \rightarrow \text{pyrophyllite} + H_2O \qquad (2)$$

produces a 2.05% increase in volume and 12.4% water by volume (11). Quartz is produced in some reactions and consumed in others, pyrophyllite is produced then consumed. Comparing the diagenetic and greenschist facies assemblages some 15% by volume of the rock was pore-fluid and a further 3.8% was generated by reactions, 3.6% H_2O and 0.2% CO_2. 35.15% by volume of the final rock consists of new phases, 15% of the quartz was produced by reactions.

Solubilities of Quartz and Calcite

The solubilities of quartz and calcite in dilute aqueous solutions vary with T, P_{fluid}, and for calcite the wt.% of CO_2 in solution (12,13), Fig. 1. Although the values for the solubilities are small, the large volumes of pore-fluid involved mean that appreciable amounts of calcite and quartz can be carried in solution (11).

Permeability and Pore-fluid Pressure

The net migration of escaping pore-fluid must be upwards. The rate at which this and hence compaction can occur depends on the permeability of the sediments. If the permeability is low P_{fluid} will rise towards P_{load} and the local compaction rate will be reduced. If dehydration reactions produce new pore-fluid rapidly in a formation below a low-permeability layer, there will be a local volume increase and high P_{fluid} maintained for a considerable time, e.g. 10^6 years in Triassic evaporites below the Jura Mountains (14). High P_{fluid} in a sediment has several effects. The solubilities of quartz and calcite are increased. It also reduces the frictional resistance to sliding (15), so facilitates grain boundary sliding, and may induce hydraulic fractures (16). If there is a net volume increase due to production of new pore-space, the rock may be partially disaggregated and grain boundary sliding further facilitated.

Rotation of Inequant Grains

Reduction in porosity of a muddy sediment during burial is largely achieved by rotation and grain boundary sliding of clay floccules, resulting in a closer-packed configuration, oriented with their long axes at lower angles to bedding (17). This process may be inhibited by precipitation of cement in pore-spaces. It is a progressively less effective process with increasing depth, though influxes of newly-created pore-fluid will facilitate it.

Pressure Solution

It is frequently observed that impinging clasts of quartz or calcite have indented each other so that locally material has been removed (Plate 1), and that overgrowths of quartz or calcite often grow in pore-spaces on clastic grains, or form in veins (18). This is the phenomena of pressure solution, in which material goes into solution at points of high normal or mean stress, and migrates to points of low normal or mean stress where it is precipitated (11,13). The scale of such transfer of material ranges from that of individual grains to several metres. Thus material **may move** into or out of a local volume of rock. A special case of deformation by pressure solution is the formation of stylolites in limestones (20). These arise where calcite is removed along bedding planes leaving a film of insoluble material behind (Plate 2). Stylolites have a characteristic zig-zag pattern in cross-section, resulting from the heterogeneous nature of the pressure solution process. Deformation by pressure solution must be accomodated by grain boundary sliding.

Plate 1. Pebbles pitted by pressure solution. Carboniferous, Asturias, Spain.

Plate 2. Stylolites in Jurassic limestones, Alpes de Haute Provence, France.

Strains Associated with Compaction

The volume loss achieved by reduction in porosity results in compactional strains. This is a flattening type of deformation, the bedding planes remainimg planes of no incremental longitudinal strain. Crystallisation of denser minerals also contributes to these strains. Pressure solution contributes heterogeneously on all scales from that of individual grains to widely spaced stylolites. Measured over several stylolites in limestones, the compactional strains achieved by their formation may be 5 to 10%.

Textures Associated with Compaction

Rotation of clay floccules towards bedding planes results in the characteristic bedding-plane textures of shales, in which the maximum pole-density for basal planes of clay minerals is normal to bedding (21, 22). This is because the shapes of clay floccules are controlled by their crystallography. Although grain-shapes may be modified by pressure solution, no textural modification results when the quartz or calcite is reprecipitated in crystallographic continuity with the clastic grains. The planar texture of shales may be enhanced by crystallisation of new minerals in a thermodynamically stable orientation relative to P_{load} (23). New minerals may grow parallel to the bedding plane anisotropy because it controls diffusion paths.

3. TECTONIC PROCESSES

Strain Path

A simple strain path to envisage is uniaxial compaction, achieved by volume loss during burial, followed by tectonic, progressive pure shear, at constant volume, with coaxial increments of shortening parallel to bedding, of extension normal to bedding, the third axis remaining one of no incremental longitudinal strain. Permian red mudstones in the Alpes Maritimes appear to have followed such a strain-path, resulting in red slates with ellipsoidal green spots flattened in the cleavage (24). Strain paths are usually more complex, with increments of triaxial strain superposed non-coaxially (25). This is probably the case in and around folds (26). The strain-path of a deformed rock is usually unknown.

Rotation of Inequant Grains

Rotation and grain boundary sliding of inequant grains is a necessary consequence of homogeneous strain. The model of deformation proposed by March (27) relates the pole-density of basal planes of platy grains to the elongation in any direction. Some workers (28,29) have found phyllosilicate textures and finite strain states in slates to be in close agreement with this model, others (30) find the relationship does not apply.

Pressure Solution

Observations on fossils, Plate 3, conglomerates and clastic grains (31,32) clearly show that pressure solution is an important deformation mechanism at low metamorphic grades, as predicted by deformation mechanism maps for quartz and calcite (13). Probably much of the quartz and calcite that occurs in tectonic veins, Plate 4, and slickenlines is a product of pressure solution.

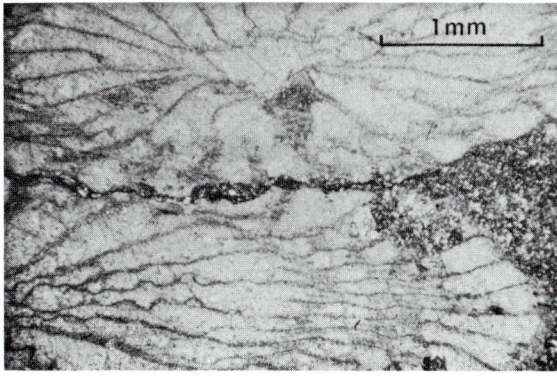

Plate 3. Corals deformed by pressure solution. Devonian, S.W. England.

Plate 4. Fibrous calcite in necks of boudinaged belemnites. Lias, Valais, Switzerland.

Crystallisation and Recrystallisation

In orogenic belts it is often found that deformation is accompanied by prograde mineral reactions (10,22) due to heating. In such cases the new phases may grow in thermodynamically stable orientations, or in diffusion path controlled orientations. Recent work (33,34) has shown that on the microscopic scale slates may have a domainal structure, in which cleavage lamellae containing highly oriented phyllosilicates anastamose around pods containing less or differently oriented phyllosilicates, Plate 5, Fig. 3. The margins of these pods are sites of crystallisation or recrystallisation, where new phases grow in thermodynamically stable orientations. Thus the cleavage lamellae grow at the expense of the pods.

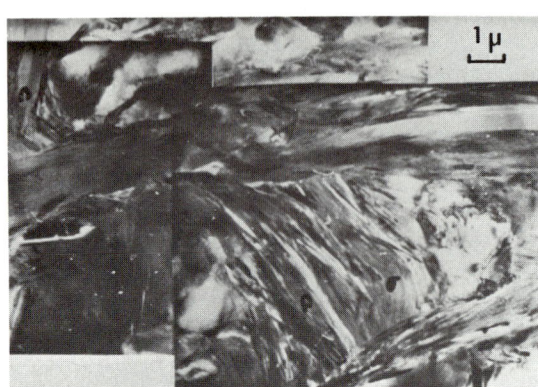
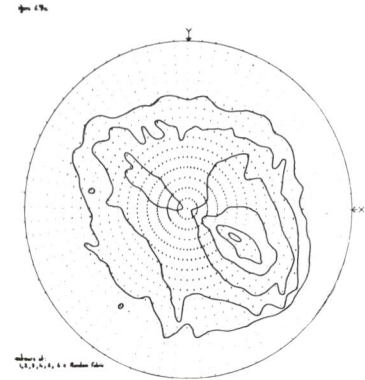

Plate 5. Domainal structure in Ordovician slate. Anglesey, Wales.

Fig. 3. Pole figure for chlorite (002) from slate shown in Plate 5. Contours at 1,2,3,4,5,6 x random.

Internal Buckling

A well-compacted shale is mechanically anisotropic due to its texture. If tectonic compression acts sub-parallel to the anisotropy internal buckling instabilities may develop into crenulations or kinks (35), Plate 6. With continued deformation these may be modified by pressure solution so that quartz or calcite migrate from the limbs into hinge regions, leaving the limbs relatively enriched in phyllosilicates (36), Plate 7. The hinge regions of kinks within pods are also sites of crystallisation or recrystallisation (34).

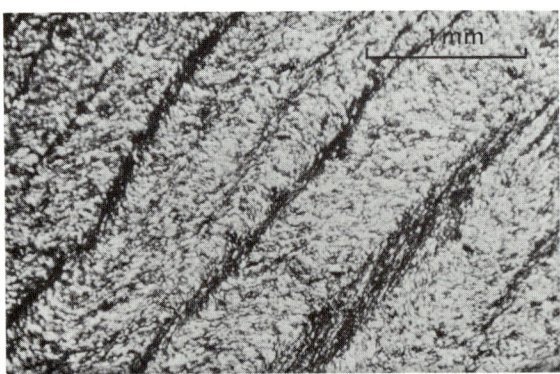

Plate 6. Crenulations in Dalradian slate. Scotland.

Plate 7. Pressure solution stripes. Cambrian. Virginia, U.S.A.

Pore-fluids

The presence of high P_{fluid} facilitates rotation and grain boundary sliding, and increases the solubility of quartz and calcite. The presence of a migrating pore-fluid facilitates transfer of material in solution, both into and out of local volumes of rock. Tectonic compression tends to reduce pore-space and prograde reactions produce new pore-fluid. Large volumes of pore-fluid and high P_{fluid} may be inherited from the burial history of the rock, if tectonic deformation closely follows a critical burial stage. It thus seems likely that the processes associated with tectonic deformation may not take place at a uniform rate or make uniform contributions to the overall deformation. Probably muddy rocks react unevenly to tectonic stresses, according to thermal events, pore-fluid production due to prograde reactions, P_{fluid} behaviour and permeability.

Textures of Low-grade Metamorphic Tectonites

The phyllosilicate textures of slaty rocks are variable. Some have orthorhombic symmetry and appear to correlate with the March model (28, 29), others have monoclinic or triclinic symmetry (22). The textures of different phyllosilicates in the same rock may themselves differ (22,30). Slates with a domainal microstructure may have bimodal phyllosilicate textures (34,37), Fig. 3. Rocks that have deformed by internal buckling have textures that vary from broad maxima to bimodal (38). If the finite strain state of the rock has a marked extension direction in the plane of flattening, this is usually reflected in elongate maxima in the phyllosilicate texture and a distinct grain in the cleavage planes of the rock. Although the shape fabric of quartz, calcite and clastic grains may be well developed, Plate 8, according to the geometry of the finite strain ellipsoid (39), they do not seem to develop a crystallographic texture (22). This probably reflects the extent to which pressure solution predominates over intracrystalline processes as the dominant deformation mechanism affecting these minerals in muddy rocks at low metamorphic grades.

Discussion

Many workers (40,41) find that the mesoscopic fabric of slaty rocks is normal to the short axis of the finite strain ellipsoid (Plate 9). This is puzzling, since the stress-controlled deformation increments that produce the fabric are usually superposed non-coaxially. It may be that the bulk of the deformation occurs during relatively short parts of the whole deformation history, controlled, perhaps, by periods of prograde reactions and high P_{fluid}. It may also be that once the tectonic fabric has developed to a certain extent, its anisotropy controls local stress fields and diffusion paths.

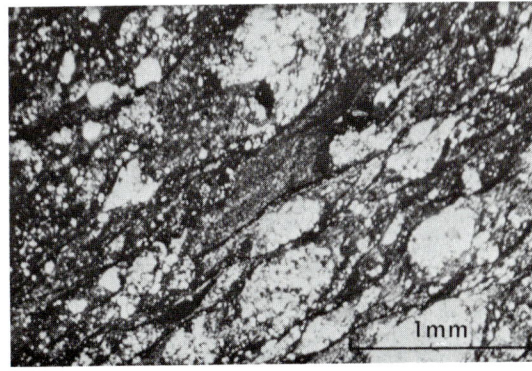

Plate 8. Sandstone deformed by pressure solution. Rio Tinto, Spain.

Plate 9. Green spots flattened in slaty cleavage plane. Cambrian, Wales.

ACKNOWLEDGEMENTS

I am grateful to Prof. J.G. Ramsay for permission to use Plate 1, to Drs. R.J. Knipe, S.H. White and The Institute of Physics for the use of Plate 5 (34, Fig. 3), and to J.S. Whalley for the use of Fig. 3 (37, Fig. 6.7a).

REFERENCES

(1) Heard H.C., Phil. Trans. R. Soc. Lond. A. 283 (1976) 173-186.
(2) Skempton A.W., Geol. Soc. Lond. Q. J. 125 (1970) 373-411.
(3) Burst J.F., Bull. Am. Ass. Petrol. Geol. 53 (1969) 73-93.
(4) Powers M.C., Bull. Am. Ass. Petrol. Geol. 51 (1967) 1240-1253.
(5) Kübler B., Bull. Centre Rech. Pau-SNPA 2 (1968) 385-397.
(6) Maxwell D.T. & Hower J., Am. Mineral. 52 (1967) 843-857.
(7) Esquevin J., Bull. Centre Rech. Pau-SNPA 3 (1968) 147-153.
(8) Curtis C.D., J. Geol. Soc. Lond. 135 (1978) 107-117.
(9) Eames T.D., in Petroleum and the Continental Shelf of NW Europe, 1. Geology, Ed. A.W. Woodland, Applied Science Publishers Ltd. (1975) 501p.
(10) Frey M., Sedimentology 15 (1970) 261-279.
(11) Fyfe W.S., Phil. Trans. R. Soc. Lond. A. 283 (1976) 221-228.
(12) Sharp W.E. & Kennedy G.C., J. Geol. 73 (1965) 391-403.
(13) Rutter E.H., Phil. Trans. R. Soc. Lond. A. 283 (1976) 203-219.
(14) Heard H.C. & Rubey W.W., Bull. Geol. Soc. Am. 77 (1966) 741-760.
(15) Hubbert M.K. & Rubey W.W., Bull. Geol. Soc. Am. 70 (1959) 115-166.
(16) Phillips W.J., J. Geol. Soc. Lond. 128 (1972) 337-359.
(17) Moon C.F., Earth-Sci. Rev. 8 (1972) 303-359.
(18) Elliott D., Bull. Geol. Soc. Am. 84 (1973) 2645-2664.
(19) de Boer R.B., Tectonophysics 39 (1977) 287-301.
(20) Bodou P., Bull. Centre Rech. Pau-SNPA 10 (1976) 627-644.
(21) Oertel G. & Curtis C.D., Bull. Geol. Soc. Am. 83 (1973) 2597-2606.
(22) Siddans A.W.B., Tectonophysics 30 (1977) 533-557.
(23) de Vore G.W., Contr. Geol. 5 (1966) 19-43.
(24) Graham R.H., Mem. B.R.G.M. (in press).
(25) Durney D.W. & Ramsay J.G., in Gravity and Tectonics, Eds. K.A. de Jong & R. Scholten, John Wiley & Sons (1973) 502p.
(26) Dieterich J.H. & Carter N.L., Am. J. Sci. 267 (1969) 129-155.
(27) March A., Z. Kristallogr. 81 (1932) 285-297.
(28) Tullis T.E. & Wood D.S., Bull. Geol. Soc. Am. 86 (1975) 632-638.
(29) Wood D.S., Oertel G., Singh J. & Bennett H.F., Phil. Trans. R. Soc. Lond. A. 283 (1976) 27-42.
(30) Siddans A.W.B., Phil. Trans. R. Soc. Lond. A. 283 (1976) 43-54.
(31) Plessman von W., Z. Deutsch. Geol. Ges. 115 (1965) 630-663.
(32) Ramsay J.G., Folding and Fracturing of Rocks, McGraw-Hill (1967) 568p.
(33) Williams P.F., Am. J. Sci. 272 (1972) 1-47.
(34) Knipe R.J. & White S.H., in Developments in Electron Microscopy and Analysis, Proc. of EMAG 1975, Ed. J.A. Venables, Institute of Physics.
(35) Cobbold P.R., Cosgrove J.W. & Summers J.M., Tectonophysics 12 (1971) 23-53.
(36) Cosgrove J.W., J. Geol. Soc. Lond. 132 (1976) 155-178.
(37) Whalley J.S., M.Sc. thesis, Univ. of London (1973) 98p.
(38) Weber K., Geol. Jb. 15 (1976) 3-98.
(39) Mitra S., Contr. Mineral. Petrol. 59 (1976) 203-226.
(40) Siddans A.W.B., Earth-Sci. Rev. 8 (1972) 205-232.
(41) Wood D.S., A. Rev. Earth and Plan. Sci. 2 (1974) 369-401.

STUDY OF FRESH FRACTURE SURFACES IN METALS BY ELECTRON SPECTROSCOPY

M. Lähdeniemi, E. Minni, L. Ranta and E. Suoninen

Department of Physical Sciences, University of Turku, SF-20500 Turku 50, Finland

ABSTRACT

Use of Auger spectroscopy for studying segregation phenomena at grain boundaries of crystalline solids is described briefly. Results of preliminary experiments with a commercial Esca-Auger spectrometer, equipped with a fracture stage to produce fresh intercrystalline metal surfaces, are presented.

1. INTRODUCTION

The range of electrons with energies in the region of a few hundred eV in solid materials is known to be only a few atomic layers, regardless of the type of material. Hence, the electron spectrum, created by irradiation with x-rays or primary electrons in this region and observed outside the target, reflects the composition and the structure of the first few atomic layers on the surface of the sample. This fact is the basis of the rapidly increasing use of electron spectroscopy in surface studies [1 - 4].

Studies of very thin impurity or segregation layers on grain boundaries are now possible by Esca or Auger spectroscopy, provided that an intercrystalline surface can be produced and investigated in ultrahigh vacuum in order to prevent its practically instantaneous contamination in atmospere. In the following will be described some preliminary experiments in studies of grain boundaries in metals using a commercial Esca-Auger spectrometer (VG Scientific Esca3 MarkII) equipped with a fracture stage.

2. EXPERIMENTAL ARRANGEMENT

The samples to be studied are made into round notched bars and loaded in the fracture stage which is situated above the specimen preparation chamber of the spectrometer. After evacuation of the chamber and of the fracture stage, the samples are cooled to liquid nitrogen temperature. One sample at a time is then moved into a special anvil and fractured by a hammer operated by a mechanical feedthrough from outside the vacuum. One half of the fractured sample is moved by a special mechanism into the sample holder of the spectrometer and further into the analyzer chamber where the electron emission from the fracture surface can be studied either by the Esca or the Auger technique.

Because of the clean vacuum and the low pressure in the spectrometer (10 - 100 nPa) the fracture surface can be considered practically free from external contamination at least for one hour from the time of fracture.

The fracture technique described above seems to work reasonably well for small grain size metals with BCC structures. For FCC materials, such as austenitic stainless steels and copper alloys, it is often difficult to obtain a completely intercrystalline fracture surface. Coarse grain materials also often require special treatment. In most cases, it is nevertheless possible to produce enough fracture surface with intercrystalline nature to allow spectroscopic analysis.

Since elemental analysis is usually the information required, Auger spectroscopy is normally used in the actual analysis. The instrument allows average analysis over a wide area of the surface, or scanning Auger analysis giving information on the distribution of different elements over the studied area. The geometric resolution obtainable in the scanning mode is for a flat sample surface about 2 - 3 μm. For most fracture samples the resolution is, however, somewhat worse because of their surface roughness and the small depth of focussing of the primary electron beam.

3. STUDIES OF FRACTURE SURFACES

Figure 1 shows the results of Auger analysis of a region of the fracture surface of an austenitic steel sample. The region studied was found to represent intercrystalline fracture on the basis of the SEM (scanning electron microscopy) picture of the fracture surface. The approximate composition of the sample is 0.05 % C, 8.5 % Ni, 18.5 % Cr, 0.1 % Mo. The sample had been annealed for 10 h at 800 $^\circ$C which treatment was suspected to cause segregation of certain impurities at the grain boundaries with subsequent embrittlement of the material. The Auger spectrum from the fresh fracture surface (curve a) shows, in addition to the major metallic elements, strong signals of carbon and sulfur. The surface was then cleaned by sputtering (bombardment by argon ions) which causes removal of successively thicker layers from the sample surface and revealing of the bulk structure. This process is found to cause important

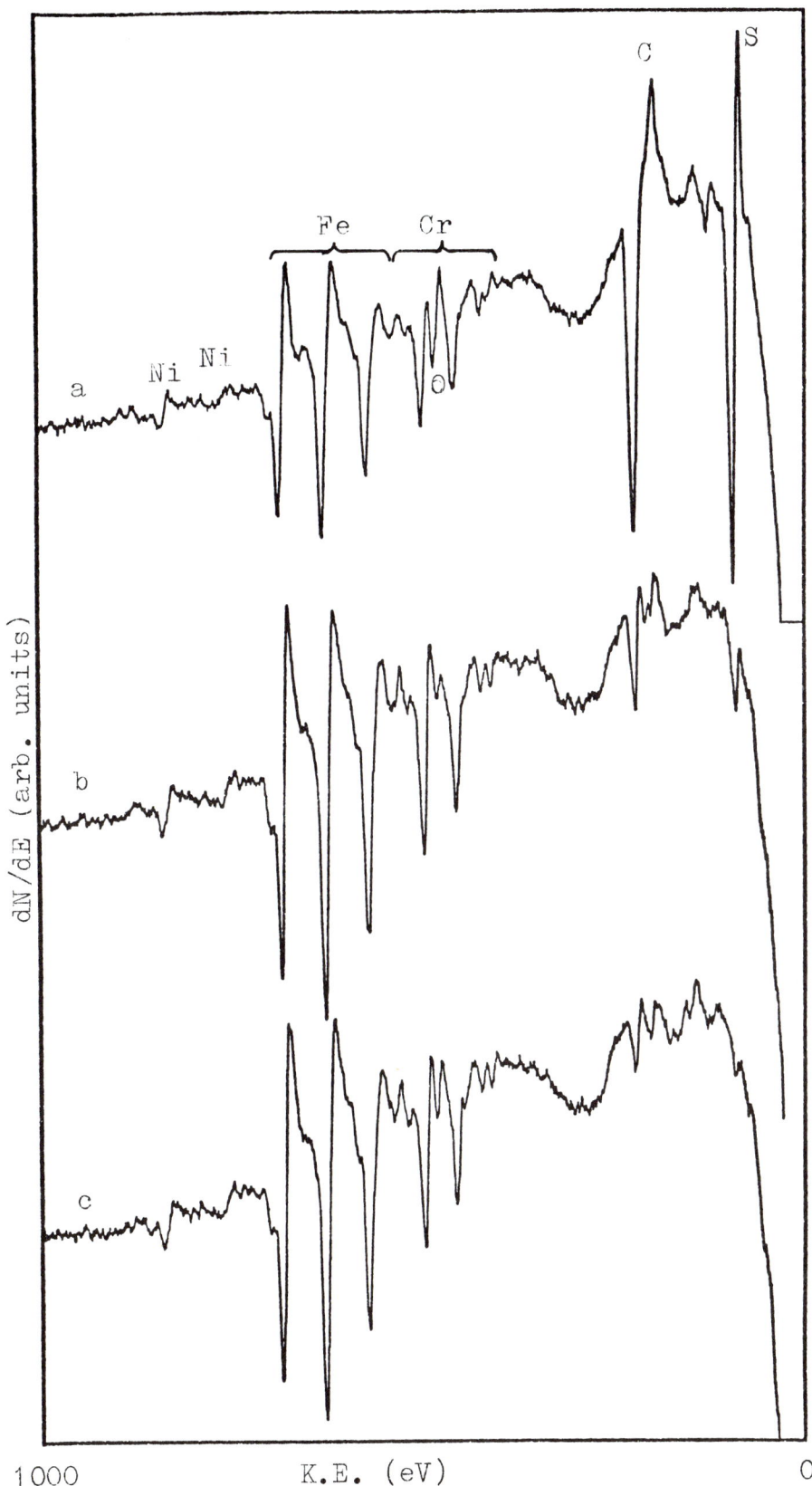

Fig. 1. Auger spectra of the fracture surface of an austenitic steel sample:
a) fresh fracture surface
b) sputtered with argon beam (5 kV, 10 μA) for 2 min.
c) sputtered 20 min.

changes in the Auger spectrum (curves b - c). The most pronounced change is virtual disappearing of the sulfur signal and strong attenuation of carbon. The result clearly indicates the presence of a thin impurity layer with a high concentration of sulfur and carbon on the grain boundaries. Analysis of the fracture surface of a sample, not sensitized by the above heat treatment, does not reveal any impurity layer. Hence, the layer is formed by the heat treatment and is probably the cause of the brittleness observed in the sensitized sample.

Figures 2 - 3 show a case in which the scanning mode of analysis was used to study a fracture surface which was only partly intercrystalline. The sample was a chromium-iron alloy (13 % Cr, 0.5 % Ni, 0.07 % C, 0.8 % Mn, 0.4 % Si, balance Fe). Figure 2 shows the Auger spectrum obtained with a wide primary beam over a large portion of the surface. Besides the main components iron and chromium, nickel, carbon and phosphorus are detected in the spectrum. Maps of the distributions of the iron and chromium signals over the same part of the surface (Figure 3) show that the regions from which these signals are obtained are almost identical. We can assume that the regions from which no signal of the main components is obtained are "shadowed" by the roughness of the surface. Hence, absence of signals from these regions probably does not imply any change in the surface composition as compared with the rest of the surface. A comparision of the distribution of the phosphorus signal with the above distributions shows, however, that in

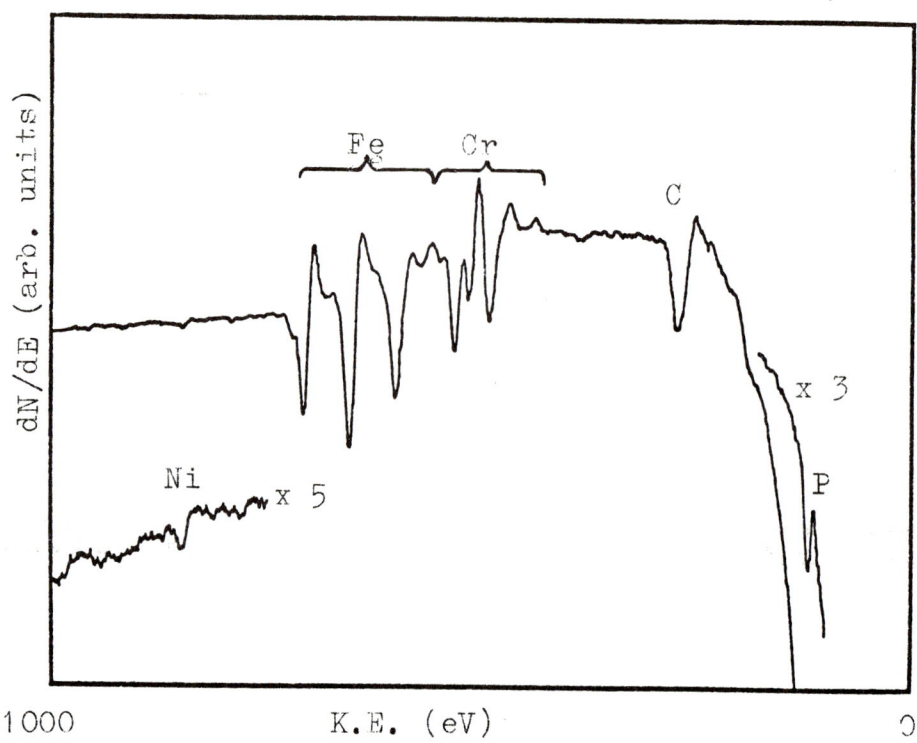

Fig. 2. Wide-area Auger spectrum of the fresh fracture surface of an iron-chromium alloy.

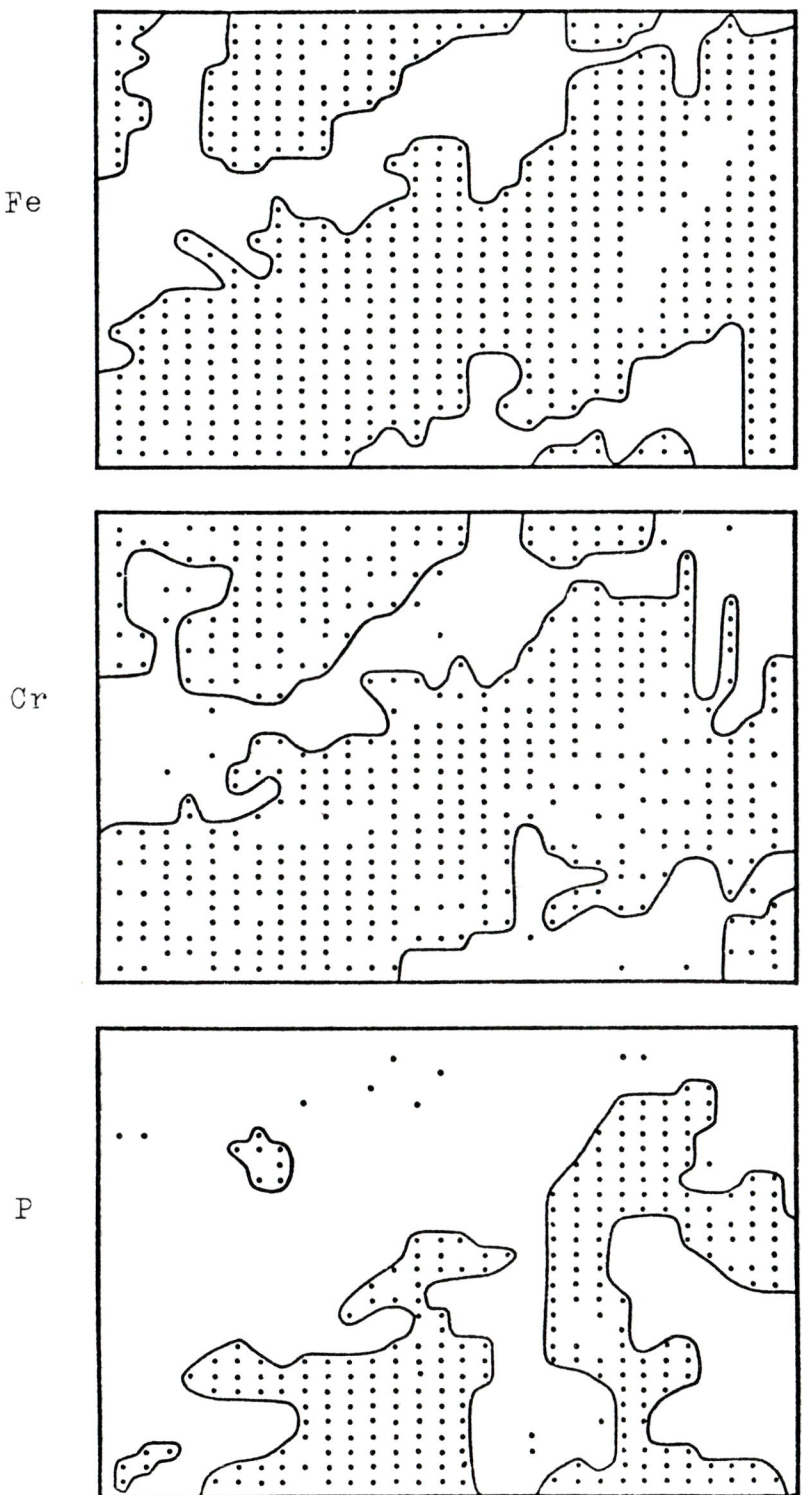

Fig. 3. Distribution of iron, chromium and phosphorus Auger signals from the fracture surface of an iron-chromium alloy. For interpretation, see text.

addition to the absence of phosphorus signal from these regions, this signal is obtained from only a part of the region giving iron and chromium signals. The regions giving a phosphorus signal are found from a comparison with the SEM picture (not shown here) to correspond to grains of δ ferrite. Hence, the conclusion is that an impurity layer with a high concentration of phosphorus covers the surfaces of the δ ferrite grains.

Other cases studied so far include, e.g., segregation of impurities on the grain boundaries in nickel copper alloys.

4. CONCLUSIONS

The above examples show that Auger spectroscopy offers interesting possibilities to detect thin impurity layers which often cause intercrystalline fracture. The special advantage of the method is its ability to detect even monatomic layers which are practically impossible to observe by other microscopic or spectrometric methods. The invividual nature of every Auger spectrometric study of a fracture surface and the subsequent need for great experimental skill in performing and interpreting such measurements must, however, be emphasized. Especially the scanning Auger analysis is still in its infancy as a tool of fracture research, and much work is needed to develop it to an easily applied technique. The same applies to the use of Esca measurements which have so far been used only very rarely in fracture research. This technique which has already been very succesfully applied in other areas of surface studies, could probably substantially add to the possibilities of electron spectrometry in explaining fracture phenomena.

REFERENCES

[1] J.H. Richardson and R.V. Peterson (ed.), Systematic Materials Analysis, Vol.I, Ch.3 (by G. Stupian) and Vol.II, Ch. 19 (by W.G. Proctor), Academic Press, New York (1974).
[2] A.W. Czanderna (ed.), Methods of Surface Analysis, Ch.4 (by W.M. Riggs and M.J. Parker) and Ch.5 (by A. Joshi, L.E. Davis and P.W. Palmberg), Elsevier, New York (1975).
[3] P.F. Kane and G.R. Larrabee (ed.), Characterization of Solid Surfaces, Ch.13 (by S.H. Hercules and D.M. Hercules) and Ch.20 (by C.C. Chang), Plenum Press, New York (1974).
[4] J.C. Eriksson and C. Leygraf, Kemisk Tidskrift, $\underline{88}$, 46 (1976).

Session II

DISCUSSION

J.F. Knott: You mention (D. Shockey) that you are able to vary the duration of the pulse by altering the thickness of the disc. How did the size of voids vary with pulse duration?

I am worried about direct comparison between disc void sizes and void sizes in tensile specimens, because the degree of triaxiality in the latter case depends on the profile of the neck. This profile could be a function of flow properties, and hence of grain size.

D. Shockey: The size of the voids increased with increasing pulse duration. This effect can be observed in a single specimen in the void size distributions measured at various distances below the impact surface (which correspond to various stress durations) as was shown in one of the slides.

Direct comparisons between void sizes in tensile bars and impacted disk specimens were not made. We only compared fracture morphology and found in both specimen types populations of roughly spherical voids that formed primarily at grain boundary triple points. Void formation in both specimens occur under high levels of triaxiality, in the disk specimens because of the uniaxial strain conditions produced by plate impact, and in the tensile bars which underwent severe necking before any voids could be observed.

G. Swan: Have you observed any damage due to the incident compressive wave?

D. Shockey: The impactor plate mounted in the leading edge of the projectile experiences a compressive wave similar to that experienced by the specimen plate, but subsequent tensions are not produced. Then we could examine polished cross sections of the impactor plate to establish whether the compressive wave produced any damage in the specimen. We saw no evidence of fracture, twinning or slip in the impactor, and therefore feel that no damage resulted in the specimen from the initial compressive wave.

P. Smart: With reference to the paper of Porter and Easterling, did you check by observation that the volume fractions of cementite are the same for "coarse" and "fine" samples?

D.A. Porter: The volume fraction of cementite was not determined but should be very similar in the two types of pearlite since the chemical analysis of both the coarse and fine, cold drawn pearlite was the same.

P. Smart: Was there a variation in grain size between the pearlites with fine and coarse lamellae? And did this affect the bulk properties?

D.A. Porter: The specimens containing finer pearlites also had smaller grains, or colonies as they are called, and this does influence the bulk mechanical properties such as toughness. However, the results that have been presented deal mainly with the mechanics of deformation within the grains.

M.S. Paterson: The structure of pearlite suggests that it would behave as an anisotropic material analogous to slaty or foliated rock. Does it similarly show kinking of this structure when compressed parallel to the layering?

D.A. Porter: Yes, such kinking behaviour is found during compression testing of both coarse and fine pearlite. In the fine pearlites we even observed kiniking in those colonies that were oriented perpendicular to the tensile axis. (See Porter et al, Acta Metall. $\underline{26}$ (1978) 1405).

O. Stephansson: The structures you have described are very common in structural geology. In nature you always find a relationship between thickness of layers and the length of the layer. Is this also the situation in deformation of pearlite?

D.A. Porter: The relationship between the length and thickness of cementite during deformation is rather complicated. If we just consider those colonies with lamellae parallel to the tensile axis then in coarse pearlite the lamellae fracture and become shorter without any appreciable thinning whereas in fine pearlite thinning can also occur. Unfortunately the thin cementite lamellae, which may be as thin as 10 nm, are too narrow to permit accurate measurement of any changes in the thickness/length ratio during deformation.

I think I should point out that there should be large differences between the behaviour I have described here and the behaviour of layered structures in rocks for two reasons. Firstly, the cementite and ferrite in pearlite can be described as two interpenetrating single crystals i.e. the grain size is much larger than the interlamellar spacing, which is not the case in the rock macrostructures. Secondly, rock deformation is taking place under high hydrostatic pressures which were not present during the deformation of pearlite in these experiments. Under such conditions of deformation the cementite ductility in both coarse and fine pearlites is greatly increased.

K. Easterling:

It should be emphasised that pearlite is (ideally) a two phase structure in which each phase is plate or disk shaped and arranged in bundles called colonies. Colonies are initially randomly oriented with respect to eachother, but after mechanical drawing operations, these colonies tend to rotate until the bundles of plates are all parallel to eachother. At a later stage of the drawing process the plates themselves thin down and the colonies become elongated into a lamellar shape. This sequence of operations, from initial growth to drawing, is illustrated in Fig "A".

D. A. Porter's experiments concerned the deformation and fracture mechanisms of pearlites of different inter-lamellar spacing. The coarse pearlite behaves differently from fine pearlite for the following reason. In coarsely spaced pearlite, dislocation pile-ups in the ferrite lamellae reach such an intensity to cause brittle failure of the Fe_3C. In fine spacing pearlite, such pile-ups are unable to form and brittle fracture is not observed. In fact since the flow stress is proportional to the inverse of the spacing, fine pearlites have very high flow stresses, this in turn induces <u>fibre loading</u> of the hard phase, causing it to plastically deform along with the soft phase. The short equisized segments of the Fe_3C phase shown e.g. in Fig 6 of Porter's paper (lower fig, top L. H. corner) are direct evidence of such fibre loading. Could you further elucidate on the similarities in deformation behaviour between pearlite and rocks?

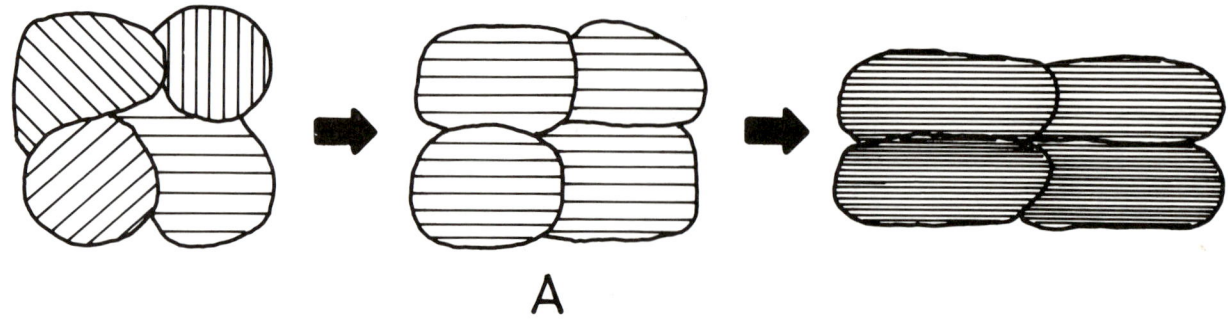

A

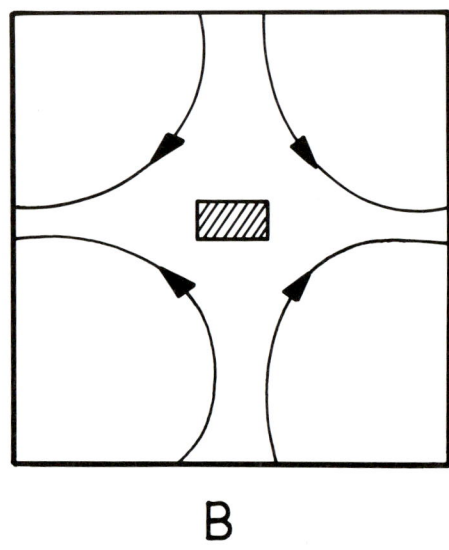

B

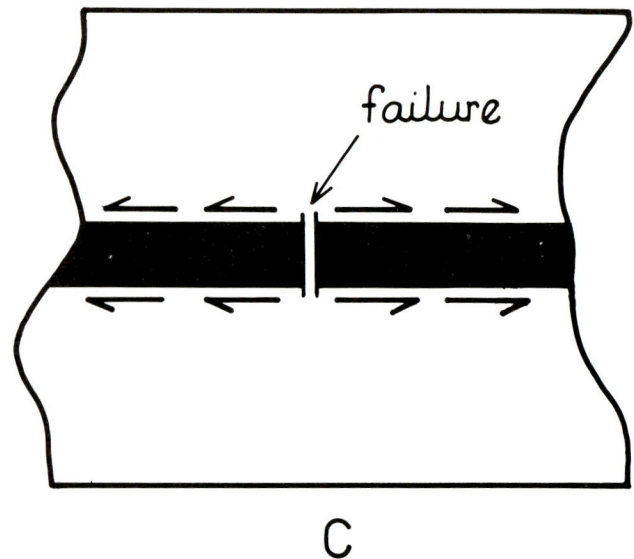

C

D

E

O. Stephansson: There seems to be many structural similarities between deformed pearlite in metallurgy and boudinage in structural geology. A boudinage structure is formed by failure and separation of a stiff layer surrounded by softer material in plastic flow. Let us consider a multilayer system of stiff and soft layer deformed in pure shear, Fig B. As the shear stress along the surface of the stiff layer exceeds the shear strength it will separate and form a boudin, Fig C. The length of the boudin is determined by the strength and thickness of the stiff layer. Hence, in nature we often find a constant length/thickness ratio for boudins of the same rock type. The boudines often have a barrel shape due to strong shear forces along the surfaces, Fig D. If the stiffness ratio between layer and surrounding is low to moderate the layer forms a structure called "pinch and swell", Fig E. The stress distribution in a boudinage structure leads to the formation of voids or low pressure areas between the boudines. These low pressure areas show segregations.

Boudinage formations and the stress distribution in a boudin and its surrounding have been studied by Stephansson and Berner 1968 in a paper entitled "The finite element method in tectonic processes", Physics of the Earth and Planetary Interiors 4, 301-321.

E. Suoninen: I would like to emphasize the fact that because of the process of pearlite formation, the crystallographic orientation relationship between the ferrite and cementite particles is fixed. This results in a definite anisotropy in the behaviour of pearlite with respect to an applied load which does not have any direct analogy in macroscopic structures.

J. Knott: In connexion with <u>Jilkén and Bäcklund's</u> paper, the technique they describe is very appealing: we have done somewhat similar work, using the signal from a potential drop system to drive a servo-hydraulic machine. One point, however, concerns the size of the crack growth increment as the load is decreased. It is important that the crack should grow well beyond the plastic zone corresponding to the higher ΔK level to remove residual stress effects? Since the increment/load decrease in your tests is governed by testpiece compliance, can you guarantee that you have eliminated these residual stress effects in your threshold determinations?

L. Jilkén,
J. Bäcklund,
H. Knutsson: The method has been compared with more elaborate ones, such as the "Proposed methods of test for constant amplitude fatigue crack growth rates below 10^{-8} m/cycle" due to ASTM. Our threshold values were slightly below those of the proposed methods.

G. Weidmann: Does "fracture" refer to complete fracture of the specimen?

L. Jilkén &
J. Bäcklund: "Fracture" refers to the onset of (stable) crack growth in fracture toughness testing (J_c) of TPB specimens.

J. Kratochvíl: Is not your method geometrically sensitive due to the demagnetization effect? During loading both geometry and stress-state are changed and both influence the flux.

L. Jilkén &
J. Bäcklund: The method is based on geometry changes on the micro-level and is hence geometrically sensitive.

J. Knott: How much do the effects depend on geometry changes and how much on processes such as dislocation movement? I am particularly interested in the case of fatigue crack propagating under applied load, where both the geometry changes and the locally yielded region around the crack tip increases in size as the crack grows?

L. Jilkén &
J. Bäcklund: Geometry changes on the micro level should not be undependent of the movement of dislocations. The method has not yet been tested on the case of a propagating crack.

O. Stephansson: Is it possible to use a tube instead of a cylinder in testing materials with your electromagnetic technique?

L. Jilkén &
J. Bäcklund: It should be possible, but the influence of the air volume in the tube has to be examined.

O. Stephansson: Do you know of any test or theoretical work where one has developed electromagnetic field due to failure?

L. Jilkén &
J. Bäcklund: No.

J. Hult: For some mechanical testing methods, such as photoelasticity, it is not fully clear whether the recorded effect depends primarily on stress or strain or some combination of the two. What can be said in this respect about the electromagnetic method which you have presented?

L. Jilkén &
J. Bäcklund: We think that stress only can be measured indirectly through its action, which may be e.g. strain. Stress is a defined quantity and not a physical one. On the atomic level the term stress is unacceptable, since it is difficult to relate a certain area to a certain force. The present method measures change of flux, which has been related to a number of physical phenomena in the paper.

J. Knott: In Dr Magnusson's paper, he has related effects of case-carburising simply to differences in the fatigue properties of the regions of different carbon contents (in strain-controlled tests). What significance do you attach to the levels of residual stress in case and core in case-hardened components in service (which are usually subjected to alternating stresses), particularly with respect to the balance of initiation between the case region and the core region?

L. Magnusson: The experimentally determined or calculated levels of residual stress in case and core must be taken into account when the probability for fatigue crack initiation is estimated. A common rule for designing of case-hardened components is to put the probabilities equal for fatigue crack initiation in case and in core.

I.J. Smalley: I should like to ask Dr Weidmann: Can crazing occur as a result of residual stresses? And is it possible (or likely) that crazing could be observed in materials other than polymers?

G.W. Weidmann: Yes, crazing certainly could occur as a result of residual stresses. I think it is unlikely that crazing, in the sence that the term is currently used, could be observed in anything other than a polymeric material.

D. Shockey: Are the slopes of the stress-strain curves you measured at different temperatures dependent upon loading rate?

G.W. Weidmann: I would expect them to be although I have not yet performed the measurements.

C. Ferguson: Is there any theoretical basis for a linear stress-strain (2 V) relationship? Your data could be equally well (or better?) fitted by a sinusoidal curve and I wonder whether there is any possiblity of "cyclic" untangling of the long molecular chains in craze zones.

G.W. Weidmann: In view of the errors associated with deriving the stress-strain data for craze zones I don't think anything more complex than a linear relationship is justified. This also fits with the rubbery nature of the crazed material. I don't think enough is yet known about detailed molecular mechanisms of deformation to comment on your last point.

A.W.B. Siddans: My question concerns Dr Smalley's paper. Many clastic quartz grains must be derived from highly deformed, high grade metamorphic terrains. To what extend do you think grain release mechanisms are related to stored elastic strains?

I.J. Smalley: I think that stored elastic strains make a very considerable contribution to the grain release mechanism. This is essentially what I proposed in the Nature paper of 1966 - but with the strain internally generated by quartz transitions rather than imposed by metamorphic events. To give a more complete picture these metamorphically produced strains should obviously be included in the collection of relevant factors.

M. Paterson: The experiments of Mr van der Molen (see our paper tomorrow) show that quartz in granite fractures very easily at high temperature in the presence of a water-rich phase - perhaps due to the presence of the water as well as the Hertzian loading conditions. This would be consistent with the observations of Tapponnier and Brau that there appear to be many microcracks in quartzes in granite that come down from very early in the history of the granite.

I.J. Smalley: This essentially agrees with the observations of Moss and his coworkers and suggests that major factors controlling the nature of elastic quartz particles in sediments are operating a long time before the particles actually appear in the sedimenting system.

O. Stephansson: What are the processes of quartz fracture in nature in your opinion?

I.J. Smalley: In the production of quartz particle clastic sediments two sets of fracture processes are important: (a) those occurring in the original quartz containing rock where fractures occur due to internal stresses of various types and imposed metamorhic stresses (possibly); and (b) those occurring as the quartz particles are released from the parant rock- and subsequently. The most important external stress is still seen to be that imposed by glacial action and glacial controls must still be ranked among the most important affecting the formation of observed sediments in the Northern Hemisphere.

M. Paterson: Concerning Dr Smart's paper, how is the shear failure initiated in the structure that is developed in the pre-peak deformation?

J.F. Knott: Is there any guaranteee that the deformation in a shear test does remain as pure shear as deformation proceeds, or might local dilatations occur as a function of end restraints, local bending etc?

P. Smart: The principal defects were: the initial anisotropy of the samples was not exactly uniform; at the start of shearing the samples, they were lightly bent and the top cap slightly tilted. During shearing: the void ratio changed slightly more at the centre than at the ends of the sample; and in most of the tests the top cap was first forced level and then kept level. Despite these factors, the stress-strain curves were reproductible before peak, compare the two sets of results in Fig 1. However, the breakaway at peak stress was more variable, see Fig 1. Many or all of the failure planes were close to a corner of the sample, and their location was probably influenced by local stress concentrations.

A.W.B. Siddans: There appears to be a close analogy between the deformation of cementite lamellae in pearlite and rutile inclusions within single grains of quartz, which boudinage or buckle according to orientation within a plastically deforming quartz grain (Mitra, 1976).

How common are pure kaolinite soils? To what extent do you think the results of your experiments would be modified by the presence of shear mineral phases, e.g. illite or montmorillonite?

P. Smart: Our experiments are on model soils. The factors deliberately excluded include (1) silt, sand and gravel, (2) organic matter, (3) cementing agents, (4) worm holes, etc, (5) structural units consisting of groups of clay plates which are packed closely but randomly within the units. Variation of the composition of the pure water may result in a dispersed clay which behaves differently (Ref 16).

Otherwise the behaviour of illite might be similar; but we should like to confirm this. We would expect montmorillonite to be different, for example work on swelling pressure and on the angle of residual friction has shown that the patterns of behaviour of illite and of montmorillonite are different (R.R. Rao, Glasgow Ph.D. thesis, in preparation).

In montmorillonite, physico-chemical effects predominate. In kaolin and illite, mechanical friction is more important. However, the immediate aim is to establish what modifications result for flocculated kaolin when sheared (1) with no volume change, and alternatively (2) when the confining pressure, $\sigma_2 = \sigma_3$, is lowered before shearing commences.

K. Easterling: Plase can you give som details of preparation techniques for EM studies of clays and clay minerals?

P. Smart: The method used by Smart and Dickson, vol 1, p 121 was to replace the pore water by successive changes of acetone + water mixtures, of increasing strength, about six changes of acetone, and a final change of aceton + Vestopal W resin + initiator + activator. The acetone evaporated, the Vestopal infiltrated the sample and hardened, after which it was cured at $60^{\circ}C$. For other clay minerals, the water must first be removed, e.g. with alcohol.

In general, for electron microscopy of clays, considerable attention has been paid by various research groups to quality control. The methods have included visual examination, measurement of dimensional changes, and X-ray diffraction at various stages. Also various comparisons have been made. Different resins have been used for impregnation. Ultra-thin sections of impregnated samples have been compared with scanning electron microscopy of fracture surfaces of dry samples. Air-drying, substitution-drying, freeze-drying, and supercriticial-drying have been compared. Electron microscopy has been compared with optical microscopy. X-ray texture diffraction, etc.

For further details see:

P. Smart; Soil structure, mechanical properties and electron microscopy; Cambridge Ph.D. thesis; 1967.

N.K. Tovey; Electron microscopy of clays; Cambridge Ph.D. thesis; 1970.

P. Smart; Electron microscope methods in soil micro-morphology; pp 190-206 in A.K. Rutherford ed.; Proc. 4 Int. Work. Mett. Soil Micromorphology; Kingston, Ontario (Limestone Press); 1973.

P. Smart and N.K. Tovey; Electron microscopy of soils; in preparation.

May I also say something concerning dislocations in soils.

Ordinary dislocations are in effect atomic sized holes which move through the atoms. A different sort of dislocation could occur in a soil composed of relatively large sand particles. It might be worth studying the movement of sand-particle-size holes through the particles. There is also a situation in which a large hole, e.g. an abandomed coal mine, can break into smaller holes, which diffuse up to the surface of the ground, spreading out to form a rich shallow depression.

P. Smart: I should like to ask a question in connexion with Professor Pusch's paper.

Is Fig 5 hypothetical?

Does it show syneresis?

R. Pusch: Yes, it is the hypothetical picture suggested by Arnold (see ref in paper). It was confirmed experimentally by the author using electron microscopy; an astonishing similarity was found. The process, which could be termed syneresis although I have a feeling that this expression, in practice at least, is commonly used to describe fissuring in gels, is best described as a phase separation.

C. Ferguson: Is it possible that the microsctructures discussed by yourself (and Dr Smart) are not characteristic of the bulk of the material, but are largely induced during specimen preparation?

R. Pusch: Many years of systematic investigations using various techniques for preparation have shown, beyond doubt, that the various structural elements discussed are original. The best evidence is furnished by a series of studies where structural changes were observed at various strains. For illitic, soft clays it has been shown that natural undisturbed specimens are practically free from "domains" while they develop in course of strain and are finally responsible for bulk failure.

I. Rosenqvist: Is the strength region of up to 15-25% of the original value over 120 days (in fig 6) a real thixotropic region, or is it partly a slow, weak cementation?

R. Pusch: Cementation may have contributed although this is hardly credible. The slow process may be due not only to a delayed regrouping of particles but also to a slow reformation of strength-contributing water lattices. These conclusions were drawn from nuclear magnetic resonance tests which showed that the water molecule order close to the particle was successively improved and that this process went on for several months.

K. Easterling: The arrangement of particles in Fig 3 of your paper could well be described as a loose network rather than as "individual agglomerates" as you interpret the micrograph. On that basis the effect of deformation (shown in your Fig 4) would be understood as a simple rotation of individual particles (not agglomerates). How sure of your "agglomerate" interpretation are you? Have you any evidence that whole <u>agglomerates</u> rotate with respect to one another during deformation - which seems a likely process for aiding alignment in the initial stages at least?

R. Pusch: The proper description of the material is: <u>a network of fairly big particle aggregates linked together by groups of small particles</u>. A large number of electron microscopical investigations have shown, beyond doubt, that most aggregates operate like strong units ("silt grains"), while the weak links between them are broken down by shear distortion at increased bulk strain. This distortion, which produces the "domains", takes place in connection with a mutual displacement of intact aggregates. The displacement involves slight aggregate rotation. I should add that strong mechanical remoulding of these clay types produces a break-down and dispersion also of the denser and stronger aggregates. The resulting large increase in specific surface area is well documented experimentally.

I.J. Smalley: Can we envisage the attraction between clay mineral particles in a clay-water system as being transmitted via the cations in the system - somewhat in the same way as cohesion was believed to be produced in the old Dude-Lorenz model of the metallic bond by attractions between positive metal ions and electrons.

R. Pusch: Interparticle attraction is produced by electrostatic forces (positive edges to negative basal planes) and by London-van der Waals' forces. Cohesion is produced by hydrogen bonds, polyvalent adsorbed cations which connect adjacent particles, and polarized inter-particle cations.

J.F. Knott: My question refers to Prof Paterson's paper. Does plastic deformation in minerals occur with no change in volume? Poisson's ratio 0.5?

M.S. Paterson: It depends on temperature. At high temperatures, the volume is constant. At intermediate temperatures, both plastic deformation and micro-fissuring occurs.

P. Smart: Was there local melting at the points of contact (similar to pressure solution)? How did differential stress vary with confining stress? Is the graph of log differential stress against void ratio straight?

M.S. Paterson: There was no evidence for any pressure melting effects. We did not investigate how the differential stress varies with confining pressure but the effect may be small since the "effective pressure" should be zero in any case. The log differential stress is not linear with melt fraction but falls off more rapidly as the melt fraction increases.

O. Alm: Will the melt film around the particles in the centre of the specimens vary much in thickness when the water added to the specimens exceeds 0.6 w %?

M.S. Paterson: The melt is distributed fairly uniformly within the specimens and so the thicknesses of melt films directly reflect the water content, except as influenced by the stress (i.e. thickening films parallel to the stress direction and thinning those normal).

I.J. Smalley: You relate your observations on granites to possible effects in sandy soils - would it not be better to compare your situation in which the particles are surrounded by melt films to that existing in a clay mineral soil in which each particle has an associated water film?

M.S. Paterson: My impression is that in our case we do not have the complicated electrical change interactions that can arise with very small particles such as clays.

C. Ferguson: Has it been possible to relate the results of your experiments to the microstructures, and inferred deformation mechanisms, of granites believed to have been naturally deformed in the partially melted state?

M.S. Paterson: The deformation mechanisms in the experiments are essentially brittle on the microscopic scale, lacking especially any effects associated with recrystallization or intergranular flow, and so may not apply to many natural situtations.

O. Stephansson: Have you noticed any reorientation of the grains due to deformation?

M.S. Paterson: No; the strains are in fact relatively small and so one would not expect to be able to see substantial rotations.

P.K. Panda: My interest is on the geodynamics aspect, and I am wondering how the theory of dislocations and deformation can be accomodated in this direction?

M.S. Paterson: Dislocation theory arises in understanding the flow of rocks on the scale of the crystal grains. There may of course be other deformation mechanisms too.

E. Suoninen: Comparing deformation of metals with nonmetals, the importance of the following aspects should be studied:

(1) The symmetry of the crystals forming the material?

(2) How high a dislocation density can the crystal accommodate?

M.S. Paterson: Some minerals such as olivine have a rather limited number of independent slip systems but others such as quartz and calcite have plenty to satisfy the von Mises strain compatibility criterion. Dislocation densities as high as 10^{12} cm^{-2} have been observed in minerals such as quartz.

I.J. Smalley: In Dr Röshoff's paper, I presume that the quartz crystals in the initial granodionite were small; could you tell me the sort of size range which is observed when quartz recrystallization takes place in the gneiss?

K. Röshoff: The quartz has a diameter of 1-3 mm in undeformed state. In the augen gneiss the quartz are linsoidal with maximum elongation of 50 mm but very thin.

M.S. Paterson: Can Dr Alm comment on what factors are responsible for the macroscopic ductility that arises in this situation where the microscopic processes are essentially brittle.

O. Alm: The solid medium Griggs' hot creep apparatus is too insensitive to allow any reliable measurement of the dilatancy. It is difficult to point out any particular factor responsible for this ductility, but I would guess that the very small grain size of the ground mass would easily be squeezed into cracks in the larger grains when they open up. The deformation rate is fairly low and the cracks may open slowly enough to prevent the specimens from failing by through-going shear faults. If this is the case the dilatancy would be very small.

G. Ranalli: The samples are anisotropic. Have you looked at the dependence, if any, of the stress-strain curve on orientation?

Also, what was the orientation of the samples with respect to schistosity?

O. Alm: The specimen used in this work were cored out normal to the schistosity of the granodiorite and normal to the foliation in the matrix samples. No experiments have so far been performed on specimens in other directions in these rocks. I would, however, be surprised if the foliation would not affect the stress-strain results.

M.S. Paterson: I have a question for Dr Swan. An interesting property of rock is the relatively slow propagation of cracking, as evidenced by the possibility of servocontrol in machines with response time of the order of microseconds. Would you not expect to need stress pulses of duration of orders of magnitude more than microseconds in order to influence the initiation of comminutive fracturing?

G. Swan: Firstly the response time you give seems to me to be higher than that quoted by manufacturers of servo-controlled machines of 1-5 milliseonds. In answer to your question I would say that although the process of coalescing cracks is a relatively slow process, the propagation rate of a microcrack before alignment and stabilisation is probably quite high. This means that a disturbance of the order of microseconds is sufficient. The slower process of coalesence is basically controlled by the static load.

G. Weidmann: Concerning Dr Siddans' paper, has the process of pressure solution been observed under laboratory conditions?

A.W.B. Siddans: Yes, see references 13 and 19 in the text of my paper.

G. Weidmann: How unambiguous is the identification of pressure solution as a process occurring in the field as opposed to, for example, material removal by mutual abrasion of rocks, or abrasion by a slurry and material deposition by crystallization out of saturated solution?

A.W.B. Siddans: Fossils provide the best evidence for pressure solution because of their three-dimensional structure - removal of material, rather than shearing, can be demonstrated. Perhaps the best evidence for the deposition part of pressure solution comes from the observations that the composition of some tectonic veins change in composition as they pass through different lithologies i.e. the source of the vein material seems to be in immediately adjacent rock. Many identifications of the pressure solution are in practise often made by analogy.

Extreme ductility in rocks is well known. The most striking examples are in ductile shear zones where rocks have become mylonites, i.e. highly deformed, banded rocks, whose grainsize has been reduced by cataclasis, sub-grain formation and dynamic recrystallisation (1). For example the shockness of the Mesozone strata forming the lower, mounted limb of the Mordes Nappe, a large flat-lying field on the Swiss/French Alps, is about 1000 m. This is reduced to about 100 m on the "root zone" of the nappe. Congranulation limestones in the invested limb show that it is itself highly deformed, probably by dislocation creep (J.G. Ramsay, personal communication). The extremely deformed rocks of the "root zone" are mylonites. Recent studies of calonite mylonites and laboratory studies on deformed limestones (2,3) show that such extreme strains may be achieved by superplastic behaviour on free-grained rocks, in which the dominant deformation mechanisms appear to be grain boundary sliding accomodated by grain boundary diffusion.

Such rocks illustrate a fundamental problem in the application of materials sciences' techniques of microstructural analysis to geologic materials. The deformation mechanisms that operate in rock may change during prolonged, slow tectonic deformation, so that microstructures produced early in the deformation history may be completely overprinted. Other deformation mechanisms that operate in some rocks at low temperatures, e.g. pressure situation accommodated by grain boundary sliding, appear to have no counterparts in materials sciences. It may thus be impossible to obtain a clear picture of the sequence of changing deformation mechanisms that have produced a deformed rock, from the microstructure of the end product alone.

1   White S., Phil. Trans. R. Soc. London A., 283 (1976) 69-86.

2   Schmid S.M. et al., Tectonophysics, 43 (1977) 257-91.

3   Casey M. et al., Proc. ICOTOM5, Aachen, March 1978 (in press).

I.J. Smalley: I have a three questions:

a) What factors would have to be involved to allow burial of a sediment without compaction?

b) Is this possible in a muddy sediment (perhaps to a relatively shallow depth)?

c) Can a "muddy sediment" be defined - must it contain a certain proportion of clay minerals?

A.W.B. Siddans:
a) High pore-fluid pressure, due to rapid burial and/or dehydration reactions, in a layer below a very low-permeability layer.

b) Yes.

c) In the context of this paper a "muddy sediment" must contain a significant proportion of clay minerals e.g. a dirty sandstone or mainly limestone.

M.S. Paterson: How do you distinguish between pressure solution and local higher permeability as the most important factor in the preferential solution phenomenon of which you give many examples?

A.W.B. Siddans: In the end-product rock I don't think this distribution is possible. However, I think neither are the most important factors in determining the rate of deformation by the pressure solution process. This is largely controlled by the rate at which material can diffuse through the solid in order to get into solution.

J.F. Knott: I should like to ask Prof Suoninen, over what area are you detecting electrons from?

E. Suoninen: Approximately 10-100 μm diameter in the scanning mode.

S. Johansson: What kind of signal do you think you would have got in this case if the surface had been broken in air instead of vacuum? This point is important if one wants to study fatigue fracture surfaces.

E. Suoninen: The surface would have been completely covered by atoms adsorbed on it from the air. In principle, they can be removed by sputtering, but this usually causes additional damage of the surface. No good alternative exists for fracturing the sample in good vacuum.

D.A. Shockey: In your Auger investigations, electron emission was stimulated by irradiation of the fracture surface with other electrons. But is there not spontaneous electron emission (exo electrons) from fracture surfaces that endures for hours after the fracture events? Cannot these electrons be analyzed to determine what elements are present? Could the exoelectrons be used to tell us various information about the fracture process?

E. Suoninen: Spontaneous emission is indeed observed, and studies such as you mention have been made. The important difference is, however, that the exoelectrons are generally less energetic than the Auger electrons used in studies which I described. The techniques used to detect them are quite different. The most difficult question is, however, the interpretation of these spectra in terms of the fracture phenomenon. For chemical analysis, Auger spectra are probably more convenient.

K. Easterling: What are the advantages and disadvantages of using this technique for non-metallic materials?

E. Suoninen: The method can be and has been applied to many non-metallic materials. The only change is in the technique of preparing the fracture surface. Cleavage or scraping of the samples is used for these materials.

K. Easterling: (Chairman) I should now like to take up a more general question. In trying to bridge the gap between Materials Science and Rock Mechanics, the main problem appears to be one of scale. How far dare we take comparisons between the deformation of metals and (say) mountains? Would anyone like to comment?

I.J. Smalley: There might be some useful comparisons to be made, on a macroscopic scale and dealing largely with creep phenomena, between rocks and concrete. In concrete we observe that the creep rate decreases as the aggregate content is increased. A simple cement paste specimen has a relatively high creep rate but a specimen with a dispersed phase (i.e. a set of rock aggregate particles) has a much lower creep rate.

G. Weidmann: Isn't there a better parallel between the creep of rocks or minerals and the creep of composite materials than with the creep of polycrystalline metals? From studies on composite materials it is known that the creep rate depends on the relative stiffnesses and concentrations of the different phases and on how good the bonding between them is.

G. Ranalli: In the steady-state creep regime, flow laws in rocks, metals and ice are similar. However, rocks consists of several phases. It is qualitatively known that the presence of a second phase decreases the strain rate by orders of magnitude. Has this problem been investigated by metallurgists in a way that could be extrapolated to rocks?

P. Feltham: The phenomenon is well known as precipitation hardening. Small additions of copper to aluminium will form precipitates which obstruct the movement of dislocations, thus strengthening the material. Similar principles hold for creep. In rocks at elevated temperatures, when the Peierls forces no longer effectively control dislocation movement similar principles would be expected to hold. Similarity of creep laws can be explained by the stochastic model of creep, to which I have referred to in my lecture.

K. Röshoff: I should like to give a comment, which may throw some light on the question of scale in rock materials and also on the links between Earth Science on one side and Materials Science and Solid Mechanics on the other. I have been working with rock material in scales from mega-to micro scale.

At low temperatures and confining pressures the fracturing of rock materials is very similar to fracture mechanics observed in Solid Mechanics. The geometry of a specific fracture is the same on every scale, that is from mega to microscale. Therefore it is also possible to use the same laws to resolve the stress field around a fracture in the large as well as in the small scale. At high temperatures and confining pressures, the deformation in geological time is mainly of creep character. This means that the deformation structures are restricted in scale to the dimension of a crystal grain. The same deformation feature can be followed continuously from the scale of a micron to macroscale (light optical scale). At larger scales the plastic deformation is only obvious in the elongation of the crystals or crystal aggregates and, a development of an anisotropy in the rock. The scale of material involved in such a deformation may, however, be as large as in the case of low temperature deformation. Thus the volume of the plastically deformed Tännäs augen gneiss as described in my contribution is 600 m in thickness and the length at least 300 km. The creep deformation in rock materials therefore is very similar to the deformation characteristics of metals.

O. Stephansson: Prof Easterling has brought up the question of the great variation in scale in the Earth Sciences and hence the difficulties to envisage the various mechanisms of deformation. I should like to mention that model experiment with the application of scale model theory has been of great importance for testing various mechanisms and modes of deformation in Earth Sciences.

SESSION 3
Theoretical Studies

CDM - CAPABILITIES, LIMITATIONS AND PROMISES

Jan Hult

Department of Solid Mechanics, Chalmers University of Technology,
S-402 20 Göteborg, Sweden

ABSTRACT

The basic ideas, concepts and governing equations in continuum damage mechanics (CDM) are presented and discussed. The stress-strain diagram för a damaging material is studied, and the influence of local variation in damage properties is examined.

1. INTRODUCTION

The science of solid mechanics occurs in a simplified version called strength of materials. This is taught in engineering schools and is used by designers and engineers for predicting the safety of structures under load.

What characterizes strength of materials are certain strongly simplified assumptions, which make design calculations easy yet accurate enough for many practical purposes. A good example is provided by the classical beam theory due to Euler and Bernoulli. The beam is modelled as a bundle of parallel thin fibres, each of which is subject to axial tensile or compressive forces. The strain field is assumed to vary linearly across the beam, and the Poisson effect is neglected. The results of this engineering beam theory are in good agreement with practical tests on loaded beams in a very wide range of shapes and loadings. This is in spite of the fact that a more accurate analysis, based on general solid mechanics theory, shows a much more complex behaviour than is assumed in the elementary theory.

Simplifying assumptions are, however, also made in general solid mechanics theory. The basic concept of the material continuum is, of course, completely artificial, since no such material exists in the real world. For the same reason the concept of stress at a point is a ficticious one, since it is defined as the limit of a ratio force/area, as the area shrinks to zero. The concept of material area loses all meaning on the atomic scale. The justification for these concepts and quantities

---

Dedicated to Professor Dr. Henry Görtler on his 70th birthday

is simply that they have been found useful in developing the
theory and that the theory leads to results, which are in
agreement with experimental observation.

Strength of materials and solid mechanics deal largely with
the deformation of loaded structures. It is known that material
properties relating to deformation, e.g. Young's modulus and
Poisson's ratio, depend only weakly on local material defects.
Hence the defect free material continuum is a suitable material
model for describing deformations and stresses in loaded
structures. Fracture, however, is often caused by the growth
of defects, which may or may not be present already in the unloaded material. Fracture analysis is necessarily more complex,
and many facets of fracture still remain to be well understood.
This paper will review some attempts, which have been made in
recent years, to develop an elementary model for fracture prediction. Like strength of materials it is based on certain
intuitive, strongly simplifying assumptions. It introduces a
new field quantity, denoted damage, which is no more artificial
the now classical concepts of stress and strain. The justification
again is purely pragmatic. This damage mechanics describes
several experimentally observed features of various different
fracture processes well enough for it to be of interest to
designers and engineers.

Whether it is also of interest to materials scientists is an
open question. To be sure it is an approach to fracture prediction
from a purely macroscopic point of view. A strong advance of
physical fracture research might very well make the damage
mechanics approach obsolete. For the time being, however, it
seems to offer a workable alternative in design offices.

2. TWO MECHANICAL APPROACHES TO FRACTURE ANALYSIS

Fracture implies separation within a material. Such separation
may be identified already on an atomic level. The atoms immediately below an edge dislocation are more apart than those in
the undisturbed lattice. Atomic distances across grain boundaries
in polycrystalline materials are strongly irregular. If the
distance is large enough over a large enough part of the
boundary, one might speak of a grain boundary crack. A number of
such micro cracks may join to form a macro crack. The macro
crack may finally extend across the whole body and cause complete
fracture.

Fracture mechanics deals with the conditions at the end of a
well defined macro crack. The basic assumption is that of
autonomy of the end regions of the crack (Broberg 1967). The
intensity of the stress and strain fields in these regions is
analyzed, and conditions are derived for the crack to extend
under various kinds of loads. Separation occurs at the moving
tip of the extending crack; the cohesive forces are there not
sufficient to hold the material together; the bonds are destroyed. The material at the crack tip has become completely
damaged.

Likewise a great number of micro cracks may be interpreted as
causing damage to the material in a great number of points.
The concept of a continuous damage field here suggests itself
as a possible working model. A precise definition of such a
damage quantity will be given below.

The two approaches to fracture analysis - fracture mechanics (FM)
and continuum damage mechanics (CDM) - correspond to two
different modes of fracturing. FM describes the conditions for
a preexisting finite crack to grow under load and cause fracture.
CDM describes the formation of continuously distributed material
damage, which eventually leads to fracture. In both cases
fracture is only the terminal event in a process, which extends
over the entire loading history. It is informative for a designer
to know something about the development of this process. Only
then may he safely judge the proximity of a final fracture.

Design against fatigue provides a good example of how the entire
process of fracture development may now be considered. The
classical method was to relate a maximum stress amplitude in the
structure to the endurance limit of the material, thus ensuring
"infinite" lifetime. Now the rate of growth of any preexisting
crack may be taken into account, and the remaining lifetime
may be estimated by use of FM technique.

FM deals with a single crack embedded in an otherwise intact,
defect free medium, whereas CDM deals with a deteriorating medium,
which contains no cracks. The two approaches may therefore be
interpreted as two extremes. A combined approach, where a crack
of finite size is embedded in a deteriorating medium, has been
found to give results closer to real observation (Janson and
Hult 1977, Janson 1978).

3. CONTINUUM DAMAGE MECHANICS - BASIC CONCEPTS

(a) Definition of damage and net stress

Microscopic and X-ray studies show that irreversible structural
changes may occur in materials under load. A variety of such
changes have been identified, and several mechanisms have been
found, which lead to those changes. The fundamental idea in
CDM is to express, collectively, these modes of material
deterioration by means of a single field variable.

The first analysis presented with this aim (Kachanov 1958) introduced a quantity ψ, denoted "continuity". For a completely
undamaged material $\psi=1$, and for a completely destroyed material,
with no remaining load carrying capacity, $\psi=0$.

Kachanov considered uniaxial tension, where ψ may be interpreted as the ratio between a "net load carrying area" A_n and
the current area A of the bar cross section (measured
externally), i.e.

$$\psi = A_n/A \qquad (1)$$

With P denoting the tensile force, the quantity

$$s = P/A_n \tag{2}$$

then suggests itself as a new measure of stress, which takes the deterioration into account. It will be denoted the "net stress" in contrast to the "current stress"

$$\sigma = P/A \tag{3}$$

From (1), (2), and (3) follows the relation

$$s = \sigma/\psi \tag{4}$$

With decreasing continuity ψ the net stress s will increase ever more over the current stress σ.

Slight modifications of the original Kachanov ideas have been introduced by Rabotnov (1969), Broberg (1975) and others. Instead of the continuity ψ a complementary quantity ω, denoted damage, has found use. It may suitably be defined by the differential relation

$$d\omega = -dA_n/A_n \tag{5}$$

which is analogous to the definition of natural strain

$$d\varepsilon = dL/L \tag{6}$$

Integration of (5) yields

$$\omega = \ln A/A_n \tag{7}$$

making use of the condition that $\omega=0$ (no damage) if $A_n=A$. The analogous expression for strain is

$$\varepsilon = \ln L/L_0 \tag{8}$$

where L_0 denotes the length in the unstrained state, cf. Fig. 1.

The definitions (1) and (7) are arbitrary, but have been chosen for mathematical convenience. The following relation is seen to exist between damage and continuity

$$\omega = -\ln \psi \tag{9}$$

From (2), (3), and (7) follows the relation between net stress, current stress, and damage

$$s = \sigma \exp \omega \tag{10}$$

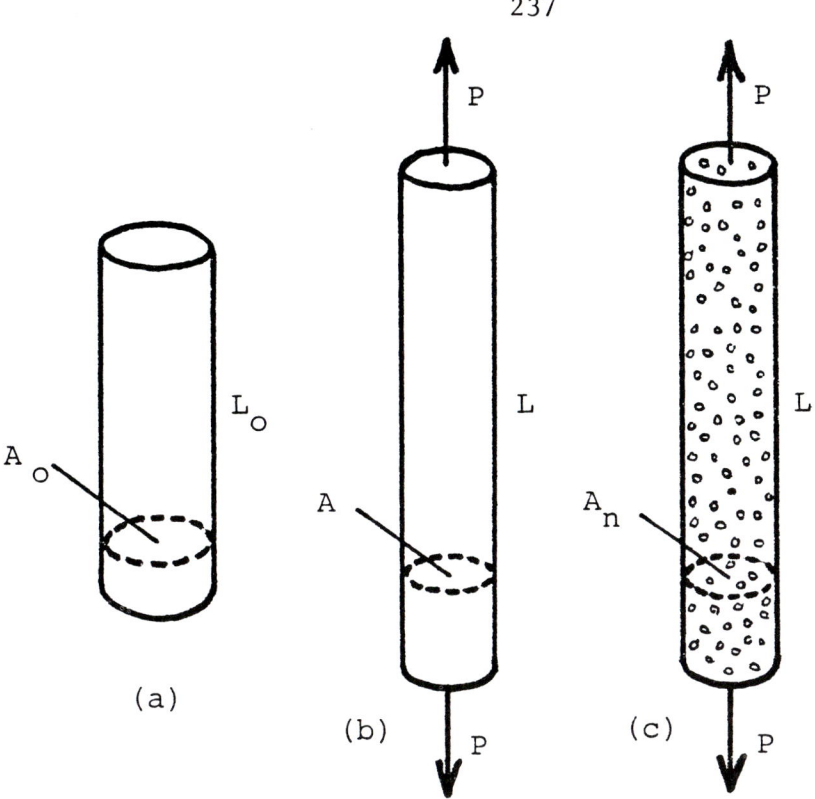

Fig. 1.
(a) Unloaded bar
(b) Strained and undamaged bar
(c) Strained and damaged bar

If the strain ε is small, the current stress σ is closely equal to the "nominal stress"

$$\sigma_0 = P/A_0 \tag{11}$$

where A_0 denotes the original cross sectional area. For large strains, however, the difference between σ and σ_0 has to be taken into account. A simple relation exists for cases of constant material volume. With

$$A_0 L_0 = AL \tag{12}$$

follows

$$\sigma = P/A = (P/A_0) \cdot (A_0/A) = (P/A_0) \cdot (L/L_0) \tag{13}$$

or, considering (8) and (11)

$$\sigma = \sigma_0 \exp \varepsilon \tag{14}$$

From (10) and (11) follows, finally

$$s = \sigma_0 \exp(\varepsilon + \omega) \tag{15}$$

which shows that the effects of strain and damage may be added to give the combined effect on the net stress.

(b) <u>Damage laws</u>

For cases of high temperature Kachanov (1958) suggested the following law for the rate of decrease of continuity

$$d\psi/dt = -Cs^\nu \qquad (16)$$

where C and ν are temperature dependent material constants. From (4) and (16) follows an expression for the lifetime t_R (reached when $\psi=0$), which has found much use in predicting brittle creep rupture in engineering components

$$t_R = 1/(1+\nu)C\sigma^\nu \qquad (17)$$

Generalized laws have later been proposed for both strain and damage (Broberg 1975, Hult 1974), which include also instantaneous effects in contrast to (16). They have been stated as

$$\begin{cases} d\varepsilon/dt = G'(s) \cdot ds/dt + F(s) & (18) \\ d\omega/dt = g'(s) \cdot ds/dt + f(s) & (19) \end{cases}$$

For instantaneous increases in net stress these laws take the forms

$$\begin{cases} d\varepsilon = G'(s) \cdot ds & (20) \\ d\omega = g'(s) \cdot ds & (21) \end{cases}$$

Starting from a virgin state, where $\varepsilon=0$, $\omega=0$, these expressions yield

$$\begin{cases} \varepsilon = G(s) & (22) \\ \omega = g(s) & (23) \end{cases}$$

as the strain and damage caused by an instantaneously applied net stress s. Similarly $F(s)$ and $f(s)$ denote the strain and damage rates which would be caused by a constant net stress s.

(c) <u>Governing equations</u>

From (15), (18), and (19) follows after elimination of ε and ω

$$[1/s - G'(s) - g'(s)]ds/dt - F(s) - f(s) = (1/\sigma_0)d\sigma_0/dt \qquad (24)$$

which is the differential equation for the net stress $s=s(t)$ as caused by a prescribed nominal stress history $\sigma_0=\sigma_0(t)$.

For instantaneous increases in σ_0 this degenerates to

$$[1/s - G'(s) - g'(s)]ds/d\sigma_0 = 1/\sigma_0 \qquad (25)$$

which is the differential equation for the net stress $s=s(\sigma_0)$.

4. STRAIN AND DAMAGE INDUCED FRACTURE

If a non-decreasing nominal stress $\sigma_0(t)$ is applied to a bar, fracture will eventually occur, cf. Fig. 2c. Two forms of loading histories $\sigma_0(t)$, shown in Fig. 2a and 2b, are of special interest, and will be analyzed first.

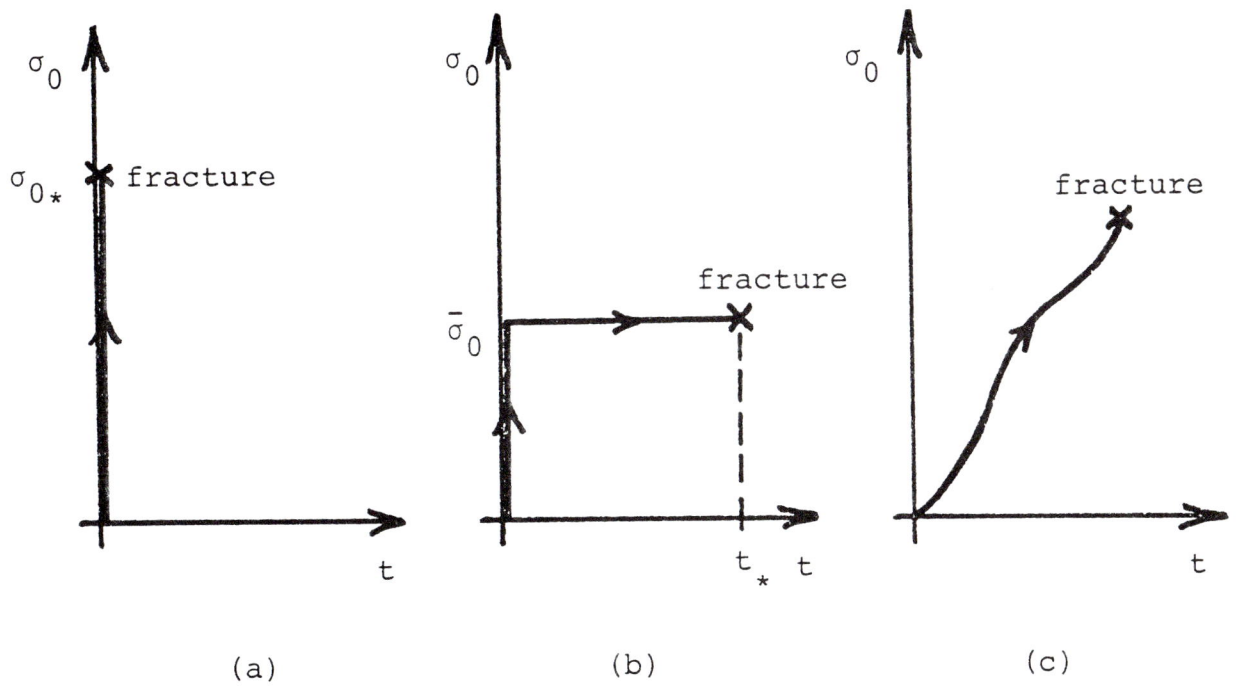

Fig. 2. (a) Rapid tensile test. (b) Constant load creep test. (c) Non-decreasing loading history.

(a) <u>Rapid tensile test</u>

If $G'(s) + g'(s)$ is non-decreasing, (25) shows that $ds/d\sigma_0$ will be monotonically increasing. A critical state, where $ds/d\sigma_0 \to \infty$, will be reached when $s \to s_*$, where

$$1/s_* - G'(s_*) - g'(s_*) = 0 \qquad (26)$$

Equilibrium is then no longer possible and fracture will occur, cf. Fig. 3 path (a). The nominal stress at fracture is denoted σ_{0*}.

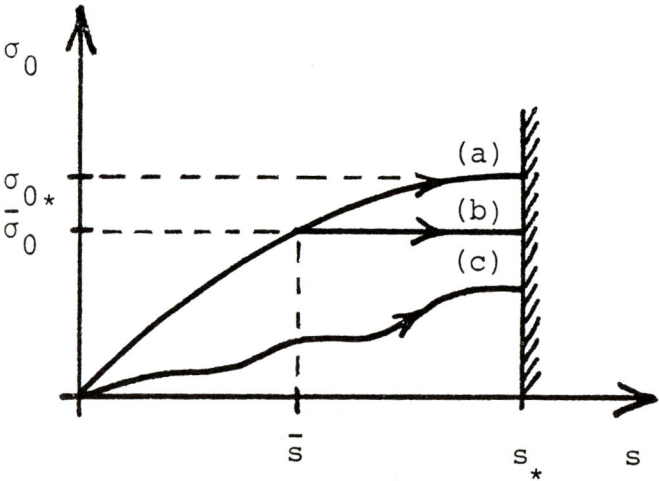

Fig. 3. Fracture occurs at a finite net stress s_*.
(a) Rapid tensile test.
(b) Constant load creep test
(c) Non-decreasing loading history.

(b) Constant load creep test

A nominal stress $\bar{\sigma}_0 < \sigma_{0*}$ is applied instantaneously and then kept constant. From (24) then follows

$$ds/dt = \frac{F(s) + f(s)}{1/s - G'(s) - g'(s)} \qquad (27)$$

If, again, $G'(s) + g'(s)$ is non-decreasing, (27) shows that ds/dt will be monotonically increasing. Hence a critical state, where $ds/dt \to \infty$, will be reached when $s \to s_*$, where s_* is again given by (26). Equilibrium is then no longer possible and fracture (creep rupture) will occur, cf. Fig. 3 path (b). The corresponding creep rupture time t_* is obtained from (27) as

$$t_* = \int_{\bar{s}}^{s_*} \frac{1/s - G'(s) - g'(s)}{F(s) + f(s)} ds \qquad (28)$$

An evaluation of this integral and comparison with experimental data has been performed by Westlund (1977).

(c) General, non-decreasing loading history

With both $G'(s) + g'(s)$ and $\sigma_0(t)$ non-decreasing it follows from (24) that fracture will occur when the net stress reaches the level s_* given by (26), cf. Fig. 3 path (c). Hence s_* has the general significance of a net fracture stress, characteristic of the particular material at the given temperature.

The material behaviour is completely characterized by the functions G', g', F, and f in (17) and (18). In the resulting governing equation (24) they appear only in pairs, viz. $G'+g'$ and $F+f$. Hence the general conclusions derived are valid also in the limiting cases of no strain ($G' \equiv 0$, $F \equiv 0$) and no damage ($g' \equiv 0$, $f \equiv 0$).

5. STRESS-STRAIN DIAGRAM FOR DAMAGING MATERIAL

Elementary strength analyses often refer to the linearly elastic material, cf. Fig. 4.

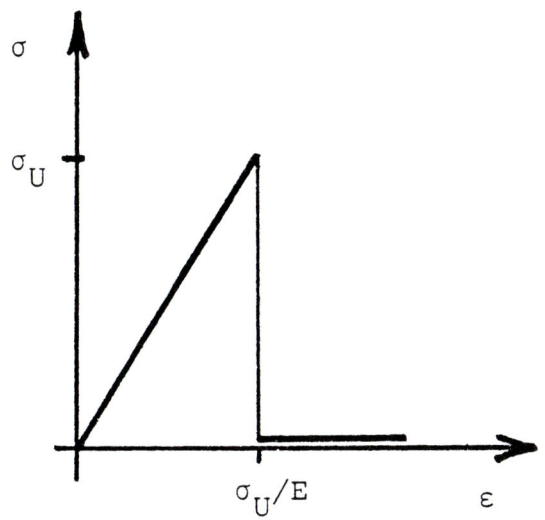

Fig. 4. Stress-strain curve for linearly elastic material with ultimate strength σ_U.

Its stress-strain curve is discontinuous at strain σ_U/E, where the stress drops from σ_U (ultimate strength) to zero.

If damage develops, the stress-strain curve will change into a continuous one. Its shape may be derived from (10), (22), and (23).

To simplify the mathematics the relation (10) will here be replaced by the linearized form

$$s = \sigma/(1-\omega) \tag{29}$$

which corresponds to the original Kachanov damage definition. For a non-damaging material $s \equiv \sigma$ and hence the following form for (22) is indicated

$$\varepsilon = s/E \tag{30}$$

Hooke's law will then be regained when $\omega \equiv 0$. The damage relation (23) is taken in the more general, non-linear form

$$\omega = (s/D)^{1/\delta} \tag{31}$$

The constant D may conveniently be denoted the damage modulus. From (29)-(31) follows the stress-strain relation

$$\sigma = E\varepsilon[1 - (E\varepsilon/D)^{1/\delta}] \tag{32}$$

which is shown in Fig. 5 for various values of δ. In the limiting case $\delta=0$ the discontinuous curve in Fig. 4 is regained.

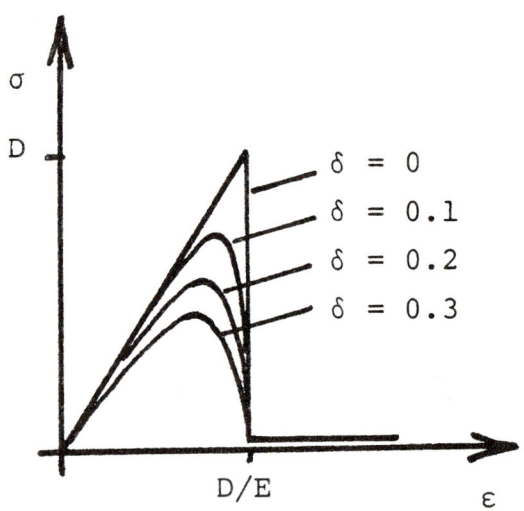

Fig. 5. Stress-strain curves for linearly elastic, damaging material.

The ultimate strength of the damaging material, i.e. the stress level at which $d\sigma/d\varepsilon=0$, is found to be

$$\sigma_U = D\, \delta^\delta/(1+\delta)^{1+\delta} \tag{33}$$

which approaches D as $\delta \to 0$.

Continuous stress-strain curves according to Fig. 5 require a stiff, deformation controlled testing machine to be experimentally recorded. Examples of such curves have been provided e.g. by Hudson and Fairhurst (1969), who also showed an alternative derivation of the theoretical stress-strain curve.

6.   LOCAL VARIATION I DAMAGE MODULUS

In the previous sections fracture was interpreted in purely mechanical terms as being a loss of equilibrium. The material properties were assumed to be uniform, and hence the non-equilibrium state was reached simultaneously in all parts of the tensile bar, with its everywhere uniform stress field.

A tensile bar will now again be considered, but the damage modulus will be assumed to vary over the cross section. The bar will be modelled as a bundle of parallel fibres, each of which has a stress-strain curve according to Fig. 5 with fixed values of E and δ. Since the shape of the cross section has no influence on the strength of the bar, it may be assumed to be onedimensional The damage moduli of the various fibres may be rearranged in monotonically increasing order as a function of coordinate x, ranging from zero to A. To simplify the mathematics this distribution will be assumed here to be linear, cf. Fig. 6.

With

$$D(x) = D_0 + \Delta(x/A - 1/2) \tag{34}$$

follows from (32) that $\sigma(x) \equiv 0$ for $x \leq c$, where

$$D_0 + \Delta(c/A - 1/2) = E\varepsilon \tag{35}$$

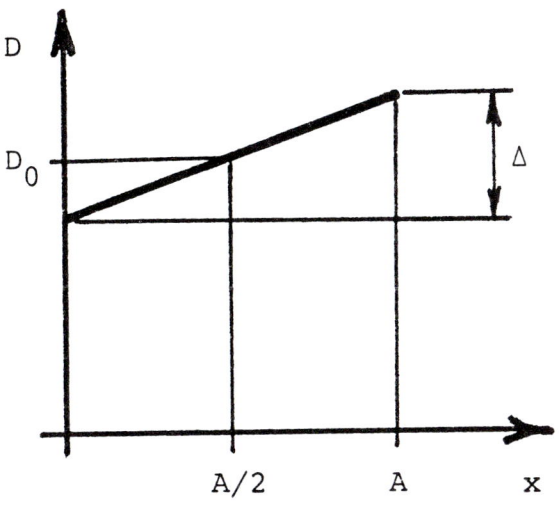

Fig. 6. Variation of damage modulus over cross section

For $c \leq x \leq A$ the stress field (32) becomes

$$\sigma(x) = E\varepsilon\{1 - [E\varepsilon/D(x)]^{1/\delta}\} \qquad (36)$$

and the tensile force is

$$P = \int_0^A \sigma(x)\,dx = \int_c^A \sigma(x)\,dx \qquad (37)$$

Introduction of (36) and (34) in (37) yields after integration

$$P = E\varepsilon(A-c) + [\delta/(1-\delta)](E\varepsilon)^{1/\delta}(A/\Delta)[(D_0+\Delta/2)^{1-1/\delta} - (E\varepsilon)^{1-1/\delta}] \qquad (38)$$

The second term is proportional to δ and may be neglected when $\delta \ll 1$. With c taken from (36) and introduced into the first term in (38) there results for the tensile force

$$P = EA\varepsilon(1/2 + D_0/\Delta - E\varepsilon/\Delta) \qquad (39)$$

This expression is valid for $c > 0$. From (35) follows that $c=0$, when $E\varepsilon = D_0 - \Delta/2$. For strains smaller than this (38) yields

$$P = EA\varepsilon \qquad (40)$$

The P-ε-relations (39) and (40) are shown in Fig. 7.

If $0 < \Delta/D_0 < 2/3$, fracture will occur instantly on reaching the load level $P'_* = A(D_0 - \Delta/2)$. If $2/3 < \Delta/D_0 < 1$, fracture will proceed in a gradual, controlled manner starting from $P = A(D_0 - \Delta/2)$.

Loss of equilibrium and instant fracture finally occurs at load level

$$P''_* = (A/\Delta)(D_0/2 + \Delta/4)^2 \qquad (41)$$

The dependence of the fracture load on Δ is shown in Fig. 8.

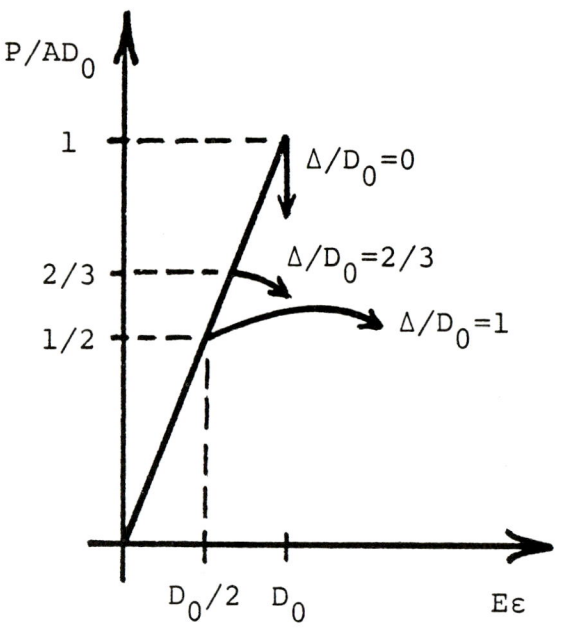

Fig. 7. P-ε-relations for tensile bar with non-uniform damage modulus

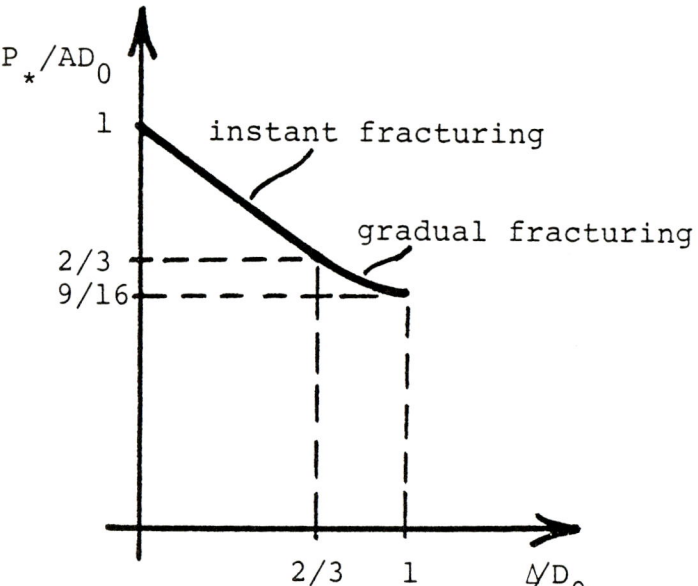

Fig. 8. The fracture load P_* decreases with increasing non-homogeneity of the damage modulus

The above considerations may be generalized to cases of arbitrary variation of the local ultimate strength. The stress-strain curve of an individual fibre is given by Fig. 4, and the ultimate strength variation over the cross section is written as

$$\sigma_U(\omega) = \bar{\sigma}_U \, \phi(\omega) \tag{42}$$

Here $\bar{\sigma}_U$ denotes the average ultimate strength, and ω the fraction of cross sectional area where fracture has occurred. As in Fig. 6 the curve $\phi(\omega)$ is monotonically increasing, and

$$\bar{\phi} = \int_0^1 \phi(\omega)\, d\omega = 1 \tag{43}$$

An increasing load P will cause ω to increase. With $(1-\omega)A$ denoting the remaining load carrying area, there results the relation

$$P/(1-\omega)A = \sigma_U(\omega) \qquad (44)$$

and hence, considering (42)

$$P/A\bar{\sigma}_U = (1-\omega)\phi(\omega) \qquad (45)$$

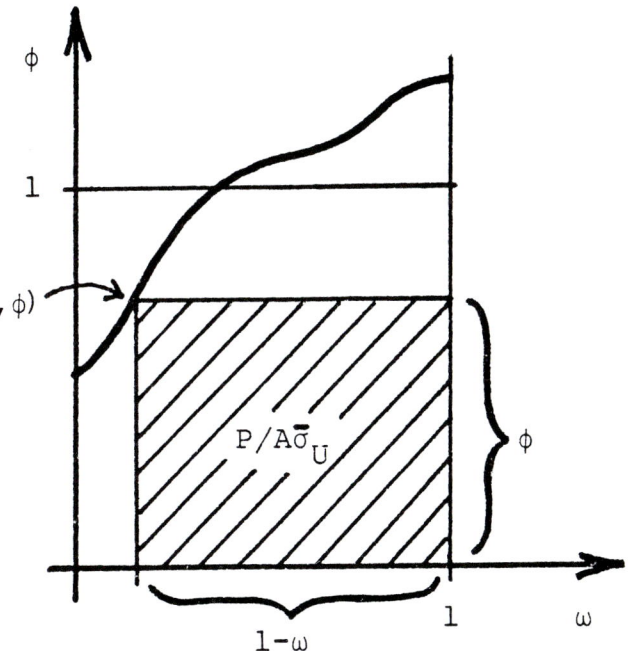

Fig. 9. Variation of ultimate strength with damaged area fraction ω

In Fig. 9 the area of the shaded rectangle equals $P/A\bar{\sigma}_U$. Since $\phi(\omega)$ is monotonically increasing, this rectangle will always lie completely below the $\phi(\omega)$ curve. Hence

$$P/A\,\bar{\sigma}_U < \int_0^1 \phi(\omega)\,d\omega \qquad (46)$$

or, considering (43)

$$P/A \leq \bar{\sigma}_U \qquad (47)$$

for all ω. Equality holds in the single point $\omega=0$ of the limiting case $\phi(\omega)\equiv 1$, i.e. the case of a completely uniform material before any local fracture has occurred. This corresponds to $\Delta=0$ in Fig. 6, 7, and 8. Hence for a bar of non-uniform material equilibrium is possible only if $P/A < \bar{\sigma}_U$, i.e. the ultimate strength of the bar is less than the average strength of the fibres.

7. DISCUSSION

Various modes of fracture in tensile bars have been considered in terms of an assumed mechanism of damaging of the material. In all cases fracture is the result of loss of equilibrium and not of the breaking of bonds with a given strength. This approach leads to a unified description of the conditions for fracture in e.g. a rapid tensile test and a creep test.

One major objection to the early damage models was that they predicted fracture only when all the material had completely lost its load carrying capacity. This deficiency disappears with the introduction of the G' and g' terms in (18) and (19). The resulting net stress at fracture, s_*, implicitly given by (26), then becomes a finite quantity.

A point of controversy is the damage variable itself. It cannot be measured directly in the same way as e.g. strain. But then the same limitation holds for stress, which can only be measured indirectly through its actions, which may be e.g. strain. Attempts to identify damage directly with the relative area of voids in a cross section have not been successful. In the present stage of development damage will only be considered as an internal state variable, and the functions appearing in the basic strain and damage relations have to be a posteriori determined from various fracture tests, cf. Bråthe (1978). Recent results by Leckie and Hayhurst (1977) do, however, indicate that the continuum damage models are essentially compatible with physical theories of nucleation and subsequent growth of voids.

Major effort is now being directed towards extending these basic relations to cases with multiaxial stress fields. Considerable advance has been made in recent years, cf. Chaboche (1978) and Leckie (1978).

CDM may well be said to offer oversimplified pictures of complex phenomena. But this simplicity is in itself important, because it enables designers and engineers to get an overview which can assist in making sound compromises in design work. Several important qualitative, and sometimes also quantitative, conclusions may be drawn by applying the CDM approach to various design problems.

8. REFERENCES

Broberg, H., Creep damage and rupture. Diss CTH, Gothenburg 1975.

Broberg, K.B., Critical review of some theories in fracture mechanics. Int. symposium on fracture mechanics, Kiruna, Sweden 7-12 August 1967. Proc. ed. by J. Carlsson, Noordhoff, Groningen 1968.

Bråthe, L., Creep deformation and deterioration. Diss CTH, Gothenburg 1978.

Chaboche, J.-L., Description thermodynamique et phenomenologique de la viscoplasticité cyclique avec endommagement. Diss Université de Paris 6, 1978.

Hudson, J.A. and Fairhurst, C., Tensile strength, Weibull's theory and a general statistical approach to rock failure. Southampton 1969 Civil Engineering Materials Conference. Proc. ed. by M. Te'emi, Wiley-Interscience, London 1971.

Hult, J., Creep in continua and structures. In Topics in Applied Continuum Mechanics, ed. by J.L. Zeman and F. Ziegler, Springer, Wien 1974.

Janson, J., Crack in material with damage formation. Diss CTH, Gothenburg 1978.

Janson, J. and Hult, J., Fracture mechanics and damage mechanics - a combined approach. J. Mech. Appliquee $\underline{1}$:1 (1977), 69-84.

Kachanov, L.M., Time of the rupture process under creep conditions (in Russian). Izv. Akad. Nauk. SSSR, Otd. Tekh. Nauk No. 8 (1958), 26-31.

Leckie, F.A., The constitutive equations of continuum creep damage mechanics. Phil. Trans. R. Soc. Lond. A. $\underline{288}$ (1978), 27-47.

Leckie, F.A. and Hayhurst, D.R., Constitutive equations for creep rupture. Acta Met. $\underline{25}$ (1977), 1059-1070.

Rabotnov, Yu. N., Creep problems in structural members. North-Holland, Amsterdam 1969.

Westlund, R., Creep rupture - deterministic and probabilistic analyses. Diss CTH, Gothenburg 1977.

SHAKEDOWN ANALYSIS IN ROLLING CONTACT PROBLEMS

G. Rydholm and B. Fredriksson

*Department of Mechanical Engineering, Linköping Institute of Technology,
S-581 83 Linköping, Sweden*

ABSTRACT

The paper presents a method of solving shakedown analysis problems in two-dimensional rolling contact problems. The elastic stress state is computed for the different load conditions during a load cycle. The "best" selfstress according to Melans theorem is then found by solving a minimax problem. Results from investigations of two-dimensional rolling contact problems are presented. The contact pressure profile is assumed to be known. Influence of different coefficient of friction and of the elasto-hydrodynamic peak are also taken into account.

1. INTRODUCTION

When a structure is loaded beyond the elastic limit plastic deformation takes place. If the load varies with time, for instance cyclically the plastic deformation will under certain conditions cease within a finite number of cycles and the structure then behaves elastically, shakedown takes place. Having analysis methods to predict when shakedown takes place will then give extended possibility for a better design.

The rolling contact problem presents a typical cyclic loading condition. Shakedown analysis of rolling contact problems have been made by for instance Johnson [1]. Johnson computes the shakedown limit for two-dimensional problems with pure Hertzian pressure loading and the material is assumed elastic-perfectly plastic. The influence of tangential forces on the shakedown limit is presented by Johnson and Jefferis [2] and by Jefferis in [3]. The methods used in papers [1, 2, 3] are different from the method presented here in the sense that the problem is not formulated as a minimax problem. Furthermore the material is assumed elastic-perfectly plastic and does not include kinematic hardening.

The results obtained in [1, 2, 3] will be compared to the results obtained by the present method.

2. SHAKE-DOWN THEOREMS

In order to introduce the terminology the shakedown theorems used will be presented very briefly

2.1 Elastic-ideally plastic material

For elastic-ideally plastic material we will make use of Melan's theorem [4] which gives a lower bound to the shakedown limit. Let σ_{ij} denote the actual stress in the solid and let $\sigma_{ij}^e(x_0,t)$ the stress state when the material is assumed to behave elastically and loaded with the same actual load. x_0 denotes a point in the solid and t the time. Further let $\rho_{ij}(x_0,t)$ denote a selfstress i. e. satisfying equilibrium conditions

$$\rho_{ij,j} = 0 \tag{1}$$

within the body and

$$\rho_{ij}n_j = 0 \tag{2}$$

on the surface where loads are applied. Yielding takes place when $\sigma_e(\sigma_{ij}) = \sigma_o$. Small strains are assumed.

The Melan theorem then states that for the structure to shakedown it is necessary and sufficient that there exists a timeindependent selfstress ρ_{ij} such that

$$\sigma_e(\sigma_{ij}^e(x_0,t) + \rho_{ij}(x_0)) < \sigma_o \tag{3}$$

at any point of the structure and for all possible load combinations.

2.2 Elastic-kinematic work hardening material

Pouter [5] has extended the Melan theorem to linear kinematic hardening. The Pouter theorem is as follows Shakedown will occur provided there exists a distribution p_{ij} throughout the solid so that

$$\sigma_e(\sigma_{ij}^e(x_0,t) + p_{ij}) < \sigma_o \tag{4}$$

p_{ij} which we may call a residual state need not satisfy any field equation and need not be continous. This theorem is thus genereally much simpler to apply than Melans theorem where ρ_{ij} has to satisfy $\rho_{ij,j} = 0$ and $\rho_{ij}n_j = 0$.

3. SHAKEDOWN FOR ROLLING CONTACT PROBLEMS

In this section we will state the problem and introduce the simplifying conditions for rolling contact problems.

3.1 Statement of the problem

We consider a load moving on a half-plane as shown in Fig 1. $q_n(\xi)$ is the pressure and 2b is the contact width. We consider Hertzian pressure distribution and pressure distribution when elasto-hydrodynamic effects are taken into account [6]. This distribution is indicated EHD in Fig 1. We also consider frictional effects. The coefficient of friction is denoted μ. When studying frictional effects we assume total sliding and the shear stress is $q_t(\xi) = \mu q_n(\xi)$.

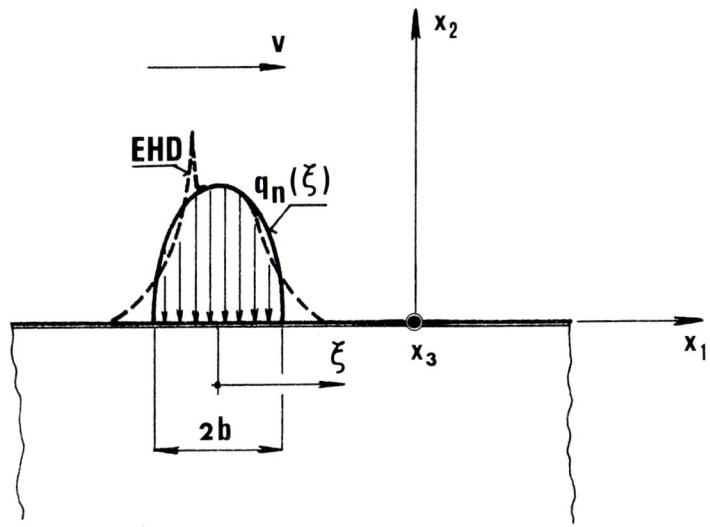

Fig 1   Load moving on a half-plane.

3.2   Problem solution

We will now examine which ρ_{ij} and p_{ij} that can prevail in the half-plane when plane deformation (PD) and plane stress (PS) are assumed.

For elastic-ideally plastic material ρ_{ij} must satisfy the equilibrium equations (1) in the half-plane and the boundary condition (2) on the surface $x_2 = 0$. ρ_{ij} is independent of time and the selfstresses in all planes perpendicular to the x_1-direction are all equal. Thus all the terms that are derivatives with respect to x_1 vanish. The assumption of plane state implies $\rho_{13} = \rho_{23} = 0$. Also terms which are derivatives with respect to x_3 vanish due to the plane state assumption. Equation (1) could then be written

$$\rho_{12,2} = 0 \; ; \quad \rho_{22,2} = 0 \tag{5}$$

Integration with respect to x_2 and making use of the boundary condition at $x_2 = 0$ implies $\rho_{12} = \rho_{22} = 0$. In addition for PS we have $\rho_{33} = 0$.

For PD the only nonzero components of ρ_{ij} are $\rho_{11}(x_2)$ and $\rho_{33}(x_2)$ and they are not coupled. For PS the only nonzero component of ρ_{ij} is $\rho_{11}(x_2)$. This simplifies the application of Melans theorem considerably.

For kinematic hardening material p_{ij} need not satisfy any field equations. The only components that due to the plane state assumption, have to be zero are p_{13} and p_{23}. This is valid both for PD and PS. It is thus shown that the nonvanishing components of both ρ_{ij} and p_{ij} may be varied independently from point to point in the half-plane.

Assuming generally that ρ_{ij} and p_{ij} may be varied independently we can use the shakedown theorems to form a mini-max problem. Let r_{ij} denote either ρ_{ij} or p_{ij}. r_{ij} is called the residual state. In absence of r_{ij} the effective stress $\bar{\sigma}_e(\sigma_{ij}^e(x_0,t))$ in a point x_0 may vary during a load-cycle as shown by the solid line in Fig 2. By superposing r_{ij}

the effective stress $\sigma_e(\sigma_{ij}^e(x_0,t) + r_{ij}(x_0))$ will vary for example as shown by the dashed line in Fig 2. The aim is now to find the best residual state, which we denote r_{ij}^*, that makes the maximum value of σ_e during a loadcycle as small as possible. This maximum value we denote σ_e^*.

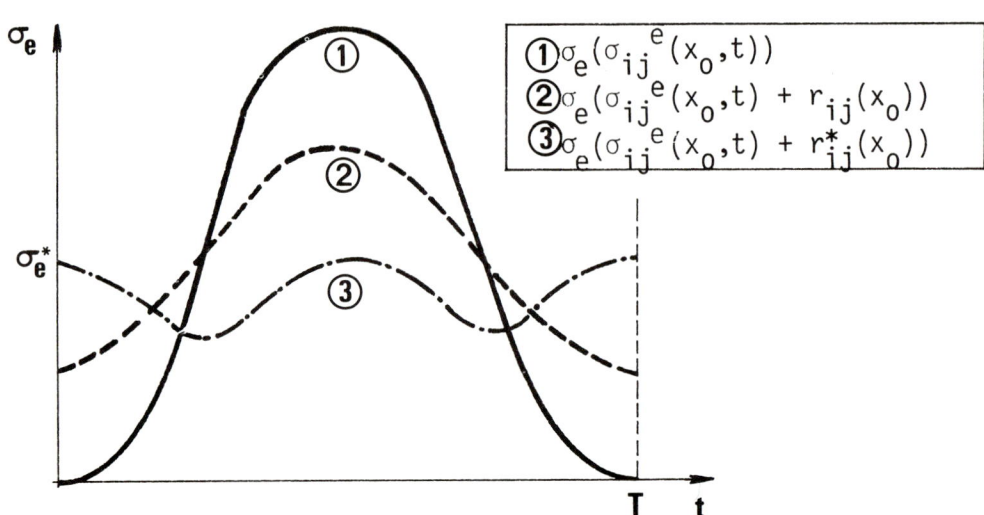

Fig 2 Effective stresses

Formally we write this

$$\text{Find } \sigma_e^*(x_0) = \underset{t, r_{ij}}{\text{Min Max}} \sigma_e(\sigma_{ij}^e(x_0,t) + r_{ij}(x_0)) \tag{6}$$

The dashed-dotted line in Fig 2 shows $\sigma_e(t)$ when $r_{ij} = r_{ij}^*$. Because r_{ij} may be varied independently from point to point we may perform the above analysis for all points and obtain

$$\sigma_e^* = \sigma^*(x), \quad r_{ij}^* = r_{ij}^*(x) \tag{7}$$

The maximum value

$$\sigma_{emax}^* = \text{Max} \sigma_e^*(x) \tag{8}$$

is then found. If σ_{emax}^* is less then the yield stress σ_0 shakedown according to Melans and Pouters theorem takes place.

Furthermore if von Mises yield condition is used in (6) it is obvious that, if α is a factor with which the load is multiplied the following holds

$$\alpha \sigma_e^*(x_0) = \underset{t, r_{ij}}{\text{Min Max}} \sigma_e(\alpha(\sigma_{ij}^e(x_0,t) + r_{ij}^*(x_0))) \tag{9}$$

By equating $\alpha \sigma_e^*$ max to σ_0 we can then compute the factor α with which the loadsystem should be multiplied to give the ultimate load for shakedown.

If we instead of using a continous time t evaluate σ_e in x_0 for discrete times t_k, $k = 1, 2,...n$. (6) takes the form

$$\text{Find } \sigma_e^* = \underset{r_{ij}}{\text{Min Max}}[\sigma_e(\sigma_{ij}^e(t_1) + r_{ij}),...., \sigma_e(\sigma_{ij}^e(t_n) + r_{ij})] \tag{10}$$

This has the advantage that $\sigma_{ij}(t_k)$ can be computed by using numerical methods. Furthermore there are computer routines [7] available to solve the minimax problem. The necessary tools for solving the problem is then a method of solving the elastic stresses for the given loading condition and a routine for solving the minimax problem.

4. RESULTS

The routines developed for solving this problem were tested on the example defined in Fig 3.

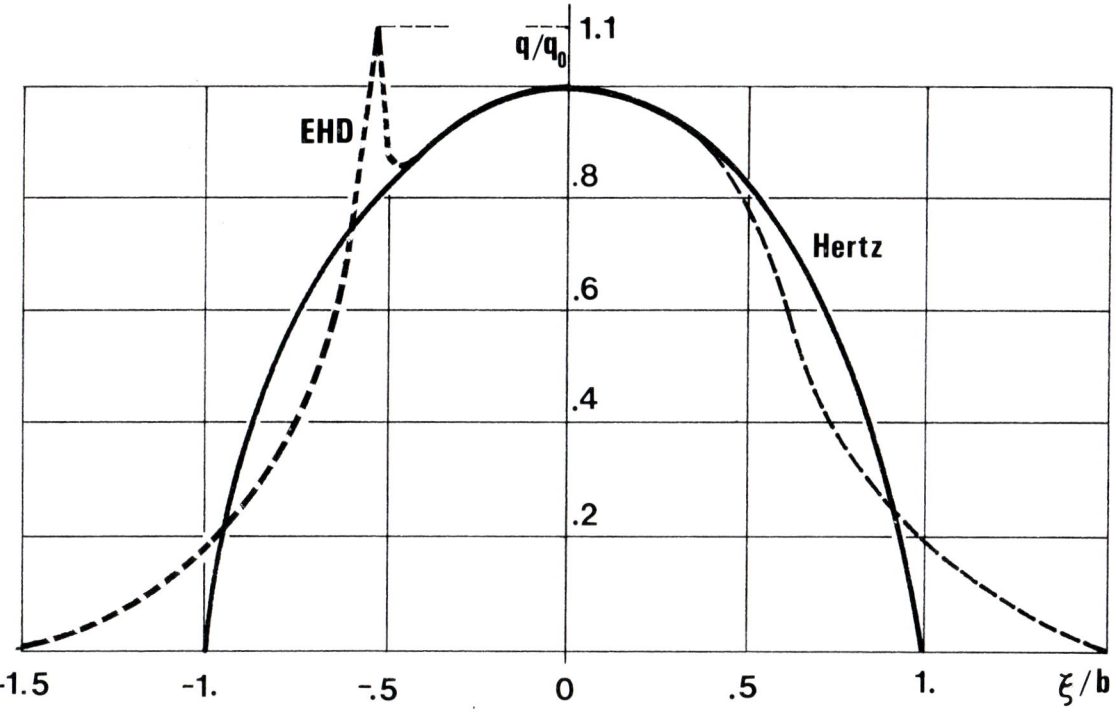

Fig 3 Pressure profiles q_0 is the maximum Hertzian pressure.

Fig 4 shows the shakedown limits expressed in maximum Hertzian pressure q_0 normalized with respect to the yield limit k in simple shear, as a function of the coefficient of friction. The result is valid for PD and elastic-perfectly plastic material. Shown in Fig 4 is also the elastic limit. It could be seen how the coefficient of friction influence the shakedown limit. The shakedown limit decreases with increasing coefficient of friction. The abrupt change in the curve comes when the point determining shakedown reaches the surface. The shakedown limit given by the solid line is also calculated by Johnson and Jefferis [2] and the results are the same. From the results it could also be seen that the EHD peak will lower the shakedown limit. Calculations were also made for PS and linear kinematic hardening. The result for frictionless case is presented in Table 1.

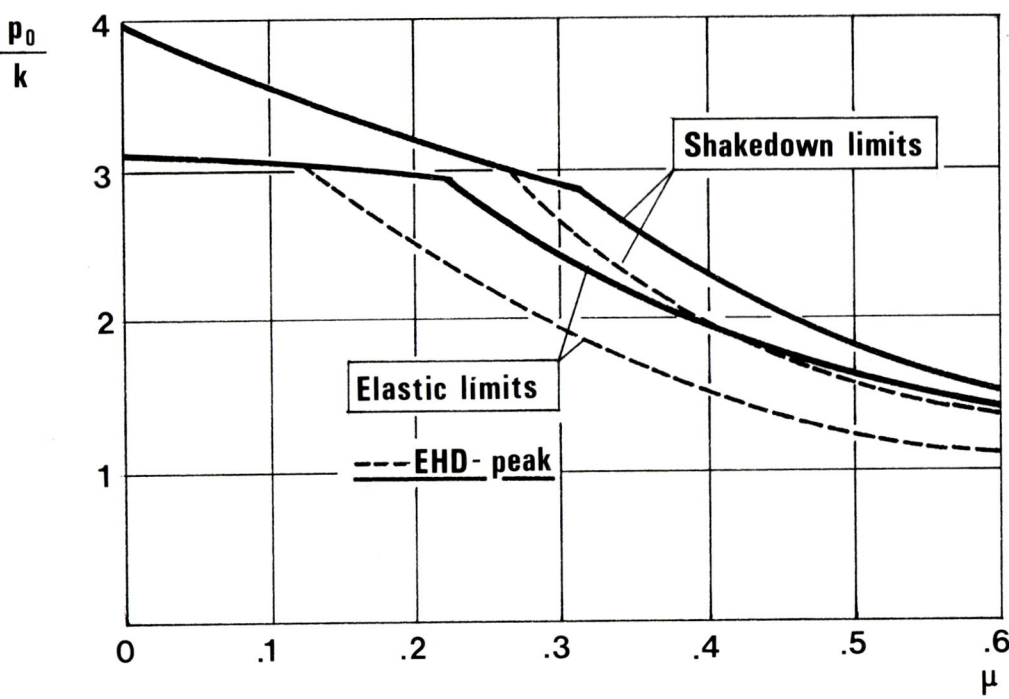

Fig 4 Shakedown limits for different loading situations.

Table 1 Comparision of shakedown limits for frictionless cases.

	Elastic-perfectly plastic q_0/k	Elastic-kinematic hardening q_0/k
PD	4	4
PS	2	3,46

For PD there is no difference between elastic-perfectly plastic and kinematic hardening. As expected the shakedown limit is lower for PS than for PD. For kinematic hardening it is however considerable higher than for perfectly plastic material in PS.

5 CONCLUSIONS

A rather general method for shakedown analysis has been developed for twodimensional rolling contact problems. Due to the uncoupled conditions for the residual states both for elastic-perfectly plastic and kinematic hardening it was possible to formulate the shakedown problem as a mini-max problem. This was then discretized and numerical methods were applied. The method is applicable to any kind of loading in a rolling situation. Testexamples were presented and compared with other methods and the result was in agreement with this. It is possible to extend this method to threedimensional problems and this work is presently in progress.

ACKNOWLEDGEMENT

The authors want to express their gratitudes to the Swedish Board for Technical Development for sponsoring this research. Gratitudes are also directed to SKF/Engineering Research Center in Holland for their cooperation.

REFERENCES

[1] Johnson, K. L., A shakedown limit in rolling contact. Proc. 4th U. S. Nat. Congr. Appl. Mech. Berkeley California 1962, pp 971-975.

[2] Johnson, K. L., Jefferis, J. A., Plastic flow and residual stresses in rolling and sliding contact. Proc. Fatigue in Rolling Contact. Institution of Mechanical Engineers, London, 1963, pp 54-65.

[3] Jefferis, J. A., Friction and deformation of rolling and sliding surfaces. Doctorial Thesis. Jesus Collage, University of Cambridge 1966.

[4] Koiter, W. T., General theorems for elastic-plastic solids. In Progress in Solid Mechanics, Vol. 2., North-Holland, Amsterdam, 1960, pp 165-221.

[5] Pouter, A. R. S., A general shakedown theorem elastic/plastic bodies with work hardening. Proc. 3rd Int. Conf. in Reactor Tehn. London, 1975.

[6] Dowson, D., Higginson, G. R., Whitaker, B. A., Stress distribution in lubricated rolling contacts. Proc. Fatigue in Rolling Contact. Institution of Mechanical Engineers, London, 1963, pp 66-75.

[7] Harwell Subroutine Library. Routine VG01A/AD

Analytical Modelling In Inelasticity

J. Kratochvíl

Institute of Solid State Physics Czechoslovak Academy of Sciences
162 53 Prague, Czechoslovakia

ABSTRACT

Analytical modelling of inelastic behavior is treated as a sequence of a constitutive analysis, identification testing and solution of the corresponding inverse problem. Models of rate-independent plasticity are studied as an example. The constitutive analysis based on the internal variables approach provides a whole spectrum of models of plastic behavior with various degrees of flexibility. Higher flexibility enables more adequate modelling. However, the complexity of the corresponding idefication testing program and solution of the inverse problem grows with increasing flexibility very rapidly. Identification tests and inverse problems for three models of the spectrum are conceptually analysed. The ranges of validity of the models are estimated. A possibility to use inhomogeneous identification tests, each of which contains far more information then a homogeneous one, is studied. Practical problems connected with actual application of the suggested procedure of modelling is briefly commented.

1. INTRODUCTION

One of the most limiting factors of the present inelastic stress analysis is the lack of analytical models which are compatible in accuracy with the currently developed numerical methods of solutions of boundary values problems. What is needed is a procedure from which concrete quantitative inelastic constitutive equations result. Such procedure can be understood as an identification process, i.e. the procedure where actual specimens of the inelastic material must be loaded and measured, and observed data used to identify explicitly the constitutive equations that mathematically characterize the material. The constitutive analysis is useful in determining the suitable format of these equations. Micromechanical structural insight from materials sciences facilitates a specification of the format.

What we mean by the identification procedure can be illustrated with a simple example of linear isotropic elasticity. The characteristic feature of an elastic material is that a current value of stress σ is a function of a current value of strain $\underline{\varepsilon}$, i.e. $\underline{\sigma} = f(\underline{\varepsilon})$. The linearity and isotropy further specify this stress-strain relation, we get

$$\underline{\sigma} = \frac{E\nu}{(1+\nu)(1-2\nu)}(\text{tr}\,\underline{\varepsilon})\,\underline{1} + \frac{E}{1+\nu}\underline{\varepsilon}. \tag{1}$$

The material function f is in (1) expressed in terms of two elastic constants E, ν. Since identifying E and ν in any strain state will produce results valid for all strain states, a simple uniaxial stress experiment will suffice to identify a linear isotropic elastic material. In a tensile test of a rod of the length l and radius r the axial load P is an input and the measured axial displacement Δl and radial contraction Δr are the output. Employing the equation (1) for the axial direction, where $\sigma_{xx} = P/\pi r^2$ and in the radial direction, where $\sigma_{yy} = 0$, determination of E and ν results in a very simple inverse problem the solution of which is $E = P\,l/\pi r^2 \Delta l$, $\nu = l\,\Delta r/r\,\Delta l$.

In the example of the linear isotropic elastic material we can trace characteristic steps of the identification procedure:

(a) At the outset of the constitutive analysis we have to choose an appropriate analytical framework which covers the class of studied materials (the relation $\underline{\sigma} = f(\underline{\varepsilon})$ in the example). Using information from materials sciences and the constitutive theory we try to specify, as much as possible, the mathematical structure of the constitutive equations (from assumption of linearity and isotropy the specified constitutive equation (1) results).

(b) The specified form of the constitutive equations achieved by the step (a) indicates the type and minimum number of identification tests needed for determination of the remaining unknown constitutive parameters or functions (in the example a single tensile test provides sufficient information).

(c) Finally, we have to formulate and solve the corresponding inverse problem, i.e. to find an algorithm for determination of the remaining unknown constitutive parameters or functions from the experimental data (determination of the constants E, ν results in solution of two linear algebraic equations).

Moreover, an inseparable step of any identification procedure should be a determination of a range of validity of the identified analytical model. This problem is usually the most difficult part of analytical modelling.

Analytical modelling of inelastic materials, which we attempt to outline in this note, follows basically the identification procedure (a) - (c). The substantional difference is that unlike the linear isotropic elasticity each of the steps (a) - (c) is more complicated due to the hereditory property and nonlinearity of inelastic response.

2. CONSTITUTIVE ANALYSIS

A current response of an inelastic material depends generally on the entire loading history to which the material has been exposed. The history dependence of the response may be expressed through internal variables. Internal variables can be understood as a convenient means to reflect the sensitivity of material response to structural changes of the material which accompany any inelastic deformation process. Here the analytical modelling will be demonstrated on rate-independent plastic materials. As an appropriate analytical framework we adopt the internal variables model of plasticity suggested in [1] (for alternative approach see [2]).

The constitutive equations of the model are generated by two potentials ϕ and F. The internal energy $\phi(\underline{e}, \underline{f})$ which is a function of elastic strain $\underline{e}$ and the internal parameter $\underline{f}$ ($\underline{f}$ can be a set of internal variables of different tensorial rank; here one second order tensor will be considered) determines the stress $\underline{\sigma}$ and the generalized force $\underline{A}$ through the relations

$$\underline{\sigma} = \frac{\partial \phi}{\partial \underline{e}} , \qquad \underline{A} = \frac{\partial \phi}{\partial \underline{f}} \qquad (2)$$

The constitutive equations for the rate of plastic strain p and the internal parameter $\underline{f}$ are expressed through the plastic potential $F(\underline{\sigma}, \underline{A})$ (superposed dot indicates time derivatives)

$$\dot{\underline{p}} = \lambda \frac{\partial F}{\partial \underline{\sigma}} , \qquad \begin{cases} \lambda \geq 0 & \text{if } F(\underline{\sigma}, \underline{A}) = 0, \\ \\ \lambda = 0 & \text{if } F(\underline{\sigma}, \underline{A}) < 0. \end{cases} \qquad (3)$$

$$\dot{\underline{f}} = - \lambda \frac{\partial F}{\partial \underline{A}} ,$$

The small strain approach is adopted, i.e. total strain $\underline{\varepsilon}$ is the sum $\underline{\varepsilon} = \underline{e} + \underline{p}$, no temperature effects are considered. It is assumed that $F(\underline{\sigma}, \underline{A}) = 0$ determines a convex surface in the $\underline{\sigma}, \underline{A}$ space called a generallized yield surface. The form of (3) guarantees the normality of $\dot{\underline{p}}$ and $\dot{\underline{f}}$ to the generalied yield surface. According to (3) the plastic deformation process can be maintained only if the values $\underline{\sigma}, \underline{A}$ remain on $F(\underline{\sigma}, \underline{A}) = 0$. Therefore the scalar plastic multiplier λ is determined by the requirement that $\dot{F}(\underline{\sigma}, \underline{A}) = 0$ during the plastic deformation process. By (2), (3)$_2$ this requirement yields

$$\text{If} \quad \underline{H} \lambda = \underline{c} , \quad \text{then} \quad \lambda \geq 0, \quad (\dot{F} = 0)$$

$$\text{If} \quad \underline{H} \lambda > \underline{c} , \quad \text{then} \quad \lambda = 0, \quad (\dot{F} < 0) \qquad (4)$$

where

$$\underline{H} = \frac{\partial F}{\partial \underline{A}} \left[\frac{\partial^2 \phi}{\partial \underline{f}^2} - \frac{\partial^2 \phi}{\partial \underline{f} \partial \underline{e}} \left(\frac{\partial^2 \phi}{\partial \underline{e}^2} \right)^{-1} \frac{\partial^2 \phi}{\partial \underline{e} \partial \underline{f}} \right] \frac{\partial F}{\partial \underline{A}} ,$$

$$\underline{c} = \left[\frac{\partial F}{\partial \underline{\sigma}} + \frac{\partial F}{\partial \underline{A}} \frac{\partial^2 \phi}{\partial \underline{f} \partial \underline{e}} \left(\frac{\partial^2 \phi}{\partial \underline{e}^2} \right)^{-1} \right] \dot{\underline{\sigma}} .$$

If $\underline{H}$ is positive definite, the system (4) has a unique solution [1] and λ is then a unique function of $\dot{\underline{\sigma}}$. The proportionality of λ and $\dot{\underline{\sigma}}$ quarantees rate-independence of the model (2), (3), i.e. a change of loading rate does not change the plastic response of the material.

To specify further the constitutive equations (2), (3) three additional experimentally resonably confirmed assumptions are introduced: validity of the linear Hooke's law for elastic strain, plastic incompressibility, and initial isotropy (more precisely the existence of an isotropic reference state). Hooke's law means that $(2)_1$ has the form $\underset{\sim}{\sigma} = \underset{\sim}{C}\, \underset{\sim}{e}$, where $\underset{\sim}{C}$ is a constant furth order tensor. This implies that $\phi = \psi(\underset{\sim}{e}) + \psi(\underset{\sim}{f})$, hence $\underset{\sim}{A} = \partial \psi(\underset{\sim}{f})/\partial \underset{\sim}{f}$, i.e. $\underset{\sim}{A}$ is independent of $\underset{\sim}{e}$. Due to the plastic incompressibility only the deviatoric stress $\underset{\sim}{s} = \underset{\sim}{\sigma} - (\mathrm{tr}\, \underset{\sim}{\sigma}/3)\, \underset{\sim}{1}$ appears in F. The initial isotropy causes that the potentials ψ and F depend on the arguments $\underset{\sim}{s}$, $\underset{\sim}{A}$ and $\underset{\sim}{f}$ only through their basic and mutual invariants (for convenience $\underset{\sim}{f}$ and $\underset{\sim}{A}$ are divided into a spherical and deviatoric parts, i.e. $\underset{\sim}{f} = \alpha\, \underset{\sim}{1} + \underset{\sim}{a}$, $\underset{\sim}{A} = k\underset{\sim}{1} + \underset{\sim}{K}$)

$$\varphi = \varphi(\alpha, \mathrm{II}_a, \mathrm{III}_a), \qquad (5)$$

$$F = F(k, \mathrm{II}_s, \mathrm{II}_K, \mathrm{II}_{sK}, \mathrm{III}_s, \mathrm{III}_K, \mathrm{III}_{sK^2}, \mathrm{III}_{s^2 K}), \qquad (6)$$

where $\mathrm{II}_a = \mathrm{tr}\, \underset{\sim}{a}^2$, $\mathrm{II}_s = \mathrm{tr}\, \underset{\sim}{s}^2$, $\mathrm{II}_{sK} = \mathrm{tr}\, \underset{\sim}{s}\, \underset{\sim}{K}$, $\mathrm{III}_a = \mathrm{tr}\, \underset{\sim}{a}^3$, etc. The number of the arguments in (5) and (6) can be easily justified. Due to isotropy the tensor $\underset{\sim}{f}$ in φ can be diagonalized by a suitable orthogonal transformation and therefore there are only three functionally independent invariants of $\underset{\sim}{f}$ in (5). If the principal directions of $\underset{\sim}{s}$ and $\underset{\sim}{A}$ do not coincide only one of the tensors can be diagonalized by an orthogonal transformation and the same transformation would only alter the magnitude of the components of the other tensor, thus only eight invariants can be functionally independent in (6) (note that $I_s = \mathrm{tr}\, \underset{\sim}{s} = 0$, $I_A = \mathrm{tr}\, \underset{\sim}{A} = 3k$). However, if the tensor $\underset{\sim}{s}$ and $\underset{\sim}{A}$ have the common principal axes, then both tensors could be diagonalized simultaneously and only five invariants in (6) would be functionally independent, i.e. in (6) e.g. III_K, $\mathrm{III}_{s\,K}$ and III_{sK^2} could be ommitted.

If only some of the invariants are retained in (5), (6) the constitutive equations (2), (3) together with (4) yield a special approximative model of a plastic material. Retaining different groups of invariants in (5), (6) the whole spectrum of plastic models of various flexibility results. To illustrate the relation of the flexibility of a model to its range of validity and complexity of the relevant identification tests and the inverse problem three models of the spectrum will be briefly analysed in the following sections.

3. CLASSICAL MODEL OF ISOTROPIC HARDENING

A very simple model arises if we retain in (5), (6) the lowest invariants, i.e. $F = F(k, \mathrm{II}_s)$, $\varphi = \varphi(\alpha)$; then from (2) – (4)

$$\underset{\sim}{\dot{p}} = \lambda d_1 \underset{\sim}{s}, \quad \begin{cases} \lambda = d_1 \dot{\mathrm{II}}_s / 6 d_2^2 (d^2\varphi/d\alpha^2) & \text{if } f = 0, \qquad (7) \\ \\ \dot{\alpha} = \lambda d_2, \quad \lambda = 0 & \text{if } f < 0, \qquad (8) \end{cases}$$

where $d_1(\alpha, \mathrm{II}_s) = \partial f/\partial \mathrm{II}_s$, $d_2(\alpha, \mathrm{II}_s) = -(1/3)(\partial f/\partial \alpha)(d^2\varphi/d\alpha^2)^{-1}$ and $f = f(\alpha, \mathrm{II}_s) = F(k(\alpha), \mathrm{II}_s)$, $k(\alpha) = d\varphi/d\alpha$. From (7) we easily get $\dot{w} = \mathrm{tr}\, \underset{\sim}{s}\, \underset{\sim}{\dot{p}} = \lambda d_1 \mathrm{II}_s$, hence, as long as $d_1 \mathrm{II}_s \neq 0$ the equation (8) yields $\dot{\alpha} = \dot{w} d_2/d_1 \mathrm{II}_s$. If II_s can be evaluated from $f(\alpha, \mathrm{II}_s) = 0$ as $\mathrm{II}_s = h(\alpha)$, from the last relation we get $\dot{\alpha} = g(\alpha)\, \dot{w}$ ($g(\alpha)$ is determined by the form of d_1, d_2, f) and by integration

$$H(\alpha) - H(\alpha_o) = w = \int_{t_o}^{t} \dot{w}\, d\tau ,\qquad(9)$$

where H is a primitive function of $(g(\alpha))^{-1}$, $\alpha_o = \alpha(t_o)$ and w is the plastic work accumulated during the time interval $\langle t_o, t \rangle$. Under the indicated restrictions the relation (9) shows that for the model (7), (8) with fixed t_o and invertible H the classical work hardening parameter w may be choosen as the internal variable α (it can be shown analogically that an equivalently suitable measure of work hardening is accumulated plastic strain). Thus from (7) and the procedure which leads to (9) it follows that the model (7), (8) can be expressed in the form

$$\dot{\underline{p}} = d(w)\, \dot{II}_s\, \underline{s}\ ,\qquad \text{if}\quad II_s = h(w) ,\qquad(10)$$

and $\dot{p} = 0$ if $II_s < h(w)$, where $d(w) = [II_s(dh/dw)]^{-1}$. The constitutive equation (10) represents the classical plasticity model of isotropic hardening with plastic work w as the hardening parameter and the yield condition $II_s = h(w)$ of von Mises type. The identification of the model (7), (8) is thus reduced to a determination of $h(w)$ in (10). The classical identification experiment is sufficient; from one-dimensional stress-plastic strain diagram $\sigma(p)$ one gets $II_s = (2/3)\sigma^2 = h(\int \sigma\, dp) = h(w)$.

To estimate at least qualitatively the range of validity of the classical model two characteristic features of the model are usually check by experiments:
(a) the overall shape of the yield surface and its changes during the deformation process, (b) the coaxiality of the plastic strain rate $\dot{p}$ and the deviatoric stress $\underline{s}$ required by (10).

(a) Yield surfaces can be represented as yield curves in e.g. a deviatoric plane of the deviatoric stress space. According to $II_s = h(w)$ in (10) the subsequent yield curves should be concentric circles. Such picture is obtained approximately from experimental data if a rough stress measure of the onset of yilding is used(stress at a strain offset of a given amount (usually 0.02) or the stress obtained by a backward extrapolation from the stress-strain curve to the elastic line)[3]. Then the measured yield curves tend to be close to the curves which are produced by isotropic expansion from the initial yield surface. Hence, the subsequent yield surfaces are concentric in agreement with $II_s = h(w)$, however, often they are not represented precisely by circles in deviatoric plane [4]. On the other hand, if for determination of the yield curve is used a procedure sensitive to the very earliest indication of onset of yielding the measured yield curves are very far from the analytical form $II_s = h(w)$, e.g. [5,6].

(b) Non-coaxiality of plastic strain rate $\dot{p}$ and deviatoric stress $\underline{s}$ which violates (10) has been intensively studied in [7-9]. It appears after a sharp change of a loading path. The direction of plastic strain rate delays from that of deviatoric stress just after the change but the delay disappears with continuing loading. It means that the non-coaxiality is sensitive to the local curvature of the loading path and does not occur if the curvature is small.

From (a) and (b) we can tentatively conclude that the model (7), (8) is suitable for rough estimate of plastic response at general type of loading or is valid for a moderately accurate modelling of plastic behavior for a restricted class of loading paths; only loading paths of small curvature are permitted (no unloading, as it requires sharp change of curvature). For a more adequate analytical modelling of the observed deviation from behavior of the classical model some of the higher invariants of $\underset{\sim}{a}$, $\underset{\sim}{s}$ and $\underset{\sim}{K}$ have to be retained in (5), (6).

4. MODELLING OF HIGHER STRESS EFFECTS

As a first example of a conceptual analysis of a more complex identification procedure we now consider a slightly generalized model. We assume that in (5), (6) are retained α, k, II_s, III_s, i.e. $\varphi = \varphi(\alpha)$, $F = F(k, II_s, III_s)$, then from (2) – (4)

$$\dot{\underset{\sim}{p}} = \lambda [b_1 \underset{\sim}{s} + b_2 \underset{\sim}{r}], \quad \left\{ \lambda = \frac{(b_1 \dot{II}_s/2) + (b_2 \dot{III}_s/3)}{3 b_3^2 (d^2\varphi/d\alpha^2)} \quad \text{if } f = 0 \right. \quad (11)$$

$$\dot{\alpha} = \lambda b_3 \quad , \quad \lambda = 0 \quad , \text{if } f < 0 \quad (12)$$

The tensor $\underset{\sim}{r}$ denotes $\underset{\sim}{r} = \underset{\sim}{s}^2 - (II_s/2) \underset{\sim}{1}$, and $b_1 = 2 \partial f/\partial II_s$, $b_2 = 3 \partial f/\partial III_s$, $b_3 = -(\partial f/\partial \alpha)/3 (d^2\varphi/d\alpha^2)$, $f = f(\alpha, II_s, III_s) = F(k(\alpha), II_s, III_s)$, $k(\alpha) = d\varphi/d\alpha$.

If f is of the type $f = g(II_s, III_s) - \alpha$ (if $\partial f/\partial \alpha \neq 0$, we get such f at least locally) the model (11), (12) can be transformed to the form

$$\dot{\underset{\sim}{p}} = c_1^2 \frac{\dot{II}_s}{2} \underset{\sim}{s} + c_1 c_2 (\frac{\dot{III}_s}{3} \underset{\sim}{s} + \frac{\dot{II}_s}{2} \underset{\sim}{r}) + c_2^2 \frac{\dot{III}_s}{3} \underset{\sim}{r}^2 , \quad (13)$$

$$\text{if} \quad g(II_s, III_s) = \alpha ,$$

$\dot{\underset{\sim}{p}} = 0$ if $g(II_s, III_s) < \alpha$, where $c_i = b_i(g(II_s, III_s), II_s, III_s) \sqrt{3 d\varphi^2/d\alpha^2}$, $i = 1,2$ (note that for $d^2\varphi/d\alpha^2$ negative c_i assume imaginary values, but the coeficients c_1^2, $c_1 c_2$, c_2^2 in (13) remain real). The evolution equation (12) for $\dot\alpha$ is then identical with an equation gained from $\alpha = g(II_s, III_s)$ by time differentiation.

Due to the fact that the flexibility of the model (11), (12) with respect to hereditary properties remains the same as for the classical model (just one internal variable) we expect that the range of validity for both models is the same. Only difference is that the generalized model is able to describe the stress dependence of plastic behavior more precisely.

The form of f of the model indicates that the subsequent yield surfaces are represented by one-parameter family of yield curves in the deviatorstress plane. As it has been already mentioned such situation occurs experimentaly, if a rough stress measure of the onset of yielding is used in measurement. The only gain is that the shape of the family of subsequent yield curves can be modelled by (11), (12) more realistically. The detailed yield curves pattern as revealed by sensitive experiments [5,6] cannot be modelled without retaining $\underset{\sim}{a}$ and $\underset{\sim}{K}$ in (5), (6).

On the other hand, if only restricted class of loading paths with small curvature is admitted, we expect that the generalized model provides an accurate description of plastic behavior. The reason is that the delay effect, description of which requires a flexibility provided by $\underset{\sim}{a}$ and $\underset{\sim}{K}$, is suppresed. Hence, $\underset{\sim}{K}$ and $\underset{\sim}{a}$ may be excluded from φ and F and (11), (12) is then the most general model (within the considered framework) for such type of loading.

Consider the specialized form of the model (13). To identify it we need first to determine the function $g(II_S, III_S)$ which describes the subsequent yield curves of the model. However, knowledge of g is not sufficient as in c_i also $d^2\varphi/d\alpha^2$ appears besides derivatives of g. The identification of $g(II_S, III_S)$ is complicated by the fact that the physical meaning of α, and hence a way of measuring it, remains undetermined. An attempt analogical to the procedure from the previous Section to identify α with plastic work w (or accumulated plastic strain) seems to fail unless some additional assumptions are introduced. Therefore practice to construct experimental subsequent yield curves as loci on loading paths with the same amount of plastic work (or accumulated plastic strain) [4] which utilizes the assumption $g(II_S, III_S) = \alpha = h(w)$ is not justified by the presented theory. Only unloading-reloading technique for determination of yield curves, e.g. [10], seems to be appropriate from the present point of view. Providing the subsequent yield curves are experimentally determined and may be understood as a one-parameter family of curves, then α is an assigned parameter which distinquishes one curve from another [11, 12] ; g is determined as a function which fits the experimental curves such as $g(II_S, III_S) = \alpha$. To specify $d^2\varphi/d\alpha^2$ as a function of α it suffices to measure one component of $\underset{\sim}{p}$ (e.g. principal value p_1) along a loading path which satisfies $s_1 = \tilde{s}_1(\alpha)$ and $dp_1/ds_1 \neq 0$, the principal deviatoric stress s_1 serves as a loading parameter. Then $II_S = II_S(s_1)$, $III_S = III_S(s_1)$ and from (13)

$$\frac{d^2\varphi}{d\alpha^2} = \frac{1}{3} \frac{B(s_1)}{dp_1/ds_1} \bigg|_{s_1 = s_1(\alpha)}, \qquad (14)$$

where $B(s_1) = 2(\partial g/\partial II_S)^2 (d II_S/d s_1) s_1 + 6(\partial g/\partial II_S)(\partial g/\partial III_S) \cdot [(d III_S/d s_1) s_1/3 + (\partial II_S/\partial s_1)(s_1 - II_S/2)] + 3(\partial g/\partial III_S) \cdot (d III_S/d s_1)(s - II_S/2)$.

The measurement of yield curves, specification of α, determination of g and the procedure (14) can be altogether avoided if the model is intended to be applied for the limited class of loading paths with small curvature. As then no unloading occurs the yield condition need not to be known and to identify the model a specification of c_1, c_2 in (13) is fully suficient. Consider a family of loading paths $s_1(s, l), s_2(s, l)$, such as $II_S(s, l)$ and $III_S(s, l)$ are invertible i.e. $s(II_S, III_S), l(II_S, III_S)$; s_1, s_2 are principal stress of $\underset{\sim}{s}$, s means an arc measure of loading path and l denotes a particular loading path within the family. Then from (13) we get, $i = 1,2$,

$$\frac{\partial p_i}{\partial s} = c_1^2 \frac{\partial II_S/\partial s}{2} s_i + \qquad (15)$$

$$+ c_1 c_2 \left(\frac{\partial III_S/\partial s}{3} s_i + \frac{\partial II_S/\partial s}{2} r_i \right) + c_2^2 \frac{\partial II_S/\partial s}{3} r_i .$$

If the system (15) has for a given s, l a solution $c_1(s, l)$, $c_2(s, l)$, the identified functions are $c_i = c_i(s(II_s, III_s), l(II_s, III_s))$.

5. MODELLING OF A DELAY EFFECT

Finally, we conceptually analyse a model which retains the lowest invariants of $\underset{\sim}{a}$ and $\underset{\sim}{K}$. We assume that in (5), (6) $\psi = \psi(\alpha, II_a)$, $F = F(k, II_{sK}, II_K)$, then from (2), (3) we get

$$\dot{\underset{\sim}{p}} = \lambda h_1 2(\partial \psi/\partial II_a) \underset{\sim}{a} \qquad (16)$$

$$\dot{\alpha} = \lambda h_2 \qquad (17)$$

$$\begin{cases} \lambda \geq 0, \text{ if } f = 0 \\ \lambda = 0, \text{ if } f < 0 \end{cases}$$

$$\dot{\underset{\sim}{a}} = \lambda(-h_1 \underset{\sim}{s} + h_3 2(\partial \psi/\partial II_a) \underset{\sim}{a}), \qquad (18)$$

where $h_1 = \partial F/\partial II_{sK}$, $h_2 = -(\partial F/\partial k)/3$, $h_3 = -2\partial F/\partial II_K$, $f = f(\alpha, II_{sa}, II_a) = F(k(\alpha, II_a), II_{sK}(\alpha, II_a, II_{sa}), II_K(\alpha, II_a))$, $k = (\partial \psi/\partial \alpha)/3$, $K = 2(\partial \psi/\partial II_a)\underset{\sim}{a}$. The formula for λ is not explicitely stated, as it is cumbersome to write.

From (16) and (18) we see that $\dot{\underset{\sim}{p}}$ and $\underset{\sim}{s}$ need not to be coaxial. Moreover within power of (18), which governs the difference between directions of $\dot{\underset{\sim}{p}}$ and $\underset{\sim}{s}$, is to simulate a „fading memory", i.e. the effect that a non-coaxiality of $\dot{\underset{\sim}{p}}$ and $\underset{\sim}{s}$ caused by a sharp change of the direction of a loading path diminishes with continuation of loading. In this way some kinds of delay effects can be modelled. Note also that the stress $\underset{\sim}{s}$ enters the constitutive equations only through II_{sa} and therefore no higher order stress effects are covered by the model.

The model (16) – (18) will be identified for loading paths where no unloading is permitted. Then sharp changes of loading direction are restricted in magnitude and one expects, despite a very limited number of retained invariants, that the model provides for this class of loading paths a resonable approximative description of delay effects. Moreover, if the model is constantly in a loading mode (i.e. f = 0 always), the yield condition need not to be explicitly known and identification of the model is substantionally simplified. Under these conditions, providing that α can be evaluated from $f(\alpha, II_{sa}, II_a) = 0$ as $\alpha = g(II_{sa}, II_a)$, we can transform (16) – (18) to the form

$$\dot{\underset{\sim}{p}} = \frac{K_2 \operatorname{tr}(\dot{\underset{\sim}{s}} \underset{\sim}{a})}{II_s + K_1} \underset{\sim}{a} \qquad (19)$$

$$\dot{\underset{\sim}{a}} = \frac{\operatorname{tr}(\dot{\underset{\sim}{s}} \underset{\sim}{a})}{II_s + K_1}(-\underset{\sim}{s} + K_3 \underset{\sim}{a}), \qquad (20)$$

where $K_1(II_{sa}, II_a) = (h_2 - 2 d_4(\partial g/\partial II_{sa})(\partial \psi/\partial II_a) II_{sa} - 2(\partial g/\partial II_a)[-h_1 II_{sa} + 2 h_3(\partial \psi/\partial II_a) II_a] / h_1(\partial g/\partial II_{sa})$, $K_2(II_{sa}, II_a) = 2(\partial \psi/\partial II_a)$, $K_3(II_{sa}, II_a) = 2(h_3/h_1)(\partial \psi/\partial II_a)$. To identify the model (19), (20) we have to determine three functions K_i, $i = 1, 2, 3$. To simplify the consideration assume that the tensors $\underset{\sim}{s}$ and $\underset{\sim}{a}$ can be simultaneously diagonalized, and therefore it suffices to represent $\underset{\sim}{s}, \underset{\sim}{a}$ as the vectors of the principal components $\underset{\sim}{s} = (s_1, s_2)$,

$\underset{\sim}{a} = (a_1, a_2)$; according to (19) the same is true for $\underset{\sim}{\dot{p}} = (\dot{p}_1, \dot{p}_2)$. The equation (19) suggests a convenient interpretation of $\underset{\sim}{a}$ as the unit vector in $\underset{\sim}{p}$ direction, i.e. $a_i = \dot{p}_i / \sqrt{\dot{p}_1^2 + \dot{p}_2^2}$. One easily verifies that the last relation requires in (20) $K_3 = s_1 a_1 + s_2 a_2$.

To cover a delay effect let us consider a family of loading paths which consists of two joined segments of small curvature and measure $\dot{p}_1$, a_1, a_2 and $\dot{a}_1$ along these paths ($\dot{p}_2$ and $\dot{a}_2$ provide identical information as $\dot{p}_1$, $\dot{a}_1$). Assume that the angle γ between the tangents of the segments at the joining point varies; γ lables a loading path within the family (alternatively we could vary the arc length s_o of the initial segment keeping γ fixed, but the model is not flexible enough to cover both variable s_o and γ). Along a loading paths $s_i = s_i(s, \gamma)$ and from measurement we know $a_i(s, \gamma)$, $\dot{p}_1(s, \gamma)$, $\dot{a}_1(s, \gamma)$; s is the arc length of loading path, i = 1, 2. The equation (19), (20) yield

$$K_1 = \frac{tr(\dot{\underset{\sim}{s}} \underset{\sim}{a})}{\dot{a}_1} (-s_1 + (s_1 a_1 + s_2 a_2) a_1) - II_s , \qquad (21)$$

$$K_2 = \frac{\dot{p}_1 (II_s + K_1)}{tr(\dot{\underset{\sim}{s}} \underset{\sim}{a}) a_1} \qquad (22)$$

To get K_i, i = 1,2,3 as functions of II_{sa}, II_a we need an invertibility of $II_{sa} = II_{sa}(s, \gamma)$, $II_a = II_a(s, \gamma)$. However, the invertibility is not easily quaranteed by a proper choice of loading paths, as a is not known a priori and is a consequence of the loading process.

6. INHOMOGENEOUS IDENTIFICATION TESTS

The identification testing is basically of two types. The classical approach is based on homogeneous identification tests. In Sections 4 and 5 we have seen that large number of homogeneous identification test would be needed even for models of modest flexibility. The other possibility is to utilize inhomogeneous identification tests each of them contains far more information. The number of such tests can be then substantionally reduced, but more sophisticated technique for collecting experimental data is required. The utilization of inhomogeneous tests is based on the following observation. For a special class of inelastic materials one can determine the stress field in an inhomogeneously deformed body from given boundary conditions and a measured strain field (such measurement is techniquelly possible for plane boundary value problems) without knowing the detail form of the constitutive equations. Each point of the inhomogeneous deformed body then gives information as an individual homogeneous test. Consider a special class of inelastic materials in which always the principal directions of the stress tensor coinside with the principal directions of the strain tensor (within this class falls whole spectrum of plastic models mentioned at the end of Section 2). Then at each point x, y, z of such body stress $\underset{\sim}{s}$ may be expressed in the form

$$\underset{\sim}{s}(x,y,z) = \underset{\sim}{Q}(x,y,z) \underset{\sim}{s}_D(x,y,z) \underset{\sim}{Q}^T(x,y,z) , \qquad (23)$$

where $\underset{\sim}{s}_D = (\sigma_1, \sigma_2, \sigma_3)$ is diagonal matrix of the principal values of stress and $\underset{\sim}{Q}$ is an orthogonal transformation. Due to the coincidence of the principal directions of $\underset{\sim}{s}$ and $\underset{\sim}{\varepsilon}$, $\underset{\sim}{Q}$ is determined by diagonalization of strain $\underset{\sim}{\varepsilon}$, which is assumed to be known from an experiment

$\underset{\sim}{\varepsilon}(x,y,z) = Q(x,y,z) \underset{\sim}{\varepsilon}_D(x,y,z) Q^T(x,y,z)$, where ε_D is diagonal. The relation (23) means that the stress field in the body is expressed through three scalar functions $\sigma_i(x,y,z)$, $i = 1,2,3$ and a known function $Q(x,y,z)$. The stress has to satisfy the equilibrium conditions and boundary conditions. The system of three equilibrium conditions may be sufficient for determination of three σ_i. The conditions for existence and uniqueness of such $\sigma_i(x,y,z)$ in the important case of plane strain in nonlinear elastic incompressible materials has been specified by Franěk [13]. A practical example of inhomogeneous identification has been studied in [14].

7. NOTE ON SOLUTION OF INVERSE PROBLEM

The described identification procedures have to be understood as a brief conceptual outline of types of needed identification tests and inverse problems. In an actual identification we encounter further difficulty. There is a temptation to solve the plastic inverse problems by a direct method of the type expressed in the formulae (14), (15), (21), (22). There is, however, the crucial question of stability of such solution with respect to small perturbation in experimental input data. There are indications that the direct method of solution does not satisfy the requirement of stability. The instability we encounter is of general nature and follows from the fact that inverse problems usually lead to ill-posed problems [15]. The way to transform the problem into a well-posed problem is by the incorporation of appropriate additional information on the functions to be determined (e.g. their sufficient smoothness). For internal variables models of plasticity a convenient way of such regularization has been described by Pister [16].

REFERENCES

[1] Halphen, B. and Nguyen Quoc Son, Sur les matériaux standards généralisés. J. Mecanique 14 (1975) 39 - 63.

[2] Tokuoka, T., Rate type plastic material with kinematical work-hardening. Acta Mechanica 27 (1977) 145 - 154.

[3] Phillips, A., C.S. Lin and J.W. Justusson, An experimental investigation of yield surfaces at elevated temperatures. Acta Mechanica 14 (1972) 119 - 146.

[4] Ohashi, Y., M. Tokuda and H. Yamashita, Effect of third invariant of stress deviator on plastic deformation of mild steel. J. Mech. Phys. Solids 23 (1975) 295 - 323.

[5] Phillips, A. and J. L. Tang, The effect of loading path on the yield surface at elevated temperatures. Int. J. Solids Structures 8 (1972) 463 - 474.

[6] Phillips, A. and M. Ricciuti, Fundamental experiments in plasticity and creep of aluminum - extension of previous results. Int. J. Solids Structures 12 (1976) 159-171.

[7] Lenskij, V.S., Experimentalnaja proverka zakonov izotropii i zapazdivanija pri sloznom nagruzenii. Izv. Akademii nauk SSSR, odd. techniceskich nauk (1958) 15 - 24.

[8] Maschkov, I. D., Zavisimosti naprjazenija – deformacii na ploskich mnogozvennych traektorijach deformacii. Mechanika tverdogo tela (1970) 191 – 195.

[9] Ohashi, Y., K. Kawashima and T. Yokochi, Anisotropy due to plastic deformation of initially isotropic mild steel and its analytical formulation. J. Mech. Phys. Solids 23 (1975) 277 – 294.

[10] Phillips, A. and H. Moon, An experimental investigation concerning yield surfaces and loading surfaces. Acta Mechanica 27 (1977) 91 – 102.

[11] Kratochvíl, J., Meaning of internal variables in plasticity. Letters Appl. Engn. Sci. (in press).

[12] Kratochvíl, J., Constitutive Modelling in Plasticity in Proc. Summer School „Thermomechanical Coupling in Non-linear Continuum Mechanics", Jablonna, Poland, May 1976 (to be published).

[13] Franěk, A., An inverse problem in non-linear continuum mechanics (in Czech), Master degree thesis, Institute of Solid State Physics, 1978.

[14] Iding, H., K. S. Pister and R. L. Taylor, Identification of nonlinear elastic solids by a finite element method. Computer Methods in Appl. Mech. and Engn. 4 (1974) 121 – 142.

[15] Distéfano, N., Nonlinear processes in engineering. Academic Press, New York – London 1974.

[16] Pister, K.S., Constitutive modelling and numerical solution of field problems. Nuclear Eng. Design 28 (1974) 137 – 146.

INITIATION OF CRACK GROWTH AT FULL PLASTICITY

H. Andersson

National Authority of Testing, Inspection and Metrology Box 857
S-501 15 Boras, Sweden

ABSTRACT

The well-known fact that a one-parameter criterion for fracture is not relevant for the description of the fracture zone at the crack tip in some general yield situations is illustrated by a case of torsion. Since the analysis is simple to perform in closed form various features of the solutions may be discussed in simple terms, both quantitatively and qualitatively.

1. GENERAL

The general procedure in fracture mechanics analysis is to choose a parameter, which is thought to characterize the condition of the fracture process at the crack tip uniquely, and to determine by experiments its value at crack growth initiation, or instability, on the laboratory specimen scale. This critical value is then used, computationally, to determine permissible crack length and stress in the full size structure. Such parameters are K_I, for linear elastic situations, the J-integral, for quasi-linear cases, and the crack-opening displacement for fully plastic situations.

Often the problem is to predict from test specimens with large cracks, and a low degree of plasticity, the conditions for fracture in a structure with a small crack and at full plasticity. In some instances also the test specimen is loaded to general yield, but in another configuration than the full size structure. This situation is relevant for example in the generalized three-point bend test specimen analysed with the J-integral criterion, proposed by Landes and Begley (1).

Evaluation of fracture mechanics tests, and predictions of the risks for fracture in new situations are increasingly requested. Various connections between the above-mentioned fracture criteria have been suggested to span differences between test situation and application. Present suggestions that the crack-opening displacement (COD) or the J-integral can be used generally have been satisfactorily tested in many case studies during the last decade. Still, exceptionally, discrepancies may occur, and as a reminder and a repetition, the connection between these macroscopic, integrated parameters, and the local fracture behaviour, should be continually discussed for various possible situations.

One such situation is the coupling between small scale tests and large scale applications, when different degrees of plasticity may lead to different boundary conditions for the fracture zone. This has been discussed in other terms by Broberg (2) and Rice and Johnson (3), among others. Further, the ideal scaling may be disturbed by the fact that the small constituents forming the fracture zone have an absolute size, and thus occupy varying parts in uniform specimens of different sizes. Finally, the material model used in theoretical analysis is connected to real material properties in a way that is very sensitive to disturbances, when the strain hardening is low, and then a theoretically found property of a fracture parameter may be invalidated in some real situations.

These effects are most pertinent in general yield situations. For example, McClintock (4) has discussed the many-valuedness of perfect plasticity crack tip solutions in plane strain. Broberg (5) has discussed the peculiar behaviour of plastic zones in sheets of a non-hardening material, where a state of plane stress can be assumed, and where the plastic zones may change drastically as a function of the relationship between crack length and sheet width. Such effects are clearly due to the fact that hyperbolic equations replace elliptic, and disturbances are propagated along characteristics.

On the other hand, it has been demonstrated in several investigations (6), (7) that when hardening, for example power-law hardening, prevails, the crack tip region shows autonomy (uniformity for different loads) in the continuum for at least some distance from the crack tip. When hardening decreases the solutions tend to be sensitive to disturbances, as may be seen from (8), and the coupling between autonomous properties in the continuum, and an autonomous fracture zone may be questioned. If, for example, the fracture zone consists of a few grains, with some degree of inhomogeneity, the properties of the fracture zone may vary as a function of the outside, continuum situation.

So at low hardening, ideally at perfect plasticity, the dependence of fracture initiation parameters as the crack opening displacement or the J-integral should be assessed for large scale yielding situations, where the situation of autonomy can not be assumed a priori. Here, the situation is discussed by means of a simple example, where closed form solutions may be obtained, a circular shaft with a longitudinal crack loaded in torsion. McClintock (4) has demonstrated that crack growth direction may depend on the shape of twisted bars with cracked rectangular cross-sections, with a strain criterion for crack growth.

It is demonstrated below that the crack opening displacement criterion, the maximum strain criterion and the J-integral criterion are not comparable in fully plastic cases, even for the same geometry with different crack lengths, much less as fracture parameters for scaling between model experiments and full size structures.

2. ANALYSIS

The case studied is sketched in Figure 1. A cylinder with circular cross section and an edge crack is exposed to a twist Θ per unit of length. The radius of the cylinder is R and the crack length is a. The elastic problem has been solved by Broberg (2), giving the relationship between the stress intensity factor K_{III} and the twist as a function of a/R. For the present purpose only the well-known relationships between K_{III}, J and COD at very small scale yielding are of interest for comparison, viz.

$$J = K_{III}^2/2G \qquad (1)$$

$$COD = 4J/\pi k, \qquad (2)$$

where G is the shear modulus and k is the yield stress in shear.

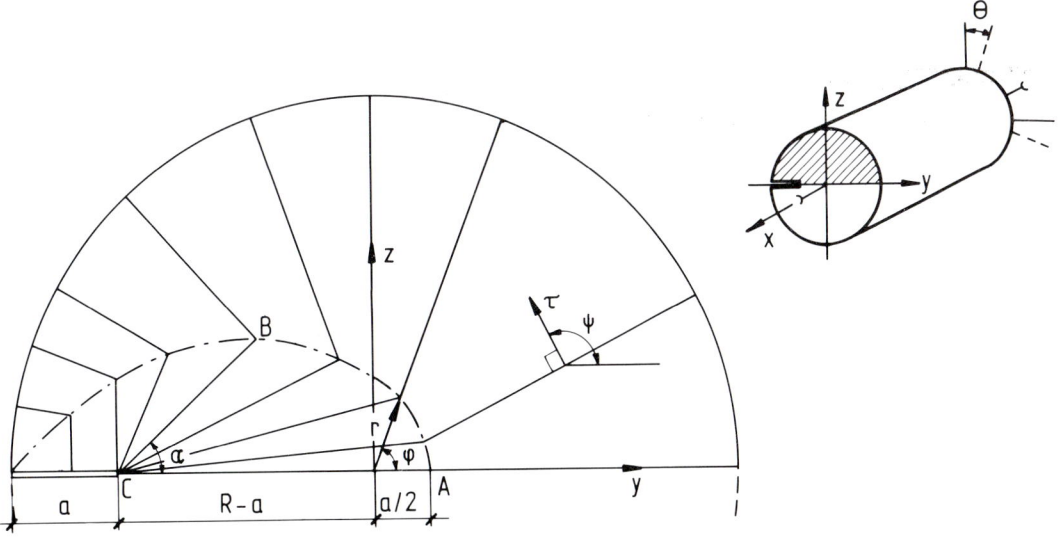

Figure 1. Sketch of the characteristics and notation

When the yield zone is no longer very small in relation to the dimensions a and R, and if perfect plasticity is assumed, the relationship (2) starts to change. Also, the factor K_{III} starts to loose its meaning. In order to demonstrate the limits of these changes, the situation at full plasticity is studied by the aid of a rigid-plastic material model. Particularly the change in (2) is of interest, as a function of the a/R-relationship, to show quantitatively what sort of errors may be introduced by using a one-parameter criterion for comparison of the fracture risks in a large variety of specimens.

The problem is formulated in terms of the sandhill analogy for a rigid-plastic material. This problem is statically determined, i.e. the stresses are obtained without any considerations of deformation. The stress characteristics, along which the stress vector is constant and perpendicular to the characteristic, then will be situated as depicted in Figure 1. (See, for example, (9) or (10)). They start at right angles from the cross section contour, and a centered fan emerges from the crack tip. The requirement of the sandhill analogy now is that characteristics meeting at the line of stress discontinuity, the crest of the sandhill, have equal length. Since the characteristics follow the gradient directions of the sandhill, and the slope is the same everywhere, the distance to reach a point of the crest must be equal from both sides.

The warping, and hence the strain distribution, of the cross section is found using the fact that the stress distribution is stationary. Then the relation between shear strains and shear stresses

$$\frac{d\gamma_{xz}}{\tau_{xz}} = \frac{d\gamma_{xy}}{\tau_{xy}} \tag{3}$$

can be integrated to

$$\frac{\gamma_{xz}}{\tau_{xz}} = \frac{\gamma_{xy}}{\tau_{xy}}. \tag{4}$$

The strains are composed of components resulting from the twist per unit length, Θ, and of the derivatives of the warping function u:

$$\left. \begin{array}{l} \gamma_{xy} = -\Theta z + \dfrac{\partial u}{\partial y} \\[2mm] \gamma_{xz} = \Theta y + \dfrac{\partial u}{\partial z} \end{array} \right\} \tag{5}$$

where x,y,z denote cartesian coordinates according to Figure 1. The origin of the coordinates may be chosen arbitrarily along the line of symmetry, the only change being a rigid-body rotation. This may be seen from eqs. (5). Here, the position shown in Figure 1 is used for computational convenience.

By integrating two differential equations, one for travel along characteristics and one for travel along the line of discontinuity, the warping function may be found.

With ψ as the angle between the stress vector and the y-axis, the equations (5) may be written, regarding (4), as

$$\left.\begin{array}{l} -\Theta z + \partial u/\partial y = \lambda(y,z)\cos\psi \\ \Theta y + \partial u/\partial z = \lambda(y,z)\sin\psi \end{array}\right\} \quad (6)$$

or, eliminating $\lambda(y,z)$

$$\sin\psi \frac{\partial u}{\partial y} - \cos\psi \frac{\partial u}{\partial z} = \Theta(y\cos\psi + z\sin\psi). \quad (7)$$

It is seen that the characteristics of this equation are the same as the stress characteristics, denoted by the length parameter c, and then

$$\frac{du}{dc} = \frac{\partial u}{\partial y} \cdot \frac{dy}{dc} + \frac{\partial u}{\partial z} \cdot \frac{dz}{dc} = \Theta(y\cos\psi + z\sin\psi) = \Theta p \quad (8)$$

where p is the length of the perpendicular to the characteristic from the origin. Since the characteristics are straight lines, p is constant, and

$$u_C = u_B - \Theta p c_{CB}. \quad (9)$$

The equation for integration of u along lines of discontinuity is found by considering these as very thin elastic regions, where rapidly varying elastic shear stresses prevail. Since the material is taken to be rigid-plastic, increments of shear strain vanish, and

$$\left.\begin{array}{l} \partial u/\partial y = \Theta z \\ \partial u/\partial z = -\Theta y \end{array}\right\} \quad (10)$$

Then the variation of u along a line of discontinuity, with length parameter ℓ, is

$$\frac{du}{d\ell} = \frac{\partial u}{\partial y} \cdot \frac{dy}{d\ell} + \frac{\partial u}{\partial z} \cdot \frac{dz}{d\ell} = \Theta\left(z\frac{dy}{d\ell} - y\frac{dz}{d\ell}\right) = \Theta p \quad (11)$$

with the direction of the perpendicular oriented in the same way, relative to $d\ell = (dy, dz)$ as in eq. (8).

With these equations, the warping function, and its discontinuity at the crack tip, may be determined. Choosing the crack tip as the origin of the coordinates, McClintock has shown that the crack opening displacement, COD, may be written generally as

$$COD = \int_{-\pi/2}^{\pi/2} \rho^2 d\alpha \quad (12)$$

where ρ is the distance from the crack tip, along the characteristic with inclination α, to the line of discontinuity.

With the use of coordinates according to Figure 1, eqs.

(8) and (11) may be integrated using the relation

$$r(\phi) = 0.5 \left[R^2 - (R-a)^2\right] / \left[R + (R-a)\cos\phi\right] \quad (13a)$$

which is obtained from the condition that characteristics from both sides of a discontinuity have equal length, viz.

$$\left[R-r(\phi)\right]^2 = \left[(R-a) + r(\phi)\cos\phi\right]^2 + \left[r(\phi)\sin\phi\right]^2 \quad (13b)$$

Then

$$(y,z) = (r(\phi)\cos\phi, r(\phi)\sin\phi) \quad (14)$$

and u may be integrated directly to an angle ϕ_0 corresponding to an angle α according to Figure 1. The result is

$$u_\alpha = -\frac{\theta}{4}\left[R^2 - (R-a)^2\right] \left\{ \frac{(R-a)\sin\phi_0}{R+(R-a)\cos\phi_0} + \frac{2R}{\sqrt{R^2-(R-a)^2}} \tan^{-1}\left[\frac{\sqrt{R^2-(R-a)^2}\tan(\phi_0/2)}{2R-a}\right] \right\} \quad (15)$$

For the case defining the crack opening displacement $\alpha = \pi/2$, the formula (15) may be simplified, using the relationship

$$r(\phi_0)\cos\phi_0 + (R-a) = 0, \quad (16)$$

to

$$COD = |2u_{\pi/2}| = \theta R^2 \left\{ 0.5(1-\frac{a}{R})\left[1-(1-\frac{a}{R})^2\right] + \sqrt{1-(1-\frac{a}{2})^2}\tan^{-1}\left[\sqrt{1-(1-\frac{a}{R})^2}/(\frac{a}{R})\right] \right\} \quad (17)$$

The calculation of the J-integral is readily performed using (15) since, in this case, the definition

$$J = \oint_\Gamma (Wdz - \tau_{xj}n_j\frac{du}{dy}ds)$$

reduces, along a semicircular path with radius δ, to

$$J = \lim_{\delta \to 0} 2k \int_0^{\pi/2} \frac{\partial u}{\partial\delta\partial\alpha}dy = \lim_{\delta \to 0} 2k \int_0^{\pi/2} \frac{\partial u}{\partial\delta\partial\alpha} \delta d\alpha \cos\alpha = \quad (18a)$$

$$= 2k \int_0^{\pi/2} \cos\alpha \frac{\partial u}{\partial\alpha} d\alpha \quad (18b)$$

where, as before, k is the yield stress in shear.

By the equations (15), (17) and (18) a case study may be performed where the possible peculiarities of various fracture criteria can be systematically assessed. In particular the relationship between J and COD is of interest.

3. RESULTS

The shape of the crack tip discontinuity, as a function of the angle, is shown in Figure 2a for several crack lengths. For a crack length $a/R \approx 0.4$ the shape is nearly coincident with the sine function obtained in the small-scale yielding solution. For shorter crack lengths the strain is more pronounced in a direction straight ahead of the crack tip, for longer cracks a more even distribution is obtained. It should be noted that for $a/R > 1$, for which value the relationship is a straight line, the strain is most severe for $\alpha = 90°$, and consequently a strain criterion would indicate crack propagation at right angles to the crack line. This effect has been shown experimentally by McClintock (4). Here it is only noted that the integrated J- and COD-criteria are not able to distinguish between directions of propagation and that very different results may be obtained when various crack lengths are used for evaluating the fracture criteria experimentally.

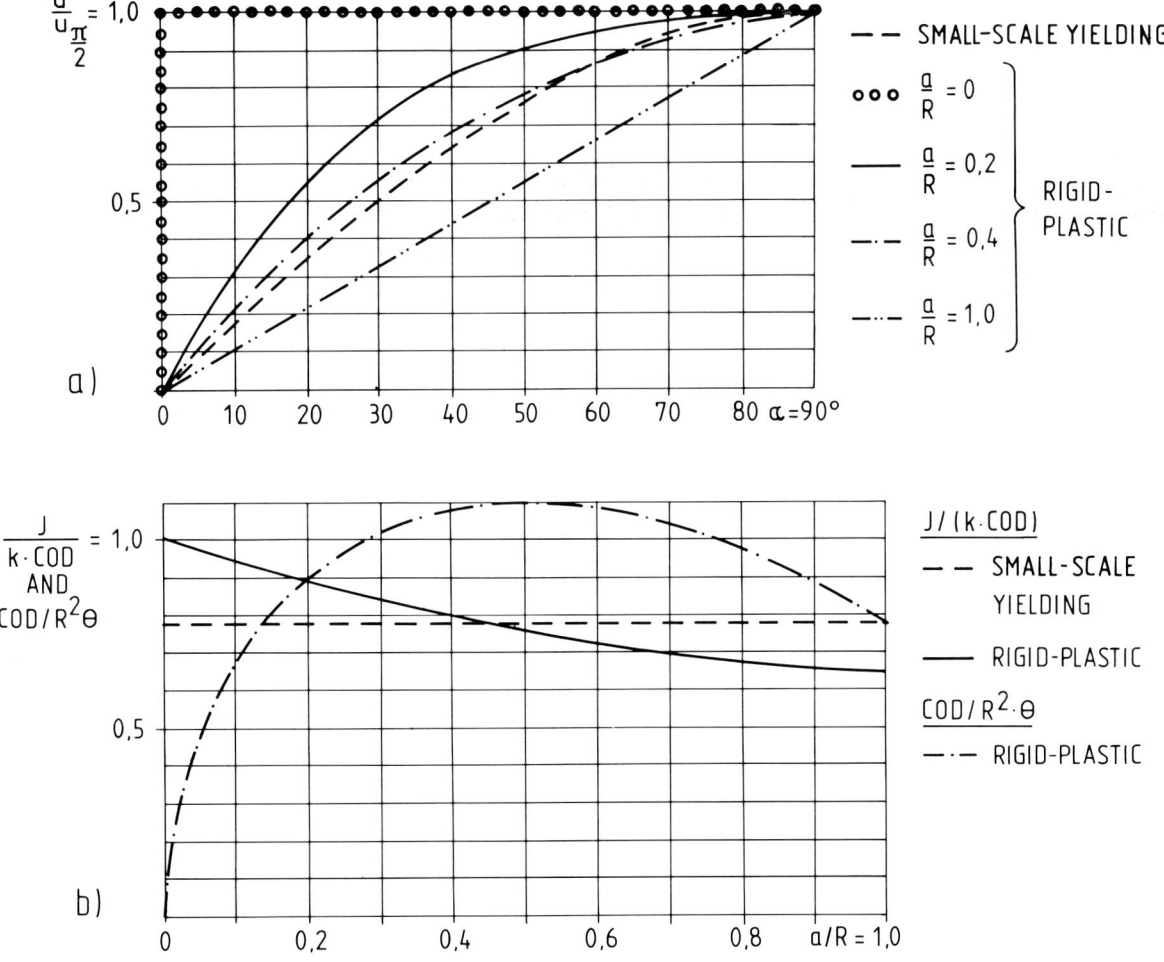

Figure 2. Shape of crack tip discontinuity (a), and relationships between crack length, and J-integral, and crack length and crack opening displacement (b).

From Figure 2b J and COD may be evaluated as functions of twist for different crack lengths. In particular it is noted that crack lengths of $a/R \sim 0.5$ lead to criticality for the smallest twists for both the J- and the COD-criterion (nearer $a/R = 0.4$ for the J-criterion). Further, the comparison between the J- and the COD-criterion for small-scale yielding, and for this rather extreme example, shows that the criteria give results of the same order but that discrepancies for more normal geometries and loadings of the order of 20-30 per cent should be expected. As is seen, here the relation between the full-plastic J/COD-value and the one for small scale yielding varies between 1.27 for $a/R = 0$ and 0.82 for $a/R = 1.0$.

This short investigation shows that care has to be taken to choose a relevant fracture criterion when scaling between structures with different degree of plasticity is necessary. Even if hardening occurs, the combination of low hardening and influence from the fracture zone may produce discrepancies of the character demonstrated here.

REFERENCES

(1) Begley, J.A. and Landes, J.D., ASTM STP 514, 1972, p 1.

(2) Broberg, K.B., private discussion.

(3) Rice, J.R. and Johnson, M.A., Inelastic Behavior of Solids, (ed. by M.F. Kanninen et al.), McGraw-Hill, 1970, p 641.

(4) McClintock, F.A., Fracture III, (ed. by H. Liebowitz), Academic Press, 1971, p 48.

(5) Broberg, K.B., J. Mech. Phys. Solids, vol 19, 1971, p 407.

(6) Hutchinson, J.W., J. Mech. Phys. Solids, vol 16, 1968, p 13.

(7) Rice, J.R. and Rosengren, G.F., J. Mech. Phys. Solids, vol 16, 1968, p 1.

(8) Amazigo, I.C., Int. J. Solids Structures, vol 11, 1975, p 1291.

(9) Kachanov, L.M., Foundations of the Theory of Plasticity, North-Holland, 1971.

(10) Freiberger, W.F., Torsion (In Handbook of Engineering Mechanics, ed. by Flügge, W.), McGraw-Hill, 1962, p 48-1.

STRESS AND STRAIN DISTRIBUTION IN TWO-PHASE SYSTEMS

Thomas Johannesson and Peter Sjöblom

Institute of Technology, S-581 83 Linköping, Sweden

ABSTRACT

Brittle two-phase materials exemplified by certain tool materials show a complex fracture behaviour. During loading elastic energy is stored in the two phases. Fracture is then initiated in the hard and brittle phase and propagates when the energy absorbtion rate becomes less than the energy dissipation rate.

In order to understand this process one need to know the stress distribution in the two-phase structure. We have made FEM-analyses of a range of geometries of idealized structures. If everything else is kept constant it is shown that contiguity influences the stress patterns to a significant extent.

1. INTRODUCTION

Toughness is a complex concept and there is many ways to measure it. These do not always vary in a congruent way (1).

The toughness concept is especially important for materials which are subject to impact as e.g. tool materials. These materials may in a simplified manner be considered as two-phase materials; one phase is hard but brittle, the other is softer but ductile. The volume fractions of the phases largely deterines the mechanical properties. On one hand we may have a large volume fraction of hard phase resulting in a hard but brittle material (e.g. cemented carbide), on the other hand we may have a large volume fraction of soft phase resulting in a softer but tougher material (e.g. high speed steels).

The development of powder metallurgy may open possibilities to produce materials with intermediate phase compositions. It is therefore important to be able to estimate the toughness for various phase combinations. In order to do this we need to study the energy balance that determines the crack propagation (2),(3) but we also need to take the crack initiation into account. In both cases the stress distribution between the phases in the composite material is of vital importance in determining the toughness behavior.

2. CRACK PROPAGATION

In a two-phase material one must consider the crack propagation events separately in the two phases. The fracture mechanics approach can be used to estimate a critical crack size, a_c

$$a_c = g\left(\frac{K_{Ic}}{\sigma}\right)^2 \quad [1]$$

where g is a constant dependent on geometry.

If we insert K_{Ic} and the fracture stress we will get the minimum critical crack size to propagate a crack.

2.1 HIGH VOLUME FRACTION OF HARD PHASE

Let us represent the material by a closed packed hexagonal arrangement of spheres of diameter d which make contact between hard spheres over a circular area with radius a. The contiguity C is then

$$C = \frac{12a^2}{d^2} \quad [2]$$

The distribution of stresses in this type of material can often be fairly well represented by the simple "parallell model" of composite materials. The hard phase carries a higher stress than the soft phase.

An initial crack will open in the hard and brittle phase when its cleavage stress is reached. Will this initial crack propagate through the rest of the material?
We shall have to distinguish between to cases.

2.1a The initial crack is supercritical

Crack propagation and complete fracture will occur immediately when the stress consentrated to the hard phase reach the cleavage stress, $\sigma_{F\alpha}$, of the hard phase. Therefore, if the stress in the hard phase is evenly distributed

$$\sigma_F = \frac{E_\alpha f_\alpha + E_\beta f_\beta}{E_\alpha} \sigma_{F,\alpha} \qquad [3]$$

where $f_{\alpha,\beta}$ is the volume fraction.

However, the stress in the hard phase is not evenly distributed but consentrated at critical points in the structure as will be discussed below.

2.1b The initial crack is subcritical

The initial crack will open at critical stress consentration points when the cleavage stress is reached there. The initial crack will spread över the contact area and thus have a radius a. It will continue propagating when the applied stress σ reach a value

$$\sigma = \sigma_F = g \frac{K_{Ic}}{\sqrt{\pi a}} \qquad [4]$$

where K_{Ic} is the critical stress intensity factor of the two phase material and g is a geometrical constant.

Inserting [2] into [4] yields

$$\sigma_F = g \frac{K_{Ic}}{d^{1/2} c^{1/4}}$$

2.2 LOW VOLUME FRACTION OF HARD PHASE

In this case the stress distribution is closer to the series model and higher applide stre-ses are needed to create a initial crack in the hard phase. Moreover as K_{Ic} is larger for this type of two-phase material the critical crack size becomes relatively large, usually larger than the grain size of the hard phase. The cracking of large inclusions in a real material could naturally result in a critical crack. However, also in this case the uneven distribution of stresses are important in determining the site of crack initiation.

3. STRESS DISTRIBUTION

As mentioned previously the detailed stress distribution is determining at what external stress the initial crack is created.

We have therefore made a finite element calculation of the stress distribution in a two-phase model structure shown in fig 1. By varying the ratio a/b the contiguity of the model structure is altered at constant volume fractions.

The model was divided into 300 triangular six-nodes finite elements. Because of the symmetry we only needed to examin a smaller part, fig 2. The model was given a total strain of 1% first in the x- and then in the y-direction. The lines of symmetry was kept straight. To do this and to keep the calculation time down we used a contact program to solve the structural equations.

The result is plotted as the tensile stress σ_x along the y-axis, fig 3. It can be seen that as the contiguity decreases at constant volume fraction the sharpness of the stress distribution increases. It follows that in this model the opening of the initial crack by cleavage should occur at lower external stresses as the contiguity decreases. In case 2.1a above the fracture stress is proportional to the cleavage stress and therefore contiguity and fracture stress should vary in a congruent way.

Another result is seen in table 1 where the effective Youngs modulus of the two-phase model at constant volume fractions is shown to vary with contiguity as foreseen in the calculations of Hashin and Shtrikman (4).

4. CONCLUSIONS

In a two-phase composite material the phase geometry has a distinct influence on the stress distribution. Therefore it should be expected that the fracture stress of the composite is dependent not only on phase properties but also on geometrical parameters like contiguity.

Table 1.

C	E_x	E_y	ν_{xy}	ν_{yx}
33 %	4.51	4.49	0.25	0.25
36 %	4.50	4.63	0.23	0.24
38 %	4.50	4.79	0.22	0.23
42 %	4.50	4.93	0.20	0.20

Volume fractions constant.

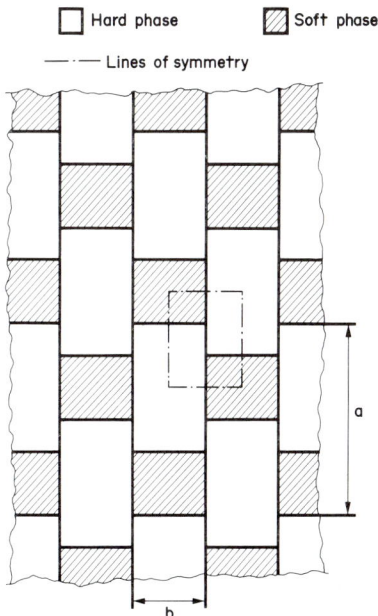

Fig. 1

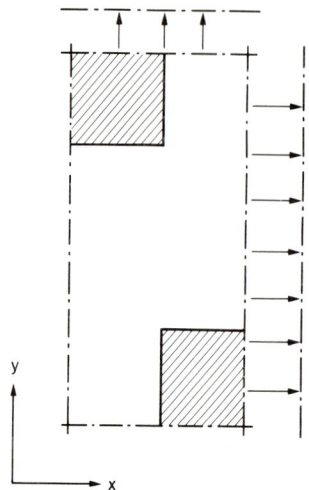

Fig. 2

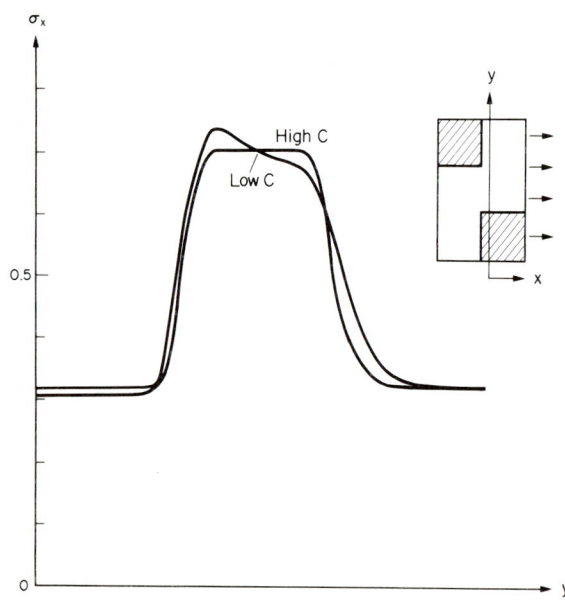

Fig. 3

REFERENCES

1. Ro Ritchie, B Francis, WL Server
   Met Trans A 1976 (7A), 831

2. T Johannesson
   4th European Symposium for Powder Metallurgy,
   Grenoble, 1975

3. C Chatfield
   5th European Symposium for Powder Metallurgy,
   Stockholm, 1978

4. Z Hashin, S Shtrikman
   J. Mech. Phys. Solids, 1963, Vol. II, pp. 127 to
   140. Pergamon Press Ltd. Printed in Great Britain

5. B Fredriksson, G Rydholm, P Sjöblom
   International Conference on Finite Elements in
   Nonlinear Solid and Structural Mechanics,
   Geilo, Norway, 1977

MICROMODELS FOR GRANULAR COMPOSITES

K. Berglund

*Department of Mechanics, The Royal Institute of Technology,
S-100 44 Stockholm, Sweden*

ABSTRACT

The microstress of a granular composite is investigated by means of a model consisting of elastic, spherical inclusions embedded in an elastic matrix and forming a cubic lattice. Both matrix and inclusions are homogeneous and isotropic. For a representative macro boundary value problem with uniaxial compressive stress the properties of both the micro and macro levels are calculated by means of an approximate expression for the displacements. The result gives of course that considerable compressive normal stress concentrations occur at those parts of the interfaces between matrix and inclusions which are perpendicular to the macro principal stress, i.e. at the polar zones of the inclusions. Moreover we obtain normal tensile stress concentrations at the parts which are parallel to the macro principal stress, i.e. at the equatorial zone of the interface. This tensile stress, which depends on the volume content of inclusions and of the elastic properties of the materials, can be of greater importance for the strength of the composite then the compressive stresses. This is in good agreement with the observed fact that cracks often start in the equatorial zone. Our result shows that this tensile stress concentration is an increasing function of the relative stiffness of the inclusions and of the Poisson's ratio of the matrix.

1. INTRODUCTION

The pure elastic properties of materials such as granular composites are rather well known, due to eg. Beran and Molyneux [1], Hashin and Strikman [2], and Kröner [3] to mention some of the investigators. In their works the macro elastic moduli are calculated from knowledge of the elastic properties of the different constituent materials, the distribution of size, form orientation of the inclusions, and so forth.

However, the connection between the strength and the microproperties of a granular composite is not well understood. Therefore it is difficult to choose the properties of the matrix and of the inclusions in some optimal manner to get as high strength as possible of the composite.

The aim of this work is to investigate the stress fields on the micro-level, from which conclusions about the initial state of crackpropagation can be drawn.

2. PRESENTATION OF THE MODEL

A model of a granular composite is considered. The model consists of spherical inclusions with radius a, arranged to form a cubic lattice with distance d between the centers of nearest neighbours and embedded in a matrix. Both the matrix and the inclusions are homogeneous, isotropic and linearly elastic. The elastic properties are characterized by the shear modulus $\mu^{(i)}$ and $\mu^{(m)}$ and the Poisson's ratios $\nu^{(i)}$ and $\nu^{(m)}$, where (i) stands for inclusion and (m) for matrix. See Figure 1.

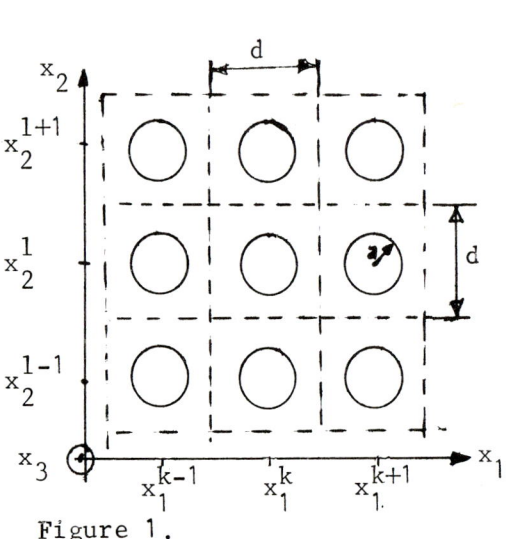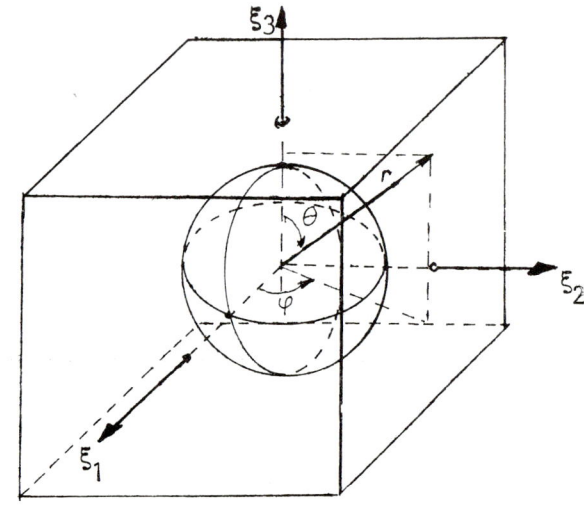

Figure 1.

In a disordered composite the micro stress field for a given macroscopic boundary value problem is impossible to solve.

Even in this simple model, it is difficult to obtain the microsolution. Thus the solution is made in an approximative manner.

We define a Cartesian coordinate system x_i with the axes parallel to the lattice. The inclusions are numbered by the indices (k,l,m). The centre of inclusion (k,l,m) has the coordinates $x_1^{(k)}$, $x_2^{(l)}$, $x_3^{(m)}$ at equilibrium, when no external forces are acting on the medium. The composite is devided into identical cells, each of them containing one inclusion at its center. These cells are denoted by the same indices (k,l,m) as the embedded inclusions.

We also define a system of local Cartesian coordinates, ξ_i, as well as a system of spherical coordinates r,θ,φ, in each cell (k,l,m). Both systems have their origin at $x_1^{(k)}$, $x_2^{(l)}$, $x_3^{(m)}$. The connections between r,θ,φ and ξ_i are

$$\xi_1 = r \sin\theta \cos\varphi$$
$$\xi_2 = r \sin\theta \sin\varphi \qquad (1)$$
$$\xi_3 = r \cos\theta$$
$$0 < \theta < \pi, \quad 0 < \varphi < 2\pi.$$

On the microscale, i.e. in the interior of the inclusion (i) or in the matrix (m), in cell (k,l,m), the displacements can be approximately expressed as:

$$u_i^{(i)(k,l,m)} = \bar{u}_i^{(k,l,m)} + \xi_j \psi_{ji}^{(i)(k,l,m)} = \bar{u}_i^{(k,l,m)}$$
$$+ r \sin\theta \cos\varphi \, \psi_{1i}^{(i)(k,l,m)} + r \sin\theta \sin\varphi \, \psi_{2i}^{(i)(k,l,m)}$$
$$+ r \cos\theta \, \psi_{3i}^{(i)(k,l,m)}, \quad r < a, \qquad (2)$$

$$u_i^{(m)(k,l,m)} = \bar{u}_i^{(k,l,m)} + a \sin\theta \cos\varphi \, \psi_{1i}^{(k,l,m)}$$
$$+ a \sin\theta \sin\varphi \, \psi_{2i}^{(i)(k,l,m)} + a \cos\theta \, \psi_{3i}^{(i)(k,l,m)}$$
$$+ (r-a) \sin\theta \cos\varphi \, \psi_{1i}^{(m)(k,l,m)} + (r-a) \sin\theta \sin\varphi \, \psi_{2i}^{(m)(k,l,m)}$$
$$+ (r-a) \cos\theta \, \psi_{3i}^{(m)(k,l,m)}, \quad r > a, \qquad (3)$$

where the quantities $\psi_{ij}^{(i)(k,l,m)}$, $\psi_{ij}^{(m)(k,l,m)}$ and the displacement $\bar{u}_i^{(k,l,m)}$ of the center of inclusion (k,l,m) are defined at the points $x_1 = x_1^{(k)}$, $x_2 = x_2^{(l)}$, $x_3 = x_3^{(m)}$. These quantities thus depend on the discrete coordinates $x_1^{(k)}$, $x_2^{(l)}$, $x_3^{(m)}$ and are consequently constants in a cell. The description guarantees that the displacements are continuous at the interface between inclusion and matrix.

On the macroscale $\bar{u}_i$ is regarded as the continuous macro displacement field.

The model and the expressions for the displacements used here are generalizations of Achenbach [4], where he used this method to describe wave-propagation and dispersion phenomena in a fiber reinforced composite. In [5] the method is used by the author to estimate the constitutive constant characterizing the antisymmetric part of the stress in a homogeneous, isotropic polar medium.

At the interfaces between the cells and between the inclusions and the matrix we shall require the displacements and the stresses to be continuous. These conditions will be discussed later.

To get some insight in the connection between macro and micro properties, e.g. between macroscopic failure criteria and micro stress concentrations, we have to solve some representative macro boundary value problems and investigate them on the micro level. Thus, as a first step the macro behaviour of the composite has to be examined.

3. CALCULATION OF THE MACRO ELASTIC MODULI

The macro elastic moduli can be calculated by first prescribing some suitable macro deformations and then solve the stress-strain relations on both levels.

According to [5] the macro stress can be obtained as the following surface average of the micro stress:

$$\Delta S_{\underline{i}} <t_{\underline{i}j}>^* = \sum_\alpha \int_{\Delta S^{(\alpha)}} t_{\underline{i}j}^{(\alpha)} dS_{\underline{i}} , \quad \Delta S^{(\alpha)} \subset \Delta S . \qquad (4)$$

(no summation over underlined indices)

$<t_{ij}>^*$ is the surface average of the micro stress t_{ij}, i.e. the macro stress $\bar{t}_{ij}$. The average is to be calculated over the collection of open surfaces $\Delta S^{(\alpha)}$ which constitutes the small macro surface ΔS. ΔS contains a statistical significant number of micro surfaces $\Delta S^{(\alpha)}$.

A composite of the kind considered here, is of course not isotropic unless the geometrical parameter a/d is much smaller then unity. However, the anisotropy is rather small even for values up to 0.5, which is the maximum value. Thus we just need two macro elastic moduli to describe the macro behaviour of the composite, eg. the macro Lamé constants $\bar{\mu}$ and $\bar{\lambda}$.

Two suitable micro displacement fields from which the macro moduli can be obtained are:

$$\bar{u}_1^{(i)(k,l,m)} = \epsilon x_2^{(l)} , \quad \bar{u}_2^{(i)(k,l,m)} = \epsilon x_1^{(k)} , \quad \bar{u}_3^{(i)(k,l,m)} = 0, \quad (5)$$

and

$$\bar{u}_1^{(i)(k,l,m)} = -n \epsilon x_1^{(k)} , \quad \bar{u}_2^{(i)(k,l,m)} = -n \epsilon x_2^{(l)} , \quad \bar{u}_3^{(i)(k,l,m)} = \epsilon x_3^{(m)} ,$$

$$\epsilon \ll 1 \qquad (6)$$

where n is an arbitrary constant. The first deformation (5) is a shear and the second (6) a tension or compression, depending on the sign of ϵ and n. For n = -1 and $\epsilon < 0$ (6) represent a hydrostatic compression. Evidently, both deformations are homogeneous on the macro level, a fact which implies the conditions

$$\psi_{ij}^{(i)(k,l,m)} = \psi_{ij}^{(i)} \quad \text{and} \quad \psi_{ij}^{(m)(k,l,m)} = \psi_{ij}^{(m)} \qquad (7)$$

for all (k,l,m), since all cells are deformed identically.

At the interfaces between the cells the displacements should be continuous. However, this is impossible in our approximation, and therefore a weaker condition is imposed, viz. that the average displacements are continuous. At the interface between the cells (k,l,m) and (k+1,l,m) we thus get the condition

$$\int_{-d/2}^{d/2} \int_{-d/2}^{d/2} \left\{ \left[u_i^{(m)(k+1,l,m)}\right]_{\xi_1=-\frac{d}{2}} - \left[u_i^{(m)(k,l,m)}\right]_{\xi_1=\frac{d}{2}} \right\} d\xi_2 d\xi_3 = 0 \qquad (8)$$

and similar for the other interfaces between cell (k,l,m) and its adjacent cells.

The strain tensor is as usual defined as

$$\varepsilon_{ij} = \frac{1}{2}\left(\frac{\partial u_i}{\partial \xi_j} + \frac{\partial u_j}{\partial \xi_i}\right) \quad \text{on the micro level, and} \qquad (9)$$

$$\bar{\varepsilon}_{ij} = \frac{1}{2}\left(\frac{\partial \bar{u}_i}{\partial x_j} + \frac{\partial \bar{u}_j}{\partial x_i}\right) \quad \text{on the macro level.} \qquad (10)$$

If (1), (2) and (3) are inserted in (9) the strain $\varepsilon_{ij}^{(i)}$ inside an inclusion and $\varepsilon_{ij}^{(m)}$ in the matrix can be expressed in terms of $\psi_{ij}^{(i)}$ and $\psi_{ij}^{(m)}$.

In addition to the condition of continuity for the displacements, (8), the solution has to give continuous stress at the interface between inclusion and matrix. However, it can be seen that it is impossible to require point to point continuity in general. Only in the case of a macroscopic hydrostatic stress state i.e. n = -1, we obtain point to point continuity. In all other cases we thus require continuity in some weaker sense.

Due to the symmetry in the deformation we just have to examine one octant of the sphere r= a, i.e. $0 < \theta < \pi/2$, $0 < \varphi < \pi/2$. On this surface we require

$$\int_{\varphi=0}^{\pi/2}\int_{\theta=0}^{\pi/2} T_j^{(i)} ds = \int_{\varphi=0}^{\pi/2}\int_{\theta=0}^{\pi/2} T_j^{(m)} ds \quad \text{for } r = a, \qquad (11)$$

where $T_j = t_{ij} n_i$ is the stress vector acting on a surface with the outward unit normal vector n_i.

We also require that the normal stress at the points $\theta = 0$, $\pi/4$ and $\pi/2$ for $\varphi = 0$ and $\pi/2$ is continuous, i.e.

$$t_{rr}^{(i)} = t_{rr}^{(m)} \quad \text{for } r = a;\ \theta = 0,\ \pi/4,\ \pi/2;\ \varphi = 0, \pi/2 \qquad (12)$$

Taking the symmetry of the macro deformation into account $\psi_{ij}^{(i)}$ and $\psi_{ij}^{(m)}$ are uniquely determined from the conditions (5) or (6) and (8), (11) and (12). When $\psi_{ij}^{(i)}$ and $\psi_{ij}^{(m)}$ are determined the micro stresses can be obtained from (9). Then, taking surface averages of the stresses t_{11} and t_{12} and comparing them with the macro description where the strain is calculated from (10) we get the macro moduli.

In Fig 2 the macro Young modulus $\bar{E}$ and the macro Poisson's ratio $\bar{\nu}$ are plotted versus $E^{(i)}/E^{(m)}$ for $\nu^{(m)} = 0.28$ and $\nu^{(i)} = 0.22$.

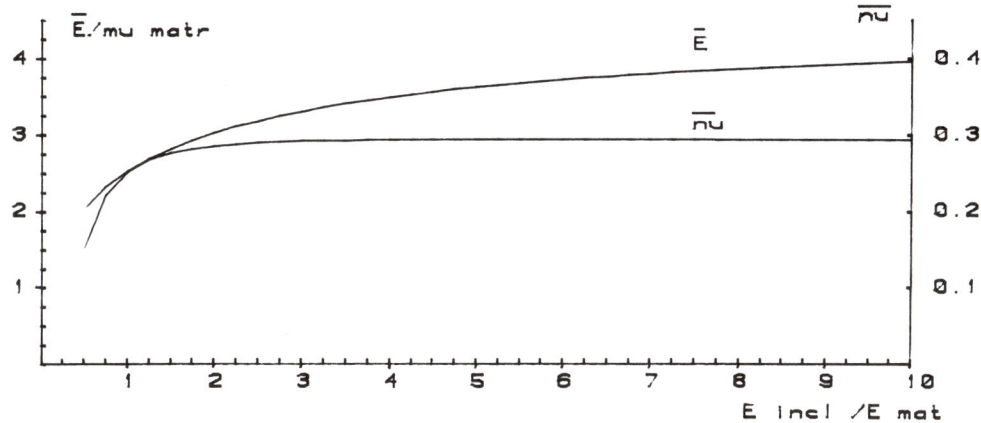

Figure 1: Youngs modulus $\bar{E}$ and Poissons ratio $\overline{nu}$ for the composite plotted versus E matr/E incl for a/d=0.40; nu matr=0.28, nu incl=0.22. E matr=2.6.

4. CALCULATION OF THE MICRO STRESS FIELDS

In a granular composite such as concrete the most common macro stress state is uniaxial compression or tension. We will therefore examine such a stress state on the micro level. Consider a cylinder oriented parallel to the x_3-axis. At the endsurfaces we apply forces which give rise to the macro boundary conditions

$$\bar{t}_{33} = -P/A, \quad \bar{t}_{31} = \bar{t}_{32} = 0 \quad \text{on the endsurfaces, and} \tag{13}$$

$$\bar{t}_{ij} n_i = 0 \quad \text{on the cylindrical surface.} \tag{14}$$

P is the total force acting on the endsurface and A is the area of the same surface.

The solution of the boundary value problem

$$\bar{t}_{ij,i} = 0, \tag{15}$$

the boundary conditions (13) and (14) and Hookes law, is well known. The relations between the components of the macro strain tensor are:

$$\bar{\epsilon}_{11} = \bar{\epsilon}_{22} = -\bar{\nu} \bar{\epsilon}_{33}, \quad \bar{\epsilon}_{12} = \bar{\epsilon}_{13} = \bar{\epsilon}_{23} = 0 \tag{16}$$

The strain $\bar{\epsilon}_{33}$ can be found as

$$\epsilon_{33} = -\frac{P}{A} \bar{E}^{-1} = \epsilon \tag{17}$$

Putting $n = \bar{\nu}$ in (6) and inserting ϵ according to (17) we get the desired macro displacement field.

$\psi^{(i)}$ and $\psi^{(m)}$ can now be determined by calculating $\bar{\mu}$ and $\bar{\lambda}$ according to Chapter 2, and then solving the set of equations composed of (6, 7, 8, 11, 12). The micro deformation fields are then known from (9).

In Fig 3 the normal micro stress $t_{rr}^{(m)}$, the micro shear stress $t_{r\theta}^{(m)}$, the stress $t_{\varphi\varphi}^{(i)}$ and $t_{\varphi\varphi}^{(m)}$, and the macro stress $\bar{t}_{33}$ are plotted versus the ratio $\mu^{(i)} = \mu^{(m)}$ for some values on a/d, $\nu^{(i)}$ and $\nu^{(m)}$.

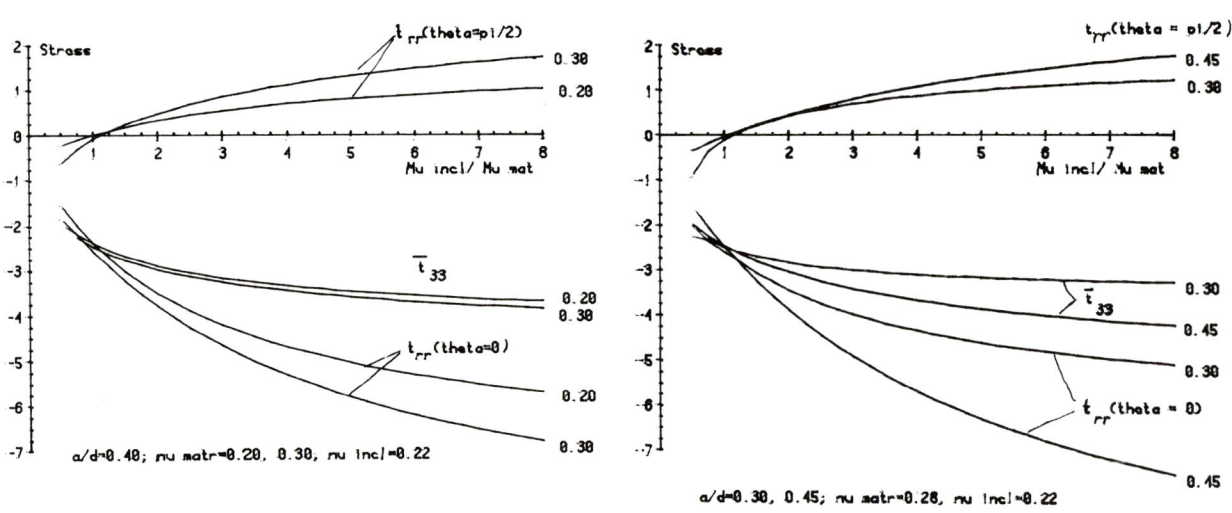

Figure 3a, b.

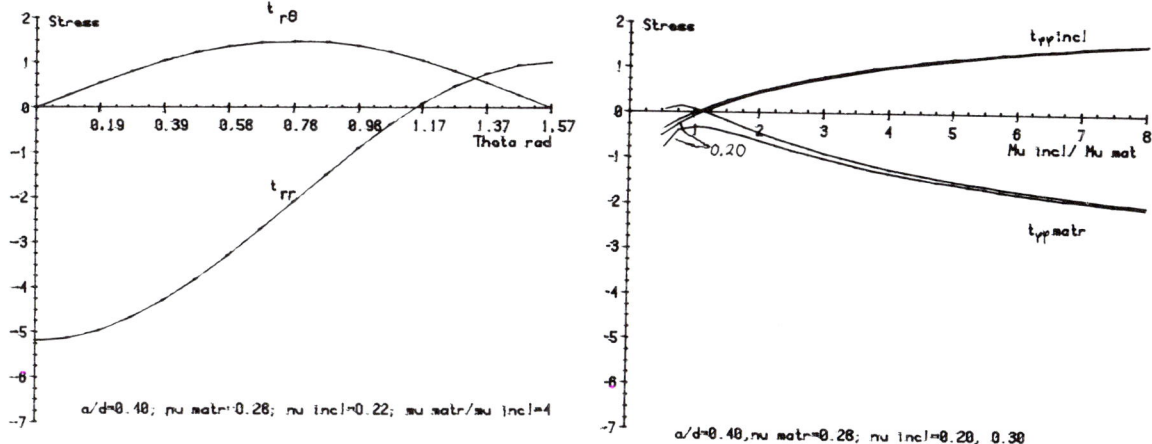

Figure 3c,d. In Figure 3 the macro deformation $\bar{\epsilon}_{33} = \epsilon$ is normalized to 1, i.e. the macro stress $\bar{t}_{33} = \bar{E}\bar{\epsilon}_{33} = \bar{E}$. In 3a and b the normal stress t_{rr} at the poles ($\theta = 0$) and at the equatorial zone ($\theta = \pi/2$) and the macro stress $\bar{t}_{33}$ is plotted versus $\mu^{(i)}/\mu^{(m)}$. In 3c the shear stress $t_{r\theta}$ and the normal stress t_{rr} are plotted as function of the angle θ. In 4d the stress $t_{\varphi\varphi}$ for $\theta = 0$ in both matrix and inclusion, is plotted versus $\mu^{(i)}/\mu^{(m)}$

5. DISCUSSION OF THE RESULTS

For the uniaxial compression here considered we see that if the inclusions are stiffer than the matrix of course we get considerable stress concentrations at the poles of the inclusions, i.e. för $\theta = 0$ and π. Dealing with materials such as concrete where the matrix consists of cement paste we have however to take the stress concentrations at the equatorial plane, $\theta = \pi/2$, into account, since in cement paste the tension strength is much lower than the compressive strength (Å.Grudemo [6]). In the equatorial plane it is seen that in a zone, $\theta = \pi/2 \pm \alpha$, where α can be of order $\pi/8$, the normal stress has changed sign from compression to tension. For $\theta = \pi/2$ the ratio of this tension stress divided by the compressive stress for $\theta = 0$ can be of the order $-1/5$. It is also seen that the tensite stress depends rather strongly on Poisson's ratio for the matrix. Thus the tensite stress concentrations in the equatorial zones can be reduced by reducing the Poisson's ratio.

An other case when the tensile stresses around the inclusions have to be taken into account is when the bounding between the inclusions and the matrix is weaker than the matrix and/or the inclusions. Then the bounds can be broken although the tensile stress is lower than the tensile strength in the matrix or in the inclusions.

If the inclusions are softer than the matrix the normal stress around the inclusions never changes sign and the zone around the midplane is never subject to tensile stresses. The critical region for fracture is then the polar zones of the interface, or perhaps the interior of the inclusions.

From Fig 2 it can be seen that the Poisson's ratio for the composite can be decreased by making the inclusions softer than the matrix. Now, consider a composite with inclusions of two types, 1, 2, embedded in a common matrix and where the diameter of type 1 is much smaller than the diameter of type 2. This composite can be regarded as 2-phase systems on succesive levels of accuracy of observation. On the higher level we consider inclusions of type 2 embedded in a matrix consisting of a composite with inclusions of type 1 in the original matrix.

If the diameters are sufficiently different the properties of these two, 2-phase composites can be investigated by means of our method. In doing so the derived macro moduli from the lover level are used as the micro elastic constants on the higher level. If now the inclusions of type 1 are softer than the original matrix we get a more favourable stress distribution around the inclusions of type 2, if these are assumed to be stiffer than the composite matrix. This simple illustration suggests the possibility of prescribing not only the elastic properties of such three phase composites, but also optimazing, to some extant, their strength properties. Further development of these possibilities is under investigation.

ACKNOWLEDGEMENT

The author is indebted to Professor Stig Hjalmars for guidance, and to him and Professor Olof Brulin for rewarding discussions. The work was financially supported by The Swedish National Board for Technical Development.

REFERENCES

[1] Beran, M. and Molyneux, J., Use of classical principles to determine bounds for the effective bulk modulus in heterogeneous media, Quart. Appl.Math. 24 (1966) pp. 107-118.

[2] Hashin, Z. and Strikman, S., A variational approach to the theory of the elastic behaviour of multiphase materials, J. Mech. Phys. Solids 11 (1963) pp. 127-140.

[3] Kröner, E., Bounds for effective elastic moduli of disordered materials, J. Mech. Phys. Solids 25 (1977) pp. 137-155.

[4] Achenbach, J. D., Generalized continuum theories for directionally, reinforced solids, Archives of Mechanics, 28 (1976) pp. 257-278.

[5] Berglund, K., The insignifficance of couple stresses in continuum models of granular composites, to be published in the Proceedings of the First Swedish-Polish Conference on the Theory of Microelastic Solids, Jablonna 1977. Also reported in TRITA-MEK-77-01, Technical reports from Department of Mechanics, Royal Institute of Technology, S-100 44 Stockholm, Sweden (obtainable on request from the department).

[6] Grudemo, Å., Strength vs. structure in cement pastes, CBI Reports 13:75. (Paper presented at Congress on the Chemistry of Cement, Moscow, September 23 to 27, 1974.)

A CALCULATION OF STATIC AND DYNAMIC ELASTICITY CONSTANTS IN AN ELASTIC INHOMOGENOUS MATERIAL

A. Bodare

Department of Solid Mechanics, Institute of Technology, Box 534,
S-751 21 Uppsala, Sweden

ABSTRACT

We show here that an inhomogenous material which consists of different perfectly elastic materials gives E_{dyn}/E_{stat} ratio which is equal or greater than unity.

1. INTRODUCTION

From a solid mechanics point of view the existence of a static Young's modulus and a dynamic Young's modulus is unsatisfactory. There are many reasons why this difference arise, [1]. The most common explanation is that the material is not perfectly elastic, but viscoelastic, so we have a Young's modulus which varies with the applied frequency. We will here pay attention to another explanation viz. that the material is inhomogenous but consists of two different but perfectly linear elastic materials. We restrict the calculation to a one-dimension rod.

2. ONE-DIMENSION ROD

We assume an one-dimension rod consisting of two different materials. The geometry, the elastic constants and the densities are given in fig. 1.

The effective static modulus for the rod is then

$$1/E_s = \alpha/E_1 + \beta/E_2, \quad (\alpha + \beta = 1) \tag{1}$$

where E_s is the static modulus and α is the volume fraction of material 1 and β is the volume fraction of material 2.

If we suppose that c_1 and c_2 is the bar-velocities of the two different materials we have

$$c_1 = \sqrt{E_1/\rho_1} \ ; \quad c_2 = \sqrt{E_2/\rho_2} \tag{2}$$

If we define the effective bar-velocity for the whole rod as the total length divided with the total travel time for a wave to pass the rod and write this as

$$c = \sqrt{E_d/\rho} \ ; \quad (\rho = \alpha\rho_1 + \beta\rho_2) \tag{3}$$

in conformity to eq. (2), we get:

$$1/E_d = \left[\alpha'/\sqrt{E_1} + \beta'/\sqrt{E_2}\right]^2 \tag{4}$$

where

$$\alpha' = \alpha\sqrt{\rho_1/\rho} \ ; \quad \beta' = \beta\sqrt{\rho_2/\rho} \tag{5}$$

E_d is the dynamic modulus and ρ is the mean density of the bar.

The ratio E_d/E_s can now easily be calculated from eq. (4) and (1). In fig. 2 we give this ratio as a function of α when $\rho_1 = \rho_2 = \rho$. For curve 1 $E_1/E_2 = 10$ for curve 2 $E_1/E_2 = 100$ and for curve 3 $E_1/E_2 = 1000$. We see from these curves that the maximum for E_d/E_s is attained for relatively large values of α.

This means that if there is a small amount of a very soft material in, for instance, a rock sample, we can expect to have a high value of the E_d/E_s ratio.

REFERENCE

[1] Dvorak, A., Seismic and Static Modulus of Rock Masses. Proc. 2nd Congress. Int. Soc. Rock Mech. Beograd 1970 pp. 313-317.

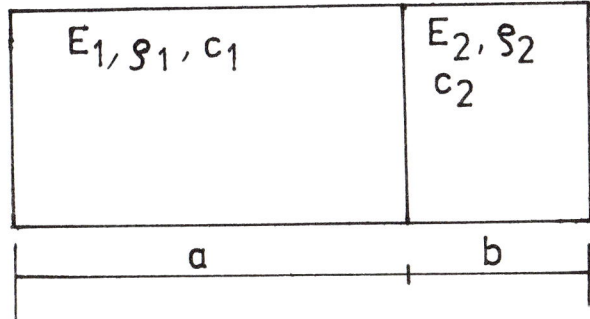

Figure 1   Geometry of the rod.

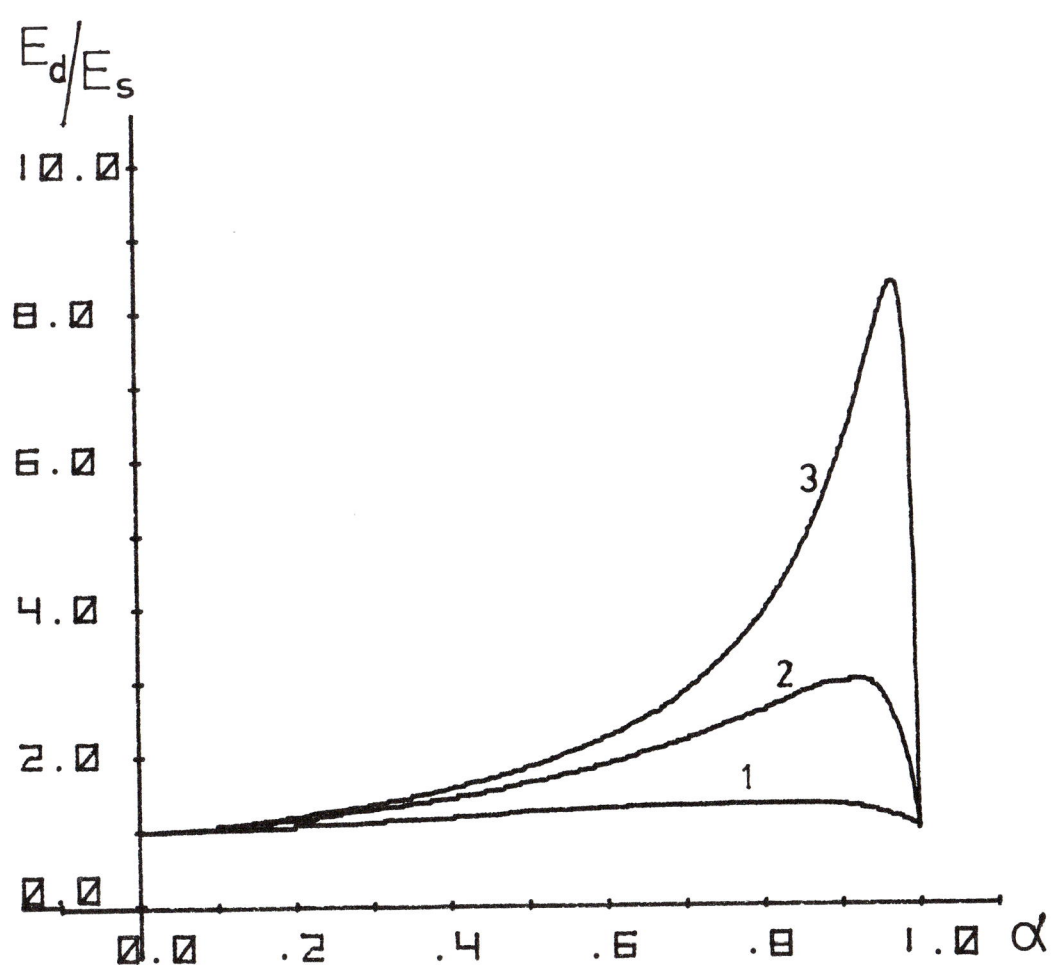

Figure 2 E_d/E_s as a function of volume ratio α
curve 1: $E_1/E_2 = 10$
curve 2: $E_1/E_2 = 100$
curve 3: $E_1/E_2 = 1000$

A COOPERATIVE MODEL OF STRESS RELAXATION KINETICS IN SOLIDS

J. Kubát, M. Rigdahl and R. Seldén

Department of Polymeric Materials, Chalmers University of Technology, S-402 20 Gothenburg, Sweden

ABSTRACT

Creep and stress relaxation phenomena in solids are usually interpreted in terms of the theory of stress-aided thermal activation. The activation volume, a key parameter of this theory, is found to be related to the stress in a unique way irrespective of the microstructure of the solid. This must be considered as a serious weakness of this theory. In this work a simple cooperative model is proposed producing the same exponential type relation between the flow rate and stress as the activation theory. However, the parameters of the model comply in a more natural way with experimental evidence. The model used is reminiscent of that used for calculating equilibria in systems obeying Bose-Einstein statistics. This type of statistics seems to represent the simplest way of describing a system of interacting flow units tending to undergo coupled transitions when activated.

1. INTRODUCTION

For many years, the time dependence of mechanical properties of solids has been an important problem for both engineers and materials scientists, and several theories have been formulated to describe the time dependent response of materials to varying conditions of stress or strain. Despite all this effort, a closer look at the state of the theoretical framework shows some serious flaws. This is, not the least, also true of such commonly accepted concepts as, for instance, the theory of stress-aided thermal activation.

The present paper intends to expose some of the weaknesses of the present theoretical picture of the physical mechanisms supposed to govern the flow of solid materials. An especially simple way to do this is to consider the kinetics of stress relaxation in samples subject to uniaxial deformation. As shown below, stress relaxation curves obtained with various solids, be it polymers, be it metals, show a far going similarity with regard to their shape in stress-log time plots, when correction is made for internal stresses. The physical implications of this similarity cannot be accomodated within the usual theoretical framework. Starting from this experimental fact we present a novel statistical approach to solid state flow based on a simple cooperative mechanism between the structural entities contributing to the macroscopic flow. Even though, for simplicity, we confine ourselves to stress relaxation in uniaxial extension, the results obtained appear to be applicable to other instances of flow as well (creep, grain boundary sliding etc.). The key feature of the statistical model to be presented is its insensitivity towards structural details of the flowing solid.

The similarity of the stress relaxation process as recorded with solids of widely differing structure and chemical composition is exemplified in Fig. 1 for cadmium (crystalline material), polyethylene (semicrystalline), polyisobutylene (amorphous, rubber-like) and cetylalcohol (crystalline). For most materials the stress approaches a non-zero level after a comparatively long time (σ_∞). In Fig. 1 only the dissipated stress, i.e. $\sigma - \sigma_\infty$ is shown. The equilibrium stress σ_∞ is equivalent to the internal stress level σ_i of the specimen. The similarity of the experimental curves is rather striking, especially when the difference in structure of these four materials are considered. This similarity in flow kinetics can be quantified using the maximum (inflexion) slope of the $\sigma(\ln t)$-curve. This slope, denoted F, is related to the total dissipated stress as [1]

$$F = (0.1 \pm 0.01)(\sigma_0 - \sigma_\infty) \tag{1}$$

where σ_0 is the initial stress of the experiments. An equivalent statement is that the stress-log time curves, when approximated by straight lines, cover about four decades of time [1].

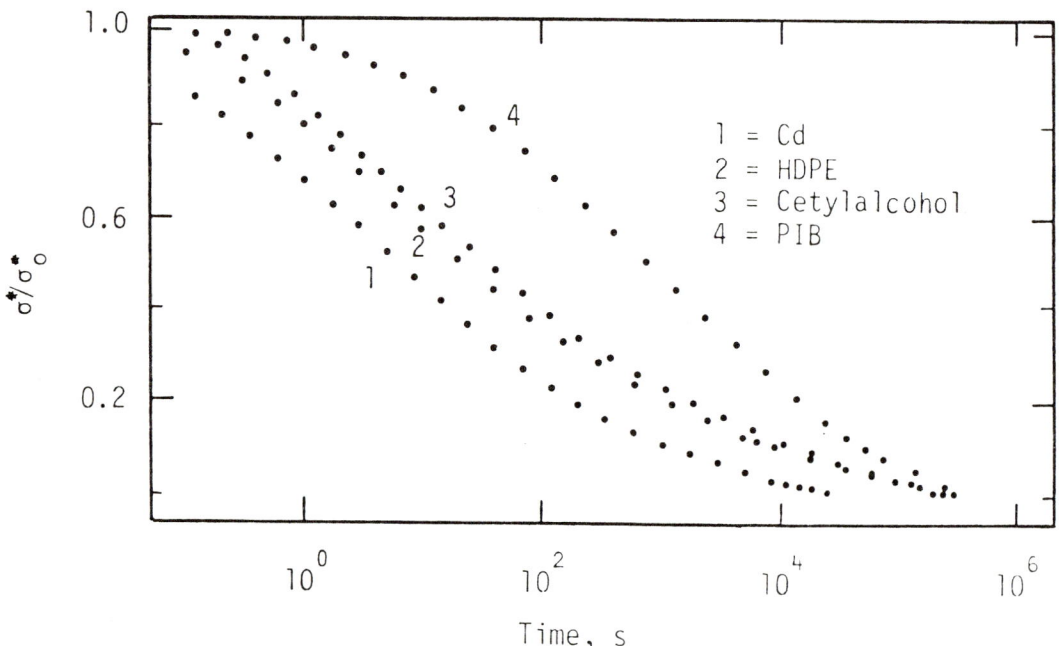

Figure 1. Stress relaxation curves for cadmium (Cd), high density polyethylene (HDPE), cetyl alcohol, and polyisobutylene (PIB). The stress concept used is the effective stress $\sigma^* = \sigma - \sigma_\infty$.

2. CURRENT THEORETICAL CONCEPTS

Basically, there are two main lines of approach in interpreting solid state flow, namely the spectral theory used in polymer physics, and the stress aided thermal activation concept used for crystalline solids.

The spectral theory, which is based on linear viscoelasticity, relates for stress relaxation the stress, strain, and time as [2].

$$\sigma(t) = \varepsilon_0 \int H(\tau) e^{-t/\tau} d\ln \tau \qquad (2)$$

Here, ε_0 is the initially applied constant strain and $H(\tau)$ the distribution of relaxation times τ. The function $H(\tau)$ is determined from experiments. Due to the formal character of the theory, it cannot provide any direct information on the mechanisms underlying eq. (1). Furthermore, the shape of the relaxation curves is rather insensitive to variations in $H(\tau)$, thus limiting the information to be extracted from experimental data [3].

The theory of stress-aided thermal activation, which is a non-linear theory, gives the following relation between stress rate ($\dot{\sigma} = d\sigma/dt$) and stress for stress relaxation [4]

$$\dot{\sigma} = A \exp(v\sigma^*/kT) \qquad (3)$$

In eq. (3), A is a pre-exponential factor including an activation energy for the process, v the activation volume, σ^* the effective stress (i.e. the difference between σ and σ_∞), k Boltzmann's constant and T the absolute temperature. The value of the activation volume is usually associated with the microstructure of the material being studied. On the other hand, combining eqs. (1) and (3) gives [1]

$$v(\sigma_0 - \sigma_\infty) \approx 10 \, kT, \qquad (4)$$

since $v = kT/F$. This relation, verified by experiments for a large number of materials [1,3], raises serious doubts about the physical significance of the activation volume. A relation similar to eq. (4) has also been reported for creep [1,5,6].

The above results strongly support the notion that the kinetics of stress relaxation may not be determined by the mobility and other physical details of the micro-units constituting the macroscopic process of flow, but rather by general interaction effects between such units. The nature of the flow units—they may be dislocations, macromolecular segments etc.—seems to play a minor role only.

3. A COOPERATIVE MODEL FOR STRESS RELAXATION

In this section, we describe a cooperative model based on the assumption that a flow unit, undergoing a successful activation in the usual sense, may induce transitions of any number of other flow units. The occurrence of multiple transitions is the main feature of this model.

Consider a volume element of a relaxing solid containing a number of flow units (mobile entities) like dislocation, segments of macromolecules etc. Each flow unit is assumed to be characterized by an average relaxation time τ. If coupling effects between the flow units were excluded the stress would decay according to

$$\sigma = \sigma_o^* \, e^{-t/\tau} \tag{5}$$

where σ_o^* is the total dissipated stress. Eq. (5) corresponds to the familiar Maxwellian behaviour, cf. also Fig. 2.

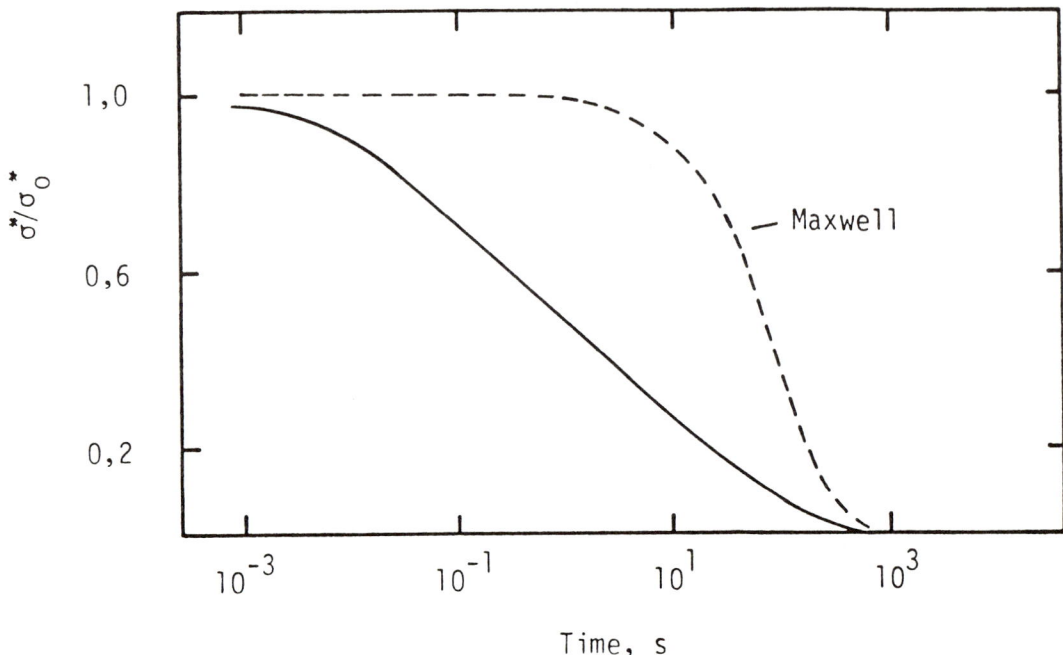

Figure 2. Theoretical stress relaxation curves according to the Maxwell model, eq. (5), and the cooperative model, eq. (8) (solid line). The relaxation time τ was chosen to 10^2 s.

We now assume that the stress rate, $\dot{\sigma}$, is proportional to the number of units undergoing a successful activation (n) at a given time t, i.e.

$$\dot{\sigma} = B n \tag{6}$$

where B is a constant. The flow units do, however, not act independently of each other but are coupled in doubles, triples etc. The corresponding relaxation time is then $\tau/2$ for a double, $\tau/3$ for a triple and, in general, for a s-multiple τ/s. It is a rather straight forward calculation to show that, for this model, the stress rate is given by [7]

$$\dot{\sigma} = - b \sum_{s=1}^{s_{max}} \lambda^{-s} \exp(-skt) \tag{7}$$

where b and λ are constants, $k = 1/\tau$ and s_{max} is the maximum multiplicity of the processes triggered by a single thermal activation. From experiments, cf. eq. (1), follows that the value of s_m is comparatively large; for t > 0 the summation in eq. (7) can thus proceed to infinity without any significant error giving

$$\dot{\sigma} = - b(\lambda e^{t/\tau} - 1)^{-1} \tag{8}$$

The stress rate can also be represented as a function of stress as

$$\dot{\sigma} = -b(e^{\sigma^*/F} - 1) \tag{9}$$

with λ contained in σ_o^*.

The parameter F, which is related to τ through

$$F = b\tau \tag{10}$$

is equal to the maximum slope of the $\sigma(\ln t)$-curve, cf. eq. (1).

Eq. (9) has the same form as the relaxation law stemming from the theory of stress-aided thermal activation, eq. (3), although the physical background is entirely different. For instance, the activation volume concept is here replaced by the parameter F (or the relaxation time τ).

From experiments we know that $F \approx 0.1 \, \sigma_o^*$. Using this value in eq. (9) results in a relaxation curve covering a much broader log time range than a relaxation of the Maxwellian type as evident from Fig. 2.

4. SOME PROPERTIES OF THE COOPERATIVE RELAXATION MODEL

We now consider some details concerning the relationship between λ, s_{max}, F and τ. For convenience we introduce two new parameters α and ε related to λ as

$$\lambda = e^\alpha = 1 + \varepsilon \tag{11}$$

since we know that ε is small (of the order $10^{-5} - 10^{-4}$)

Integrating eq. (7) gives

$$\sigma^* = F \sum_{s=1}^{s_{max}} \frac{e^{-\alpha s} e^{-skt}}{s} \tag{12}$$

since $F = b\tau$. For $t = 0$ the initial effective stress then is obtained as

$$\sigma_o^* \approx F \ln(s_{max} + \gamma) \tag{13}$$

where γ is Euler's constant (= 0.5772...). Using the experimental result $F = 0.1 \, \sigma_o^*$ it is from eq. (13) found that the maximum multiplicity number s_{max} is of the order 10^4. To get a fair approximation of the relation between the relaxation time and s_{max} we proceed as follows.

Integration of eq. (8) yields to a first approximation [7]

$$\sigma^* = F \ln \tau - F \ln(t + 1) \tag{14}$$

Here $\sigma_o^* = F \ln \tau$, at least approximately, and since τ and s_{max} are large in comparison with γ we obtain

$$\ln s_{max} \approx \ln \tau \qquad (15)$$

Here $\log \tau$ is defined as the difference between $\log \tau_{max}$ and $\log \tau_{min}$; τ is then a reduced time, cf. also Fig. 3.

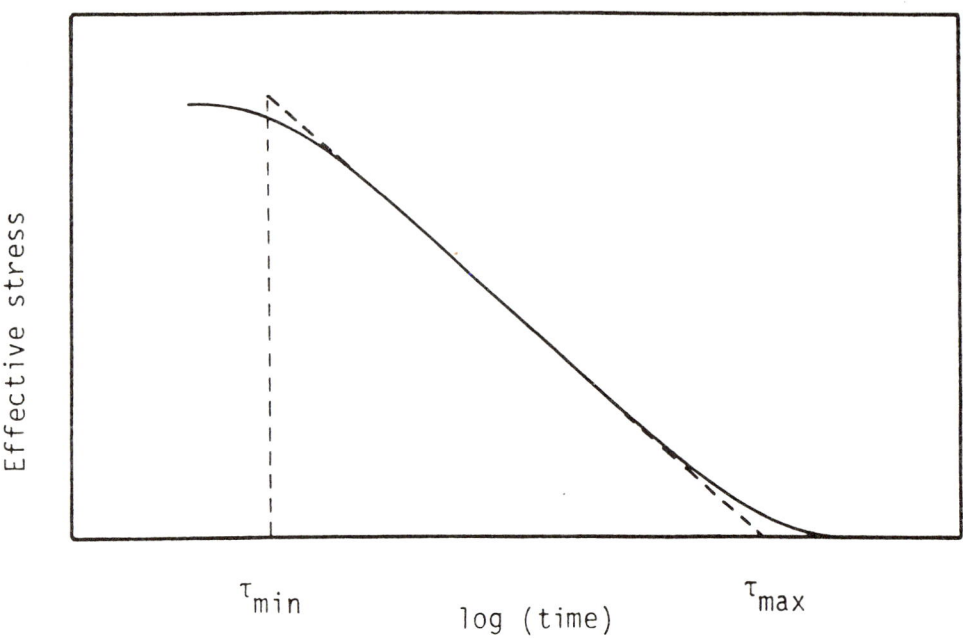

Figure 3. Definition of reduced relaxation time (τ) as $\log \tau = \log \tau_{max} - \log \tau_{min}$.

From eqs. (8) and (15) now follows

$$\exp(\alpha) = (1 + 1/\tau) = \exp(1/\tau) = \exp(1/s_{max}) = (1 + 1/s_{max}) \qquad (16)$$

The relaxation time τ is here of the order 10^4 which is in good agreement with experiments.

So far we have only discussed minimum (τ_{min}) and maximum relaxation times (τ_{max}). It is now of interest to define an average relaxation time characterizing the main relaxation behaviour. One way of doing this may be as follows.

The area under a $\sigma(t)$-curve can be calculated as

$$\sigma_o^* \bar{\tau} = \int_0^\infty \dot\sigma \, t \, dt \qquad (17)$$

where $\bar{\tau}$ is an average relaxation time. Using eq. (8) we then obtain

$$\sigma_o^* \bar{\tau} = \frac{\pi^2}{6} F\tau \qquad (18)$$

Since $F \approx 0.1 \, \sigma_o^*$, the average relaxation is related to the (maximum) relaxation time as

$$\bar{\tau} = \tau/6 \qquad (19)$$

It may be worthwhile to note that according to this analysis stress relaxation in solids can be regarded as being represented by a number of relaxators which due to interaction can relax with the relaxation times τ, $\tau/2$, $\tau/3$, ..., respectively. This is analogous to the procedure used by Leibfried [8] where the Bose-Einstein distribution is treated as a result of coupling of Boltzmann factors. There is, at present, no plausible explanation to the fact that the average relaxation time $\bar{\tau}$ is 1/6 of the time characterizing the transition of single flow units, even though it could be tempting to resort to a speculation starting from six degrees of freedom of an element embedded in a solid matrix. As a matter of fact, a distribution of the type shown in eq. (7) could then be easily obtained on the assumption that the various degrees of freedom are distributed, with the bounds given by eqs. (7) and (18), on the successful activations of such elements. This would then produce a quantitative agreement with the kinetics as recorded experimentally.

5. FINAL REMARKS

The formula derived in this paper, eq. (9) is formally similar to the exponential law, eq. (3), based on the theory of stress-aided thermal activation. Contrary to the latter, however, it does not contain the notion of an activation volume which, as ample experimental evidence shows, depends only on the initial effective stress and not on the structural details of the flowing solid. Needless to say, the concept of an activation volume is used here in a formal sense only, since there appears to be no direct proof of the activation energy being influenced by the stress in a relaxation experiment. On the other hand the present theory is not in conflict with the theory of stress-aided thermal activation, cf. eq. (20) below.

The fact that the exponential law, eq. (3), describes a substantial part of normally-observed relaxation processes certainly is no proof of the validity of the theory of stress-aided activation. As shown in this paper, an exponential law may be obtained on grounds of a physically entirely different assumption. On the other hand, this does not mean that external stress cannot influence the rate of flow. In fact, this apparent contradiction can be resolved in the following way. Let us consider the following equation

$$\dot{\sigma} = - D \exp[-(U - v\sigma_0^*)/kT][\exp(\sigma^*/F) - 1] \qquad (20)$$

where D is a constant and U the activation energy. Such a formula has the advantage of accomodating both the mechanisms in question. Here the role of the stress as a factor reducing the activation energy, thereby increasing the rate of flow, is confined to shifting the relaxation curves along the log time-axis. One may note, that this shift is related to the value of σ_0^*, i.e. the initial effective stress. The corresponding term in eq. (20) is $\exp[-(U - v\sigma_0^*)/kT]$. On the other hand, the slope of the $\sigma(\log t)$-curves is determined solely by the cooperation effect described above. It may be remarked that the exponential character of the term $[\exp(\sigma^*/F) - 1]$ is apparent only. Due to eq. (1), the slope of the $\sigma(\log t)$-curves, normalized with respect to σ_o^*, is constant and independent of the nature of the solid under study, (cf. eq. (9)). As eq. (1) shows, the initial value of the term $\exp(\sigma^*/F)$ is $\exp(10)$. The well documented linearity of the relaxation process, as reflected for instance in the linear variation of the slope of the $\sigma(\log t)$-curves with σ_o^* cannot be contained in this term. Instead, it is easy to show that this linearity is contained in the parameter D, which is proportional to σ_o^*.

In summarizing, eq. (20) keeps apart the position of the $\sigma(\log t)$-curves along the log t-axis from the slope of such curves. The position, being given by a rate determining factor, is assumed to be related to the classical notion of stress-aided thermal activation, whereas the slope $d\sigma/d \log t$ is interpreted in terms of the novel cooperative model. The linearity of the process with respect to σ_o^* is contained in the pre-exponential factor. Even though similar formulae have been formulated in the past [3,4], the invariance of the shape of the $\sigma(\log t)$ curves, as expressed by eq. (1) and illustrated in Fig. 1, has remained unexplained.

REFERENCES

[1] Kubát, J. and M. Rigdahl, Activation volumes for flow processes in solids. Mater. Sci. Eng. 24 (1976) 223-232.

[2] Ward, I. M., Mechanical properties of solid polymers. Wiley, London 1971 p. 94.

[3] Kubát, J., A similarity in the stress relaxation behaviour of high polymers and metals. Diss. Stockholm University 1965 p. 27.

[4] Kambour, R. P. and R. E. Robertson, Polymer Science, (Ed. A. D. Jenkins). North-Holland Publishing Company, London 1972 p. 772.

[5] Balasubramanian, N. and J. C. M. Li, The activation areas for creep deformation. J. Mater. Sci. 5 (1970) 434-444.

[6] Struik, L. C. E., Proc. VIIth Int. Congr. Rheology. Gothenburg 1976 p. 134

[7] Kubát, J. and M. Rigdahl, to be published.

[8] Leibfried, G. and F. Kaempffer, Über ein einfaches Verfahren zur Berechnung von Zustandssummen. Z. Physik 124 (1947-48) 441-449.

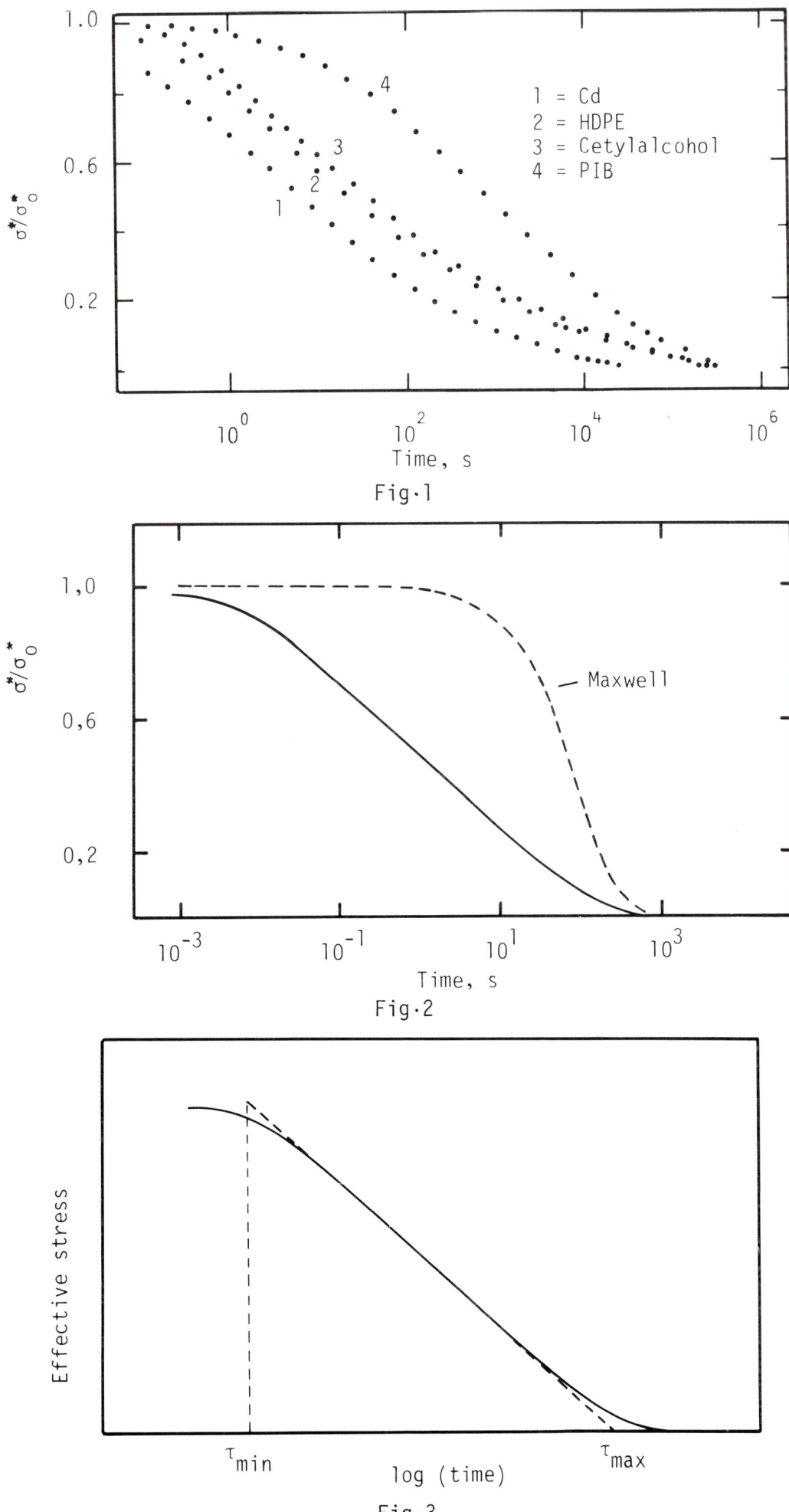

Fig.1

Fig.2

Fig.3

DAMAGE IN A PRESSURE TRANSDUCER DUE TO A HIGH SPEED LOAD

G. L. Anderson

Institut CERAC SA, CH-1024 Ecublens, Switzerland

ABSTRACT

An attempt is made to explain the failure in a longitudinal splitting mode of the quartz elements of a pressure transducer subjected to a high speed pressure front. The quartz elements are treated as a pair of parallel slender elastic cantilevers that are connected at their free tips by a rigid platform. The pressure front translates at constant speed along the length of the platform, thereby inducing coupled longitudinal and flexural deformations in the cantilevers. The motion of the system is described by four linear partial differential equations with appropriate boundary and initial conditions. The forced motion problem is solved by the method of separation of variables, modified to account for the presence of time derivatives and time-dependent loads in the boundary conditions. Numerical calculations reveal that rapidly moving pressure fronts can produce elevated compressive stresses in the quartz bars.

1. INTRODUCTION

The present study was undertaken as a result of a series of pressure transducer failures that occurred at pressures believed to be less than the maximum tolerable pressure claimed by the manufacturer. When the transducer was mounted such that a pressure front impinged normally onto its active face, no failures occurred if pressures were less than the manufacturer's critical value. However, when a pressure front passed over and moved parallel to the face of the device, failures occurred as the front's speed became sufficiently great even though the pressure levels were less than the critical value. Examination of the quartz components of the transducer revealed that one or more of them had been broken in a longitudinal splitting mode.

The objective of this study is to devise a simple mathematical model of the transducer and then to determine how the stresses in the quartz elements depend upon the speed and intensity of the pressure front. The transducer is assumed to consist, in essence, of a pair of parallel, slender eleastic (quartz) cantilevers that are connected at their free ends by a rigid platform. Coupled flexural and extensional deflections in a plane are considered, these deformations being described by the elementary theories of extension and flexure.

2. THE EQUATIONS OF MOTION

Consider a pair of identical elastic cantilevers of length ℓ, density ρ, and Young's modulus E joined at their tips by a rigid platform of mass m, length ℓ_0, thickness 2c, and depth h, as shown in Figure 1. An external force per unit area, $\underset{\sim}{F}$, acts on the upper surface of the platform. The bars undergo flexural and extensional deformations $w_s(x_1,t)$ and $u_s(x_1,t)$, respectively, s = 1,2, in the plane of the figure, where t denotes time and x_1 the axial coordinate. Each bar has the moment of inertia I, whereas the moment of inertia of the platform is

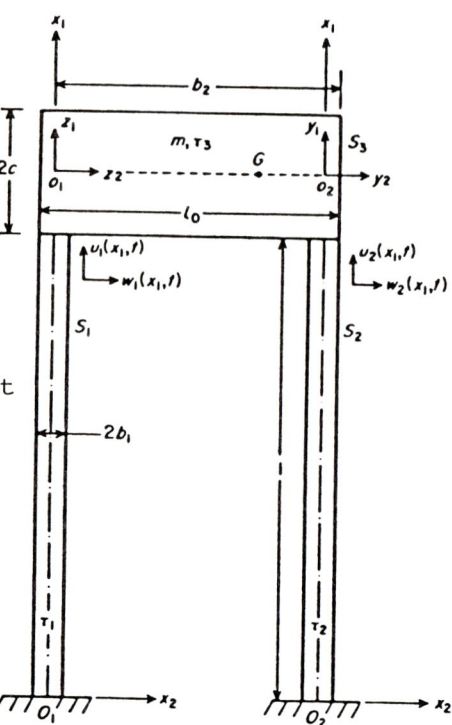

Figure 1. Co-ordinate systems and dimensions.

$$J = \int_{\tau_3} (z_1^2 + z_1^2) \, dm = m(4c^2 + 12r_1^2 + \ell_0^2)/12,$$

where $r_1 = (b_2 - b_1)/2$, b_1 and b_2 being the distances from o_1 to the left and right walls, respectively, of the platform. For the external force applied on the upper surface of the platform, we assume that
$\underset{\sim}{F} = -F(z_2,t)\underset{\sim}{i}$ and that the pressure front appears at t = 0 at $z_2 = -b_1$, moving at uniform velocity v in the positive z_2-direction. This front traverses the length of the platform, $\ell_0 = b_1 + b_2$, in time $t_1 = \ell_0/v$, leaving behind a constant pressure p at every point of the surface, i.e.,

$$F(z_2,t) = \begin{cases} p[H(z_2+b_1) - H(z_2+b_1-vt)], & \text{if } 0 < t < t_1, \\ p, & \text{if } t_1 < t, \end{cases}$$

where H(z) denotes the Heaviside step function. We can show that the rate of work done by this force is

$$\int_S \underset{\sim}{\dot{r}} \cdot \underset{\sim}{F} \, dS = -G_1(t)\dot{u}_1(\ell,t) + G_2(t)\dot{w}_{1,1}(\ell,t),$$

where, with $A_0 = h\ell_0$,

$$G_1(t) = phvt[1 - H(t-t_1)] + pA_0 H(t-t_1),$$

$$G_2(t) = \tfrac{1}{2}phvt(vt-2b_1)[1 - H(t-t_1)] + pr_1 A_0 H(t-t_1).$$

(1)

The equations of motion and boundary conditions can be derived from the principle of virtual work. If we introduce the following definitions:

$$x_1 = \ell x, \quad t = \ell \tau (\rho/E)^{\frac{1}{2}}, \quad u_1(x_1,t) = \ell \tilde{u}_1(x,\tau), \text{ etc.},$$

$$g_s(\tau) = G_s(t)/EA\ell^{s-1}, \quad \mu = m/\rho A\ell, \quad r = r_1/\ell, \quad \alpha^2 = I/A\ell^2,$$

$$\xi = c/\ell, \quad a = a_o/\ell, \quad a_o^2 = (4c^2 + 12r_1^2 + \ell_o^2)/12, \quad \sigma = pA_o/EA,$$

$$\tau_1 = \ell_o v_1/\ell v, \quad v_1 = (E/\rho)^{\frac{1}{2}}, \quad \eta = \ell_o/\ell, \quad \gamma = 2b_1/\ell_o,$$

(2)

we find (neglecting the tildes on the dimensionless deflections) that the dimensionless forms of the fundamental equations are

$$u_s''(x,\tau) = \ddot{u}_s(x,\tau), \quad \alpha^2 w_s^{IV}(x,\tau) + \ddot{w}_s(x,\tau) = 0, \quad 0 < x < 1, \quad 0 < \tau, \quad (3)$$

$$u_s(0,\tau) = w_s(0,\tau) = w_s'(0,\tau) = 0 \quad (4)$$

$$u_1'(1,\tau) + u_2'(1,\tau) + \mu \ddot{u}_1(1,\tau) - \mu r \ddot{w}_1'(1,\tau) = -g_1(\tau), \quad (5)$$

$$\alpha^2 w_1''(1,\tau) + \alpha^2 w_2''(1,\tau) - 2r u_2'(1,\tau) + \mu(\xi^2 + a^2)\ddot{w}_1'(1,\tau) +$$

$$+ \mu \xi \ddot{w}_1(1,\tau) - \mu r \ddot{u}_1(1,\tau) = g_2(\tau), \quad (6)$$

$$\alpha^2 w_1'''(1,\tau) + \alpha^2 w_2'''(1,\tau) - \mu \ddot{w}_1(1,\tau) - \mu \xi \ddot{w}_1'(1,\tau) = 0, \quad (7)$$

$$w_1(1,\tau) = w_2(1,\tau), \quad w_1'(1,\tau) = w_2'(1,\tau), \quad (8)$$

$$u_1(1,\tau) - u_2(1,\tau) = 2r w_1'(1,\tau), \quad (9)$$

where $w' = \partial w/\partial x$, $\dot{w} = \partial w/\partial \tau$, etc., and

$$g_1(\tau) = \sigma(\tau/\tau_1)[1 - H(\tau - \tau_1)] + \sigma H(\tau - \tau_1),$$

$$g_2(\tau) = \tfrac{1}{2}\sigma\eta(\tau/\tau_1)(\tau/\tau_1 - \gamma)[1 - H(\tau - \tau_1)] + \sigma r H(\tau - \tau_1).$$

The initial conditions are assumed to be

$$u_s(x,0) = \dot{u}_s(x,0) = w_s(x,0) = \dot{w}_s(x,0) = 0. \quad (10)$$

3. EIGENVALUES AND EIGENFUNCTIONS

We now write

$$u_s(x,\tau) = u_s(x)\cos \omega\tau, \quad w_s(x,\tau) = w_s(x)\cos \omega\tau, \quad (11)$$

where ω denotes a dimensionless frequency. Substitution of (11) into (3) to (9) with $g_s(\tau) = 0$ yields an eigenvalue problem from which the eigenvalues and eigenfunctions of the system can be derived. The details of this process are given in Ref. [1]; here we present only the results. The general frequency equation is factorable and leads to two independent frequency equations, namely,

$$2 \cot \tilde{\omega} = \mu\tilde{\omega}, \quad (12)$$

$$r^2\lambda(2\cos\omega + \mu\omega\sin\omega)[2(\sin\lambda\cosh\lambda + \cos\lambda\sinh\lambda) - \mu\lambda(1-\cos\lambda\cosh\lambda)]$$

$$+ \alpha\sin\omega[4(1+\cos\lambda\cosh\lambda) - 2\mu\lambda(\sin\lambda\cosh\lambda - \cos\lambda\sinh\lambda) - 4\mu\xi\lambda^2\sin\lambda\sinh\lambda$$

$$- 2\mu\lambda^3(\xi^2+a^2)(\sin\lambda\cosh\lambda + \cos\lambda\sinh\lambda) + (\mu a\lambda^2)^2(1-\cos\lambda\cosh\lambda)] = 0, \quad (13)$$

where $\lambda^2 = \omega/\alpha$. Eq. (12) is the frequency equation for the system depicted in Figure 1 when it undergoes longitudinal deformation only. Eq. (13) provides the frequencies associated with the coupled flexural-extensional deformations, in which the extensions in the individual cantilevers are unequal due to bending. The eigenfunctions associated with (12) are found to be

$$\tilde{u}_{sn}(x) = \sin \tilde{\omega}_n x, \quad (14)$$

whereas those associated with (13) are

$$u_{sn}(x) = (-1)^{s-1} B_n \sin \omega_n x, \quad (15)$$

$$w_{sn}(x) = w_n(x) = C_n(\cos\lambda_n x - \cosh\lambda_n x) + D_n(\sin\lambda_n x - \sinh\lambda_n x), \quad (16)$$

where

$$B_n = -2r\lambda_n \csc\omega_n[\mu\omega_n(1-\cos\lambda_n\cosh\lambda_n) - 2\alpha\lambda_n(\cos\lambda_n\sinh\lambda_n + \sin\lambda_n\cosh\lambda_n)],$$

$$C_n = 2\alpha\lambda_n(\cos\lambda_n + \cosh\lambda_n) + \mu\omega_n(\sin\lambda_n - \sinh\lambda_n) +$$
$$+ \mu\xi\omega_n\lambda_n(\cos\lambda_n - \cosh\lambda_n),$$

$$D_n = -2\alpha\lambda_n(\sin\lambda_n - \sinh\lambda_n) - \mu\omega_n(\cos\lambda_n - \cosh\lambda_n) +$$
$$+ \mu\xi\lambda_n\omega_n(\sin\lambda_n + \sinh\lambda_n).$$

4. SOLUTION OF THE FORCED MOTION PROBLEM

Regarding the boundary conditions in (3) to (9), we observe that two are non-homogeneous and three contain time derivatives. In view of this and the existence of two independent sets of eigenfunctions in (14) to (16), we represent the deflections as follows (see Ref. [2]):

$$u_s(x,\tau) = -\tfrac{1}{4}x^3(2-x)g_1(x) + \sum_{n=1}^{\infty}\tilde{u}_{sn}(x)q_n(\tau) + \sum_{n=1}^{\infty}u_{sn}(x)Q_n(\tau), \qquad (17)$$

$$w(x,\tau) = \varphi(x)T(\tau) + \sum_{n=1}^{\infty}w_n(x)Q_n(\tau), \qquad (18)$$

where

$$\varphi(x) = -(x^2/30)(1-x)^2(70x^5 - 187x^4 + 108x^3 + 53x^2 - 2x - 57),$$

$$T(\tau) = (\sigma\eta/4\alpha^2)(\tau/\tau_1)(\tau/\tau_1 - 1)[1 - H(\tau-\tau_1)].$$

Making use of the orthogonality properties of the eigenfunctions (see Ref. [1]), we find the functions $q_n(\tau)$ and $Q_n(\tau)$ to be given by

$$q_n(\tau) = \begin{cases} (\sigma/\tau_1\tilde{\omega}_n^3)[K_n\tilde{\omega}_n\tau + (L_n\tilde{\omega}_n^2 - K_n)\sin\tilde{\omega}_n\tau], & 0 < \tau < \tau_1, \\ (\sigma/\tau_1\tilde{\omega}_n^3)\{K_n\tilde{\omega}_n\tau_1 + (L_n\tilde{\omega}_n^2 - K_n)[\sin\tilde{\omega}_n\tau - \sin\tilde{\omega}_n(\tau-\tau_1)]\}, & \tau_1 < \tau, \end{cases} \qquad (19)$$

$$Q_n(\tau) = \frac{\sigma\eta(\alpha^2 J_n - \omega_n^2 I_n)}{(\alpha\tau_1 N_n\omega_n^2)^2}[1 - \cos\omega_n\tau - (\omega_n\tau_1/2)\sin\omega_n\tau] +$$
$$+ \frac{\sigma\eta J_n}{2(\omega_n N_n)^2}(\tau/\tau_1)(1-\tau/\tau_1), \quad \text{for} \quad 0 < \tau < \tau_1, \qquad (20)$$

$$Q_n(\tau) = \frac{\sigma\eta(\alpha^2 J_n - \omega_n^2 I_n)}{(\alpha\tau_1 N_n \omega_n^2)^2} \{\cos \omega_n(\tau-\tau_1) - \cos \omega_n \tau$$

$$- (\omega_n \tau_1/2)[\sin \omega_n \tau + \sin \omega_n(\tau-\tau_1)]\}, \quad \text{for} \quad \tau_1 < \tau, \tag{20}$$

with

$$K_n = 6[(1+\mu)\cos \tilde{\omega}_n - \mu]/(\mu\tilde{\omega}_n^3 \tilde{N}_n^2),$$

$$L_n = \{[6(1+\mu) + \tilde{\omega}_n^2]\cos \tilde{\omega}_n - 6\mu\}/(\mu\tilde{\omega}_n^5 \tilde{N}_n^2),$$

$$\tilde{N}_n^2 = \tfrac{1}{2}(1+\tfrac{1}{2}\mu \sin^2 \tilde{\omega}_n),$$

$$J_n = \int_0^1 \varphi^{IV}(x) w_n(x)\,dx, \quad I_n = \int_0^1 \varphi(x) w_n(x)\,dx,$$

$$N_n^2 = \text{see Ref. [1]}.$$

Since the longitudinal and transverse deflections of the system are now completely determined, we may compute the axial and shear stresses from

$$\sigma_{11}^{(s)} = E[N_s(x,\tau) - \alpha\sqrt{3}yM(x,\tau)], \quad \sigma_{12} = \tfrac{3}{2} E\alpha^2(1-y^2)V(x,\tau),$$

respectively, where bars of square cross section of side $2b_1$ have been assumed, $y = x_2/b_1$, and

$$N_s(x,\tau) = u_s'(x,\tau), \quad M(x,\tau) = w''(x,\tau), \quad V(x,\tau) = w'''(x,\tau)$$

are the axial forces, bending moment, and transverse shear force, respectively. Due to the linear dependence of $\sigma_{11}^{(s)}$ on y, the greatest axial stresses are to be found on the left and right faces of the vertical crystals. Dimensionless forms of these stresses may be defined as

$$S_s(x,\tau) = \sigma_{11}^{(s)}/E\,\big|_{y=-1} = N_s(x,\tau) + \alpha\sqrt{3}\, M(x,\tau),$$

$$S_s^*(x,\tau) = \sigma_{11}^{(s)}/E\,\big|_{y=+1} = N_s(x,\tau) - \alpha\sqrt{3}\, M(x,\tau). \tag{21}$$

The maximum shear stress occurs on the neutral axes of the bars:

$$S(x,\tau) = \sigma_{12}/E\,\big|_{y=0} = (3\alpha^2/2)V(x,\tau). \tag{22}$$

We now describe briefly the nature of the moving pressure fronts. A pressure transducer is mounted in the nozzle of a water cannon in which a water packet of density ρ_* is struck by a piston moving at speed u_o. A shock appears in the water and translates over the sensitive face of the transducer at velocity v. The pressure in the water behind the shock is $p = \rho_* u_o v$. Assuming that the compressibility of water may be described by the Tait equation of state [3], we can show that, to a satisfactory degree of approximation,

$$v = (4a_*^2 + 11a_* u_o + 3u_o^2)/(4a_* + 3u_o),$$

where $a_* = 1500$ m/sec is the speed of sound in water. It is evident that the dimensionless pressure parameter σ defined in (2) depends strongly upon the impact speed of the piston u_o.

5. NUMERICAL RESULTS AND CONCLUSIONS

The following values for the parameters have been used:

$\rho_o = 7800$ kg/m^3, $\rho = 2650$ kg/m^3, $\rho_* = 1000$ kg/m^3, $\ell = 5$ mm, $b_1 = 0.25$ mm, $\ell_o = 3.5$ mm, $c = 0.25$ mm, $r_1 = 1.5$ mm, $h = 0.5$ mm, $E = 78.31 \times 10^9$ N/m^2.

In performing the numerical calculations, we have examined the dynamic response of the system in two situations: (i) the system is subjected to a pressure front of intensity p translating at speed v, which gives rise to coupled extensional-flexural deformations, and (ii) it is subjected to a non-moving pressure also of intensity p but distributed uniformly over the active face of the transducer, which gives rise to purely longitudinal deformations. For brevity, we discuss the results only for the case of $u_o = 300$ m/sec, in which p = 6.18 kbar and v = 2061 m/sec. Examination of the numerical data revealed that the most severe (dimensionless) compressive axial stress is $S_2^*(0,\tau)$, which occurs at the point x = 0, y = +1 at the built-in base on the right side of surface S_2 of the right-hand bar in Figure 1. We have plotted in Figure 2 the variation of $S_2^*(0,\tau)$ over the interval $0 \leq \tau \leq 20$ for situations (i), the solid curve, and (ii), the dashed curve. It is clear that the most severe compressive stress occurs in the quartz bar loaded by the moving pressure front. Indeed, max $|S_2^*(0,\tau)|$ for case (i) is 31.5% greater in magnitude than the corresponding maximum in case (ii). In addition, the dotted horizontal line in Figure 2 denotes

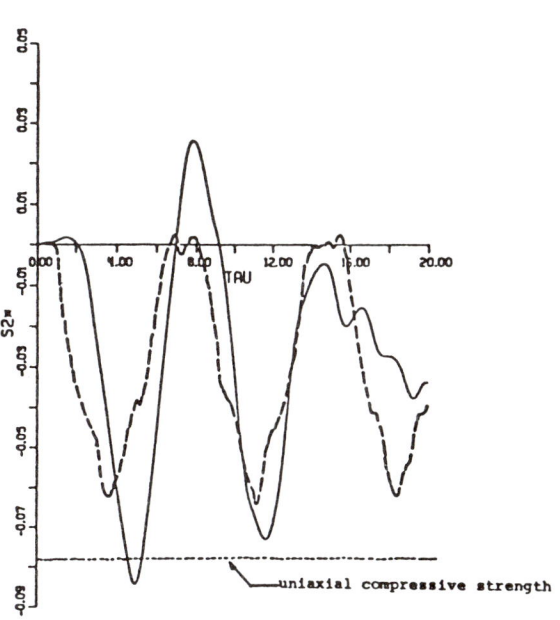

FIGURE 2 - VARIATION OF $S_2^*(0,\tau)$ FOR $u_o = 300$ m/s
(--- DIRECTLY APPLIED PRESSURE;
— MOVING PRESSURE)

the dimensionless uniaxial compressive strength of the quartz bar. The dashed curve does not intersect this line, but the solid one does. Thus, in case (ii) the quartz bar will not be damaged because the maximum compressive stress is less than the uniaxial compressive strength. Damage is predicted to occur, however, in the case of the moving pressure front even though the pressure intensities are the same in both cases, viz., p = 6.18 kbar.

Therefore, the present results clearly indicate that, for moving and non-moving pressures of the same intensity, the dynamic stresses in the system subjected to the moving front will be significantly more severe. If the pressure is sufficiently high (and perhaps less than the manufacturer's declared tolerable level), an axial stress greater than the uniaxial compressive strength of the quartz bar will occur. This may very plausibly lead to the longitudinal splitting of the crystal in the transducer. According to the present theory, the maximum pressure that the transducer subjected to the moving front can withstand is 5.7 kbar. This corresponds to a piston impact speed of u_o = 281 m/sec.

REFERENCES

[1] Anderson, G.L., Natural frequencies of two cantilevers joined by a rigid connector at their free ends. J. Sound Vib. 57 (1978) 403-412.

[2] Anderson, G.L. and Thomas, C.R., A forced vibration problem involving time derivatives in the boundary conditions. J. Sound Vib. 14 (1971) 193-214.

[3] Tait, P.G., Report on some of the physical properties of fresh water and sea water. Phys. Chem. 2 (1888) 1-71.

A GENERAL THEORY FOR THE SENSITIVITY OF CLAYS (A SUMMARY)

I. Th. Rosenqvist

Institute of Geology, University of Oslo, N-Oslo 3, Norway

The British term "sensitivity" used in soil mechanics represents a more exact expression for the older Swedish term "håldfasthetskvoten" or H-Ratio. This latter term expressed the relative strength of a soil material in its undisturbed state divided by the relative strength after complete remoulding or structure disturbance.

The ultra-sensitive type of clay called quick-clay, which has caused severe landslides in Scandinavia and on the west and east coast of Canada, represents material where the sensitivity is higher than 32 and may approach infinity. This means that after complete remoulding the material behave as a liquid.

The sheer strength of an undisturbed soil is generally expressed by the Coulomb-Terzagi-expression

$$S = (\sigma - u) \cdot tg\phi + C$$

where C is the sum of cementation and cohesion.

In uncemented clay-material the cohesional term may be expressed as a function of attractive and repulsive interpartical forces multiplied by an internal frictional coefficient. Thus, the traditional frictional term and the cohesional term may both be expressed as frictions, though it is not certain that the angel of internal friction acting in the cohesional term is identical with the angel of internal friction in the frictional term.

It may be considered an advantage to consider the sheer strength of a clay material by the following terms:

$$(\sigma + A - u - R)tg\phi$$

(σ = external normal stress, A = van der Waals and electrostatic attraction, u = pore-water pressure, R = electrostatic repulsive forces, i.e., diffuse layer repulsion, $tg\phi$ = frictional coefficient.

In the so-called quick-clays, mechanical disturbance will act upon several of the factors involved in the expressions above. The attractive and repulsive forces may change, and the cementation may be broken up.

Up to the end of the 19th century it was generally thought that the so-called quick-clay slides occurred because it was believed that under a crust of stiff clays, basins of liquid clay slurry existed, and when for one reason or another a crust failed, the underlying liquid slurry broke through, and the crust fell down. Such slides may progress retrogressively into extremely gentle slopes of only 2-3°, and slide material is known to have progressed on a slope of less than 1/4° (Aasrum-slide in Norway, 1942). Hans Reusch (1901) was the first scientist to demonstrate on the basis on fairly undisturbed samples that the clays in which these slides occur possessed ultra high sensitive values.

The transition from solid to liquid state due to remoulding may be called a gel-sol transformation, but in contrast to normal thixotropic systems this gel-sol transformation is irreversible as long as all components are held constant. The minerals of the Norwegian quick-clays belong to a great extent to the groups of hydrous micas of the illitic type (Rosenqvist, 1942). This hold for all subarctic quick-clays. Thus, the mineralogy does not differ qualitatively from the clays of high sesitivity and those of low sensitivity. The natural high sensitive clays are moreover often more coarse grained than the clays of low sensitivity. There exists, however, a complete overlap in grain size and mineralogy. It is furthermore found that all clays of non-expanding minerals and of high sensitivity has low content of dissolved electrolytes in the pore-water in spite of the fact that most of these clays are of marine origin. The salinity of the clays was in 1946 thought to be the decisive factor and the cause of the sensitivity. The original quick-clay theory as it was presented at that time, represents, however, only a special case of the present more general theory.

The most characteristic common feature in all quick-clays is that the phyllosilicate minerals belong to the non-expanding type, i.e. illitic, cloritic minerals. No real quick-clay has been found in sediments containing more than few percent or traces of expanding minerals of the smectite group.

The other characteristic feature is that the clays have a __flocculated__ structure. This means that the flaky minerals are touching each other by corner or edge to plane contact. Such clays may be of high as well as moderate sensitivity. In this respect there is no difference between clays of high and low salinity.

Floculated structures are found in sediments formed from suspensions where the flaky minerals have low double-layer repulsion. This may be either because the suspension medium has high ionic consentration, e.g. in marine environments, or because the clay minerals have low surface charge density. This may be caused by adsorbtion of highly charged ions as iron or aluminium to the clay minerals surfaces previous to erosion, transport and sedimentation.

Thus, salt-water clays as well as fresh-water clays may originate with
flocculated structures. Fresh-water clays may, however, contrary to
marine clays also be formed with disperged structure, i.e. with
parallel or subparallel arrangements of the layered silicate minerals.
This seems to be the most common fabric in lacustrine clays. This
arrangement is most stable when the mutual diffuse layer repulsion is
high, whereas the flocculated structure with corner to plane contacts
has the lowest free energy when the mutual repulsion is low. In this
case the attractive forces will be overwhelming over the repulsive
forces at greater distances than if the repulsion is high.

An increase in double-layer repulsions subsequent to depositions and
consolidation may be caused by any disperging agent in clays with low
salinity in the pore-water provided that most of the exchangeable ions
are of low valency and polarizability such as sodium. In such soil
it is sufficient only to reduce the salinity of the pore-water by per-
colation or diffusion in order to cause dispergation of the clay
minerals in the pore-fluid after moderate mechanical disturbance.

In sediments where a great part of the exchangeable ions have higher
valency, such as calcium, magnesium, iron or aluminium, dispergation
will not take place even at low salinity. Addition of disperging
agents such as sodium silicates, sodium meta-phosphate or organic
polarizable anions may counteract the neutralizing effect of these
high valency cat-ions and change the clays into sodium clays. In this
state dispergation in the remoulded state is possible provided the
salinity is sufficiently low. The Swedish scientist Rolf Söderblom
has worked considerably on this problem, e.g. (1966 and 1974) he has
demonstrated that a long series of organic substances may be present
in many quick-clays, and many of them are able to act as disperging
agents.

There is in fact no real contradiction between the two seemingly opposite
types of observation, i.e. that salt-water clays turn into high sensity
material by reduction of pore-water salinity, and that fresh-water
clays of flocculated structures turn into high sensitive material by
suitable chemical changes by the organic content.

By laboratory experiment it is easy to demonstrate that the water-content
of clay sediments depend on the salinity of the suspension medium provided the minerals have low charged counter ions, e.g. natrium, whereas
the water content will be independent upon the salinity of the suspension
medium if the counter ions have high-charged as Fe^{+++} and Al^{+++}.
Subsequent changes in the chemistry will in all clays of high porosity
(i.e. is high water-content) change the material into high sensitive
clays of the quick-clay type. A general theory for quick-clay properties
which satisfies the Canadian fresh-water quick-clays, the Baltic as well
as the marine clays is developed:

1. The clay particles of non-expanding minerals are sedimented in a
    flocculated state, because the diffuse layer potential is low
    either due to salinity in the water or due to adsorption of strong-
    ly held counter ions as Fe^{+++} to Al^{+++}.

2. After deposition and slight consolidation the diffuse layer
    potential of the minerals is increased. This may be caused by

leaching, reduction of Fe^{+++} to Fe^{++} or by complexing of the counter ions. The original disordered structure of the floculated clay remains.

3. By mechanical deformation an alignment of the minerals takes place. Due to the high mutual repulsion, refloculation is inhibited and only after considerable reduction in water content it is possible for the minerals to form new contacts.

Young clays which have developed so-called cementing bonds after consolidation, may have obtained strength properties (but not porosity as by overconsolidation) of more than 500 KPa. In such cases leaching with EDTA solution will attack the cement and bring the clay back to strength of 200 KPa corresponding to the real set of consolidation stresses. This was reported by KENNEY, MOUM and BERRE (1967).

Thus we can see that the study of sensitive clays are touching a field of fundamental importance for the understanding of the shear strength of normally consolidated clay sediments at a given water content. We will also understand why quick clays are restricted to geologically young formations and, furthermore, why they are not to be found in areas where the percolating ground water phase has a high content of calcium, magnesium or other highly charged cations. Only by typically soft waters it is possible to increase the effective charge of the particles, unless complexing molecules or ions are present in sufficient amounts.

Results of leaching with fresh water in clays of non-expanding minerals will bring these minerals out of chemical equilibrium with the surrounding aqueous phase. The resulting weathering processes will slowly change illites into mixed layer minerals and even into smectites. In such cases the initially obtained high sensitivity value will be counteracted by the expansion of the mineral layers and the quick properties will disappear.

The quick-clays properties are as demonstrated based upon collodial-chemical phenomena and are <u>not only caused by the specially open structure</u> of the grain fabrics which is the same in other loose-clay deposits. In both cases, however, deformation and fracture will cause a pore-water overpressure. These materials behave as typical pseudoplastic materials, and their stress-strain relations deviate from a Bingham body in the opposite direction of the more densely packed material which behaves as dilatent.

A deformation in the loose packed material may increase the pore-water pressure u even up to the value of σ so that the total internal friction will be reduced to zero. This is one side of the picture.

A hypothesis which was earlier proposed for the quick-clay property was to explain it as a simple tixotropic phenomenon. This is unsatisfactory because the quick-clays will not return to their original strength after remoulding, unless a considerable amount of water reduction takes place or if chemical changes of the pore-water take place.

Recently a large scale engineering operation was successfully performed in Oslo, where the arterial high-way from the north is built to cross and divert into the great "Ring-road". In this case the underground contained more than a 15 m thick, very soft quick-clay layer which was not strong enough to carry any high ramps. Thus, an alternative for the construction work involved extensive piling down to bedrock. In this case, however, the planning and the geotechnical investigations were performed so well in advance that it was time enough to press down a net of salt wells and leave them filled with potassium chloride. This material was left for diffusion for three years. <u>This increased the shear strength of the material around four times</u>, and the ground is now in the salted state <u>at the same vold ratio</u> sufficiently strong for the planned engineering work.

(This paper represents an updated summary of original material presented by the present author 1975 and 1977. The relevant illustrations are to be found in the original papers.)

REFERENCES

[1] Reusch, H., <u>Nogle optegnelser fra Verdalen</u>. Norges Geol. Undersøk. Kristiania 1901, Vol. 32 pp. 1-32.

[2] Kenney, T.C., J. Moum and T. Berre, <u>An Experimental Study of Bonds in a natural Clay</u>. Proc. of Geotech. Conf., Oslo 1967, Vol. 1, pp. 65-69.

[3] Rosenqvist, I.Th., <u>Angående norske leirers petrografi</u>. Medd. Vegdirektøren, Oslo 1942 nr. 3, pp. 24-30.

[4] Rosenqvist, I.Th., <u>Om leirers kvikkaktighet</u>. Medd. Vegdirektøren, Oslo 1946 nr. 3, pp. 29-36.

[5] Rosenqvist, I.Th., <u>Clay Mineralogy applied to Mechanics of Landslides</u>. Geol. Applicata e Idrogeol., Bari 1975, Vol. X (II), pp. 21-32.

[6] Rosenqvist, I.Th., <u>A general theory for Quick Clay Properties</u>. Proc. III Meeting Eur. Clay Groups, Oslo 1977, pp. 215-228.

[7] Söderblom, R., <u>Chemical aspects on quick-clay formation</u>. Engng. Geol., 1966, Vol. 1 no. 6, pp. 415-431.

[8] Söderblom, R. <u>Organic Matter in Swedish Clays and its Importance for Quick Clay Formation</u>. Swedish Geotechn. Inst. Proc., 1974, No. 26, 90 p.

MECHANICAL BEHAVIOUR OF CLAY EXPLAINED IN MICROSTRUCTURAL TERMS

Sven Hansbo

*Department of Geotechnical Engineering, Chalmers University of Technology,
S-402 20 Göteborg, Sweden*

ABSTRACT

The research on clay microstructure carried out in the last few years, especially by Prof Pusch, Sweden, has been of great importance for a better understanding of the geotechnical properties of clay. In this paper, a hypothesis is presented in which the mechanical behaviour of illite clay is related to its microstructure. It is shown that no difference exists in the microstructural mechanisms between soil deformations caused by consolidation on the one hand and by failure on the other.

The hypothesis gives a basic explanation of the concept of true cohesion and true friction and also of the creep behaviour and of strength anisotropy.

1. MICROSTRUCTURAL FEATURES OF CLAY

In clay the number of particles per unit volume is extremely large, on the order of 10^{15} to 10^{16} per cm^3. The total particle surface area is about 50 to 100 m^2 per cm^3 of clay. Particle size (maximum width) varies from about 2nm (= 20 Å) to 2 µm, mostly however from about 10 nm to 0.2 µm.

For these clays it is significant that the particles form a network of aggregates separated by pores with very small dimensions. Lacustrine clays have a fairly even distribution of particles while marine clays have large, dense aggregates separated by micropores which are considerably larger than those in the lacustrine clays (Pusch, 1970). The particles are flakelike - in lacustrine clays razor-edged and very thin, in marine clays fairly thick (cf. Hansbo, 1975).

The most important feature from the geotechnical point of view is that the aggregates are coupled together in a three-dimensional network via bridges built up of extremely small particles (Pusch, 1970), Fig. 1.

This network of aggregates forms the clay skeleton. The microstructural pattern just described has a decisive influence on the deformation and failure characteristics of clay.

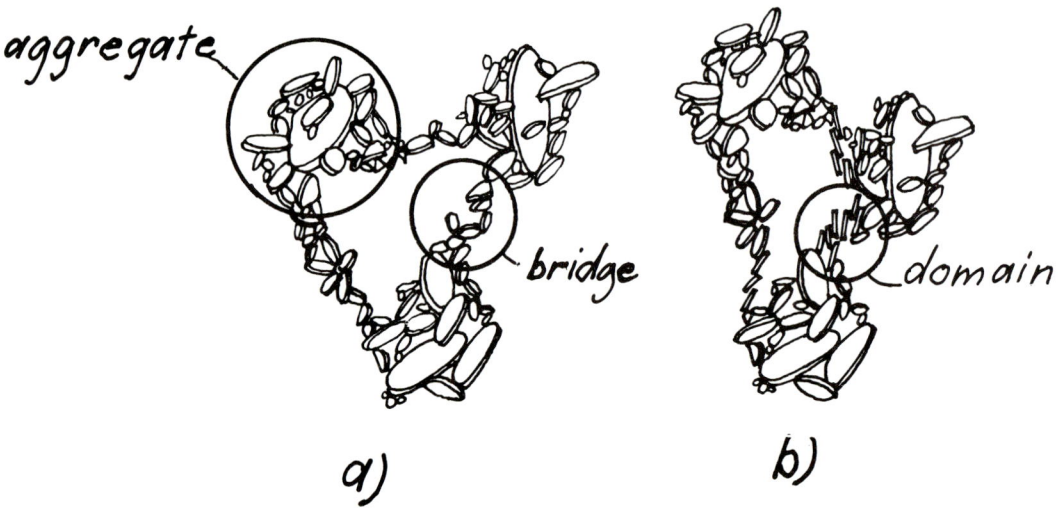

Fig. 1. Schematic picture of clay skeleton.

    a) natural microstructural pattern

    b) formation of domains

2. DEFORMATION CHARACTERISTICS

Due to extremely small spacing between the particles in the aggregates and to the influence of sorption forces, the pore water in the aggregates has av very high viscosity. This gives the aggregates a considerable rigidity. In deformation, the aggregates will therefore act as micrograins in a similar way to that in which the grains in a noncohesive soil act. In loading, shear deformations will start to take place in the bridges binding the aggregates together whereas the aggregates themselves - after the "break-down" of the bridges - will only change their positions relative to each other without internal shear deformations.

Since no contact exists between the particles - unless the clay contains a large amount of coarse materia such as silt grains - the clay skeleton is very susceptible to creep. The creep will continue until it is eventually stopped by the stress relaxation in the bridges. The deformation will be time-dependent (structural viscosity), Fig. 2.

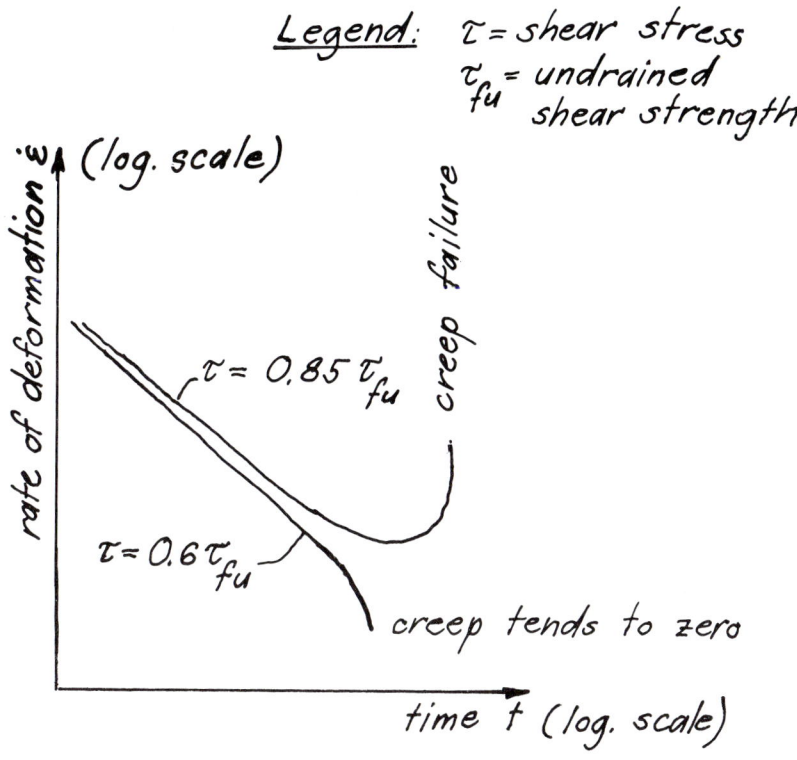

Fig. 2. Creep observed in undrained triaxial tests at two different stress levels.

In clays having a high preconsolidation pressure, groups of parallell-oriented particles, so-called domains, can be noticed in between aggregates in which the particles have a more or less random orientation, cf. Fig. 1. These domains have been created by the deformation of the bridges when, during the consolidation process, the aggregates have been forced to move closer together. In clays that have consolidated under a very high overburden pressure the aggregates will also be deformed and the particles forming the aggregates will thus be more or less parallell-oriented. In these heavily consolidated clays the resistance to deformation will be very large. The deformation will mainly take place by shear in zones of weakness and by closure of open fissures. The deformation characteristics will be volume-dependent.

While the preconsolidation pressur is being exceeded, progressive failure will take place in the bridges as just mentioned. The aggregates will start to move closer together, thereby forming a more stable clay skeleton able to resist the increased pressure. The formation of domains caused by consolidation cannot be distinguished from that caused by shear. The consolidation process is no doubt phenomenologically a continuously proceeding internal shear failure (cf. Hansbo, 1960).

At effective stresses that are low in comparison to the preconsolidation pressure, swelling will cause a decrease in the strength of the bridges. The swelling causing the weakening process in the bridges will start to increase considerably when the effective stress decreases below 0,5 times the preconsolidation pressure (Larsson, 1977), Fig. 3.

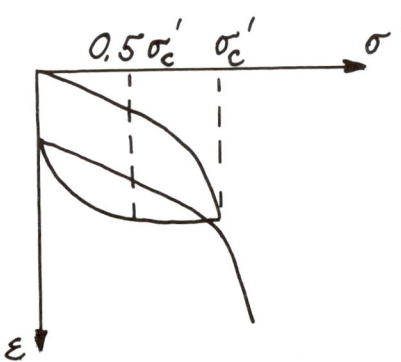

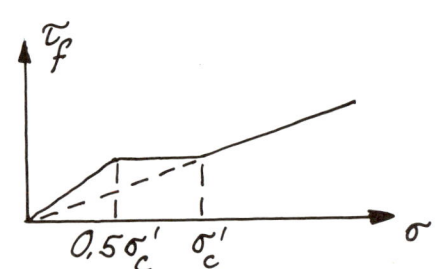

Fig. 3. Result of oedometer tests (upper figure) and drained shear tests on clay.

Legend: σ' = effective vertical stress
σ'_c = consolidation pressure

3. FAILURE CHARACTERISTICS

It is obvious from the above that the cohesion of a clay in its true sense depends on the strength of the bridges linking the aggregates together while the "true" friction depends on the energy required to move the aggregates (the micrograins) relative to eachother. As stated above the preconsolidation pressure is a critical stress level, representing the bearing capacity of the bridges. Therefore the "true" cohesion I_ε is a function of the preconsolidation pressure ($I_\varepsilon = i_\varepsilon \sigma'_c$), while the "true" friction is a function of the effective stress σ' ($D_\varepsilon = d_\varepsilon \sigma'$). The shear strength τ_f of the clay can thus be expressed as

$$\tau_f = i_\varepsilon \sigma'_c + d_\varepsilon \sigma'$$

In shear deformation the "true" cohesion will first be mobilized while the "true" friction will not be fully mobilized until after failure in the bridges has occurred, Fig. 4.

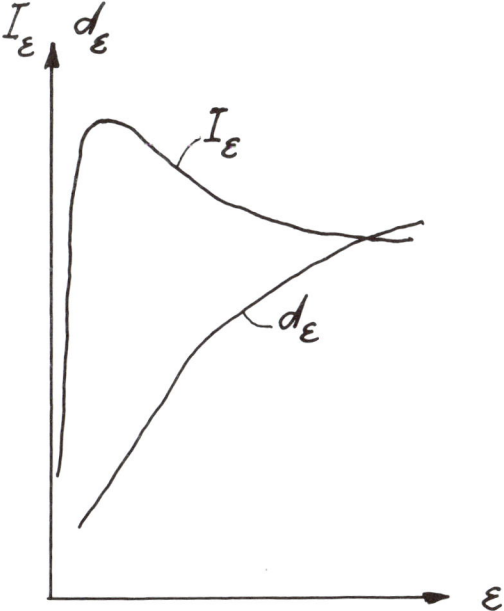

Fig. 4. "True" cohesion and "true" friction vs. axial deformation ε in an undrained triaxial test (after Schmertmann & Osterberg, 1961)

With time, creep in the bridges will lead to a stress transfer from the bridges to the aggregates. If the creep deformation comes to a stop, all the load will eventually be carried by the aggregates and the bridges will be unloaded (stress relaxation). If the aggregates are not able to withstand the load (the shear forces) the creep will eventually lead to failure, cf. Fig. 2. In other words, creep failure will occur if the "true" friction is not high enough to resist the shear stresses in the soil.

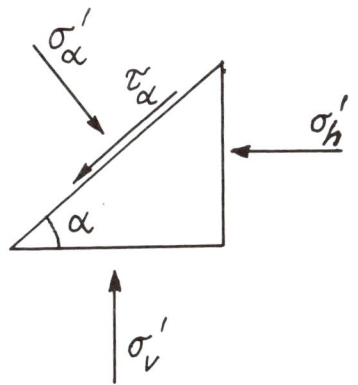

Fig. 5. Stresses acting on a soil in situ.

In a soil in situ, submitted to different horizontal and vertical stresses, shear stresses act on all inclined planes through the soil, Fig. 5. These shear stresses are carried by the aggregates while, as just stated, the bridges are unloaded due to stress relaxation. If additional shear stresses are added to those already existing the bridges will again be mobilized. Later, with increasing deformation, the remaining shearing resistance of the aggregates will supplement the resistance of the bridges.

We find

$$\tau_f = \tau_\alpha + (d_\varepsilon \sigma'_\alpha - \tau_\alpha) D_M + i_\varepsilon \sigma'_c$$

where D_M = parameter which indirectly describes how much of the shear stress at failur $\tau_{f\alpha}$ is carried by D_ε (mobilized value of D_ε when I_ε reaches maximum). This relation corresponds to the concept of strength anisotropy presented by Bjerrum (1973) at the ICSMFE in Moscow, Fig. 6

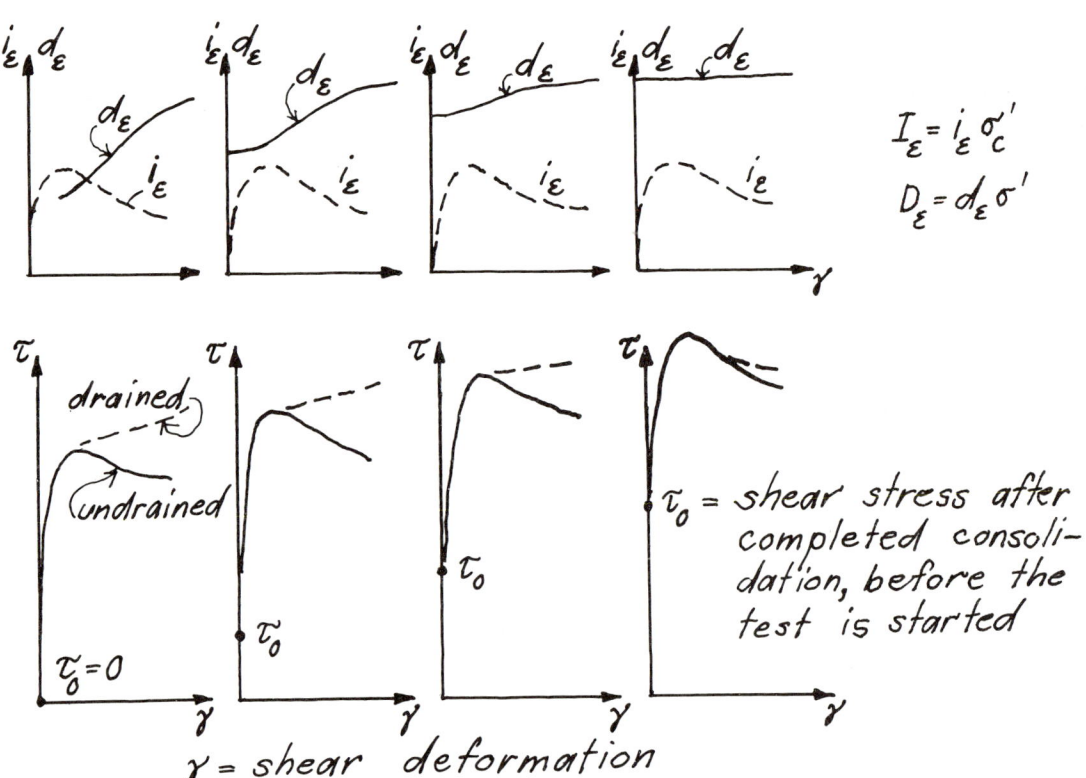

Fig. 6. Drained and undrained stress/deformation curves obtained for four samples. The mobilization of "true" friction increases with increasing shear stresses after completed consolidation.

REFERENCES

Bjerrum, L., 1973: "Problems of soil mechanics and construction on soft clays and structurally unstable soils (collapsible, expansive and others). Proc. 8th Int. Conf. Soil Mech. and Found. Engng., Moscow, Vol. 3 (p. 122-123).

Hansbo, S., 1960: "Consolidation of clay with special reference to influence of vertical sand drains", Swedish Geotechnical Institute, Proc. No. 18 (p. 26).

Hansbo, S., 1975: "Jordmateriallära", AWE/Gebers and J&W, Stockholm (Textbook, p. 17).

Larsson, R., 1977: "Basic behaviour of Scandinavian soft clays", Swedish Geotechnical Institute, Report No. 4, (p. 15 - 27).

Pusch, R., 1970: "Clay microstructure", Swedish Council for Building Research, Document D8:1970 (p. 56 and 67-68).

SHEAR STRENGTH OF SOILS

R. Larsson

Laboratory Department, Swedish Geotechnical Institute, S-581 01 Linköping, Sweden

ABSTRACT

A generalized presentation of the strength properties of the most common uniform Swedish soils is given. The soils are grouped into granular and fibrous materials. The influence of particle size and density in granular material and the influence of humification in fibrous materials is described.

1. INTRODUCTION

From strength criteria Swedish soils are usually divided into three main groups; friction soils, intermediate soils and cohesive soils. Friction soils comprise sand, gravel and coarser materials. Intermediate soils are silts. Cohesive soils usually comprise clay, mud (gyttja), muck (dy) and mucky peat (dytorv) while peat is regarded as a special material.

This grouping is somewhat unfortunate as the strength in silts and inorganic clays as well as in the classic friction soils is mainly made up by friction. The word cohesive suggests some sort of attraction forces and tensile strength at zero effective normal stress in the soil. This cannot normally be counted upon except for in peats.

An alternative way to subdivide soils with regard to strength properties is to divide them into granular materials and fibrous materials.

2. DRAINED SHEAR STRENGTH IN GRANULAR MATERIALS

The shear strength that can be mobilized in a granular material depends on the interparticle friction, the stress level in the material and the deformation properties of the material.

The shear strength can be expressed

$$\tau = \sigma' \tan \phi' \qquad (1)$$

where τ = shear strength
 σ' = effective normal stress
 ϕ' = effective angle of friction

2.1 Interparticle Friction

The interparticle friction depends on particle mineral and particle shape. Shear tests on plane smooth ground particles have given angles of friction between 6° for dry quartz and 37° for water-soaked feldspar. As these kinds of surfaces hardly exist in nature and a certain amount of moisture is usually present in soils a basic angle of friction of about 30° can normally be counted upon. The more irregular shaped and sharp edged the particles are the higher the interparticle friction becomes. At an increasing stress level the particles start to break and crush so that the influence of particle shape decreases with increasing stress level.

2.2 Deformation Properties

The deformations can be subdivided into; elastic deformations in particles and points of contact when the stresses are changed, plastic deformations in the form of rearrangement of particles as the particles are displaced in relation to each other and plastic deformations due to crushing of particles. The sum of these deformations determines whether the volume of the material will increase or decrease during shear.

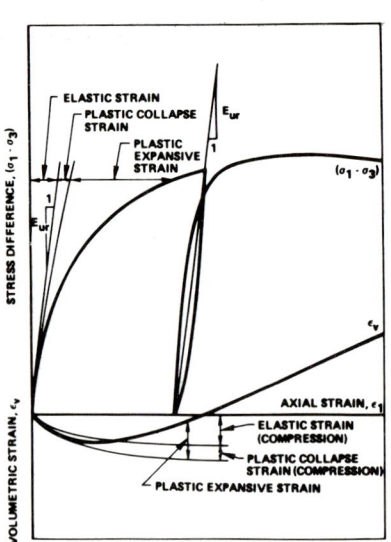

Fig 1. Schematic illustration of deformations in drained triaxial compression test. From Lade 1977.

At low stress levels the elastic deformations are small and crushing non-existent but the influence of these two factors increases with increasing stress level.

In dense materials the particles tend to climb over each other during shear and thereby increase the volume of the material. In loose materials the particles fall into hollows and the volume is decreased.

If the stress level in the material is increased over a certain level, which depends on the relative density, plastic compression occurs in the material. This plastic compression is relatively small in coarse materials but increases with decreasing particle size and becomes very large for loose materials with very small particles.

If the sum of the deformations causes the volume to increase or decrease the material has a positive or negative dilatation. If the volume during shear remains constant the material is considered to have a critical void ratio. It is clear that the critical void ratio is dependent on the stress level and on whether the principal stresses are increased or decreased.

2.3 Influence of Dilatancy

If the volume of the material increases during shear the shear strength increases. This increase in shear strength can be evaluated by energy considerations where the increase in shear strength is the energy required to expand the material (Rowe, 1962 (17)). The increase in shear strength can also be evaluated from a static consideration where the particles are displaced along a plane having the inclination α to the direction of the shear force. The inclination α is determined by the relation between volume change and shear displacement in the direction of the shear force. The effective angle of friction that can be mobilized is then the basic angle of friction + shape factor + α.

If the volume of the material decreases during shear a corresponding decrease in shear strength occurs.

2.4 Influence of Particle Size

The maximum positive dilatancy is dependent on the particle size. In coarse materials the contact points are few and the elastic deformations are small. At a large positive dilatancy failure occurs along one single shear plane as the material after failure is heavily strain-softening. The distance the particles in this shear plane are lifted perpendicular to the shear direction is dependent on the particle size. Thus the maximum angles of friction measured in standard triaxial tests are in the order of $50°$ for gravel, $45°$ for sand, $40°$ for silt and $30°$ for clay. No significant positive dilatancy can be counted upon in clay which is reasonable as the particle size is very small.

2.5 Influence of Intermediate Stress

The dilatancy is also dependent on the intermediate stress acting in the shear plane perpendicular to the shear direction. If this stress is low the particles are able to move sideways instead of on top of each other. If the intermediate stress is increased this possibility diminishes. Thus an increased intermediate stress results in an increased angle of friction. This increase of friction angle is most pronounced when the intermediate stress is increased up to the stress required to prevent lateral yield in the intermediate direction. A further increase of the intermediate stress has less significance.

The influence of the intermediate stress seems to be largest in coarse materials and decreases with decreasing particle size.

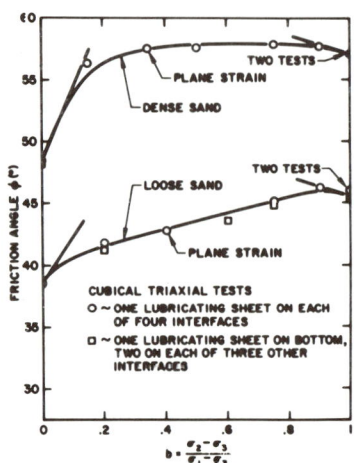

Fig 2. Variation of measured friction angles versus intermediate stress level. From Lade and Duncan 1973.

Reported tests with different intermediate stresses on clay show no clear trend and the influence is so small that it is within the measuring accuracy.

2.6 Anisotropy

The shear strength that can be mobilized in a granular soil is dependent on the direction in which the shear stress is applied. During the geological process the material is deposited and subjected to stresses. Normally the vertical stress caused by overlaying materials is the major principal stress. The magnitude of the horizontal stress depends on what shear stress the material can take during consolidation for the vertical stress. In coarse materials vibrations have a significant influence on the final density and the horizontal stress as a small vibration can compact the material much more than a small extra load. Most granular soils have a much higher in situ density than would have been caused by the previous maximum vertical load alone if the material was originally deposited in a very loose state.

This is not the case for the most fine grained materials.

A clay normally has a density almost exactly corresponding to the density caused by previous maximum stresses. A certain influence of time and load cycles exists. Vibrations do not seem to have any significance. This can be explained by the fact that in a saturated clay all sudden stress changes are compensated by pore pressure changes and the stress has to act for so long a period that the pore water can be squeezed out of the microscopic pores to have an influence on the density. Thus for clays the density is not normally used to describe the deformation characteristics. Instead the preconsolidation pressure, which is the maximum previous effective vertical stress, is used.

Owing to the maximum previous stresses being different in different directions the deformation characteristics also vary. Normally the major principal stress has been acting vertically and the minor horizontally.

The relation between maximum previous horizontal stress and maximum previous vertical stress caused by evenly distributed overlaying vertical loads is denoted K_c'. As the horizontal stress depends on what shear stress is mobilized during consolidation K_c' becomes small for a material with high angles of friction and low compressibility and increases with decreasing angle of friction and increasing compressibility. The most commonly used formula to calculate K_c' for coarse granular materials is Jaky's formula $K_c' = 1 - \sin \phi'$.

For clays it is more common to express K_c' as a function of the plasticity index or liquid limit, which are easy to determine.

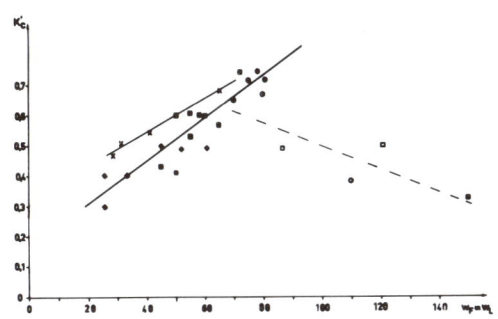

Fig 3. K_c' versus liquid limit for clay. Open symbols represent organic clay.

As high-plastic clays with a high liquid limit and a high water content are more compressible than low-plastic clays with a low water content this relation is quite reasonable.

If the maximum previous vertical stress is σ_c' and the maximum previous horizontal stress is $K_c' \sigma_c'$ the maximum previous stress in a direction with the inclination α from the vertical plane can be evaluated from

$$\sigma_\alpha' = \sigma_c'(\cos^2\alpha + K_c' \sin^2\alpha) \qquad (2)$$

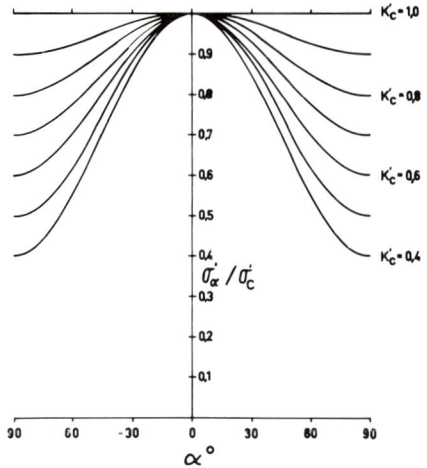

Fig 4. Effective prestress in planes with different orientation.

As a preloading of the material in one direction makes it stiffer to load in the same direction the material will become less compressible and more dilative at shear if the vertical stress is increased than if the horizontal stress is increased correspondingly.

This difference is relatively small in coarse granular materials but has recently been possible to measure with a refined technique.

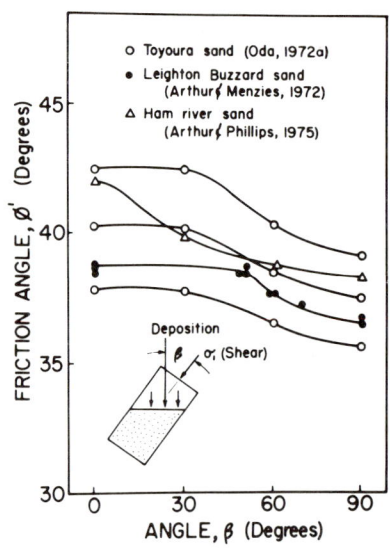

Fig 5. Effect of direction of loading on the friction angle of three sands. From Ladd et al 1977.

As the compressibility increases the influence of anisotropy increases. In clays where an exceeding of the previous maximum stresses causes very large compressions the shear strength that can be mobilized for practical purposes is often directly dependent on the previous maximum stress in the loading direction. This is valid for normally consolidated and slightly overconsolidated clays in cases where the stresses are increased. In heavily overconsolidated clays and in other cases of loading where the previous maximum stresses are not exceeded before failure occurs the influence of anisotropy is small.

2.7 Clay Structure

Electron-micrographs have shown that clay consists of particles and aggregates of particles. The contact points often consist of small particle aggregates. The large aggregates seem to function as separate, easily crushed particles and the small aggregates in the contact points seem to make these points more elastic. No other fundamental difference from other granular materials as regards the shear strength seems to exist in normal Scandinavian clays apart from the particle size and the permeability.

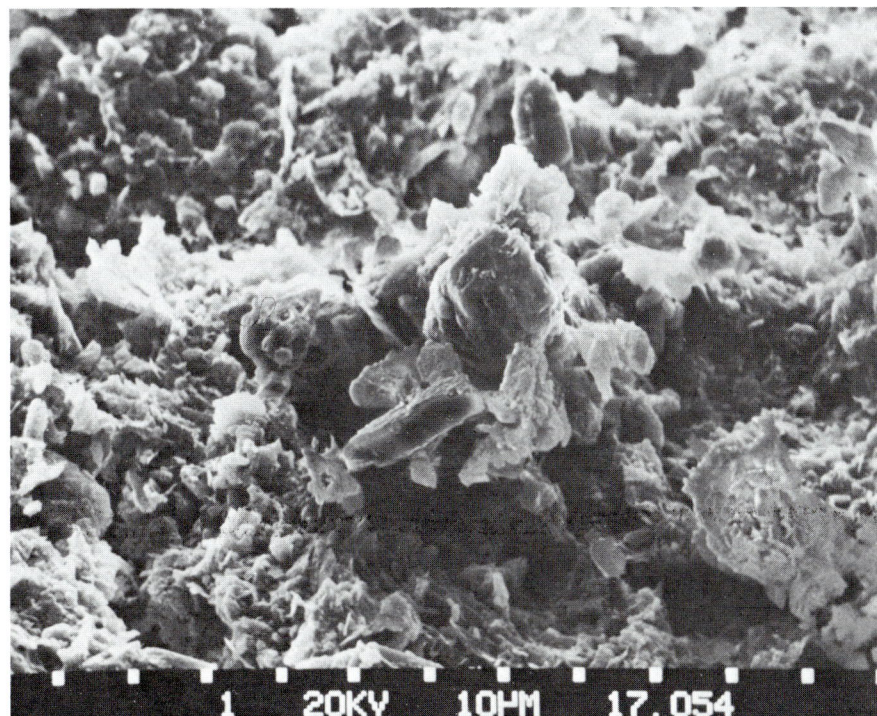

Fig 6. Micrographs of Lilla Mellösa clay.

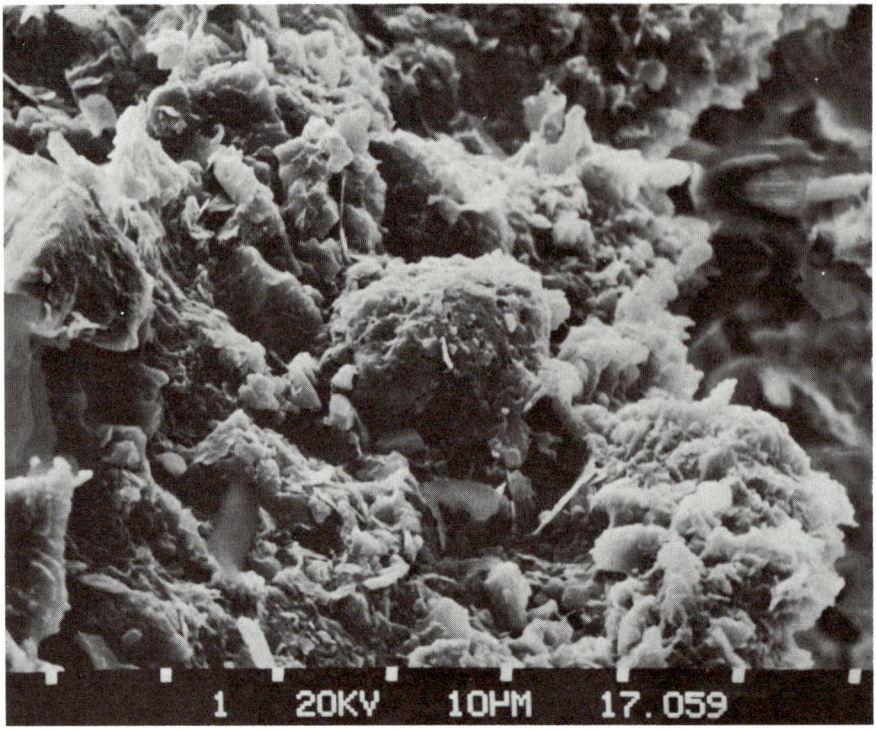

A certain type of anisotropy, particle orientation, exists in some foreign clays but if it exists in Swedish clays at all it is rare.

3. UNDRAINED SHEAR STRENGTH IN GRANULAR MATERIALS

3.1 Pore Pressure Changes at Loading in Granular Materials

If a saturated granular material is subjected to a swift loading the pore water pressure in the material changes. If particles and water can be considered as incompressible this pore water change will momentarily be of such magnitude that the mean effective stress in the material is unaltered. If shear deformations occur due to the stress change the pore water pressure decreases or increases depending on whether the material dilates positively or negatively. The duration of the pore water change depends on the permeability of the material and the magnitude of the drained volume change for the actual change in stresses. A low permeability and a large volume change increases the duration of the pore pressure change. Pore pressure changes occur momentarily in all saturated materials as the volume is momentarily constant but for coarse granular materials with a high permeability it has importance only in a few very special loading cases.

3.2 Undrained Shear Strength in Granular Materials with Coarse and Medium Sized Particles

In materials from the gravel size and coarser hardly any pore pressure changes of any duration occur. In sands the pore pressure change can be important at very swift stress changes such as earth-quakes, wave forces or vibrating foundations. If the material undergoes a positive dilatation when sheared this causes no trouble as the pore pressures in that case decrease and the effective stresses increase. On the other hand if the material undergoes a negative dilatation pore pressure increases may be generated faster than the pore pressure decreases due to the flow of water out of the material. This pore pressure increase might cause failure of the material as the effective stresses decrease.

In silts the pore pressure changes must always be considered as the permeability is low. Most cases of loading can be performed and calculated on the basis of drained shear strength provided the stress change is brought on slow enough. In some cases such as pile driving severe problems can occur.

3.3 Undrained Shear Strength in Clay

In clay the permeability is very low and the volume changes at loading large. Therefore the undrained shear strength is of great importance.

When an overconsolidated clay is subjected to a stress change and shear deformations occur the pore pressure decreases. It can continue to decrease until the effective stress in any direction amounts to about 80 per cent of the previous maximum stress in that direction. At this effective stress the dilatation is no longer positive so that the pore pressure ceases to decrease and failure occurs. As the previous maximum effective stress is different for different directions the undrained shear strength is dependent on the direction in which the major effective stress acts.

As mentioned in part 2.6 the maximum previous effective stress acting vertically is normally the preconsolidation pressure σ_c' and the maximum previous effective stress acting horizontally $K_c' \sigma_c'$.

The undrained shear strength for an overconsolidated clay where the effective angle of friction is $30°$ and where the major effective stress acts in a direction deviating by the angle α from the vertical becomes

$$\tau_\alpha = 0{,}8\ \sigma_c'(\cos^2\alpha + K_c'\ \sin^2\alpha)/3 \tag{3}$$

The deformations and thereby the pore pressures in normally consolidated clays exhibit time dependent creep effects. Therefore the effective stresses for a shorter period can amount to the previous maximum effective stresses and the short-term undrained shear strength becomes 25% higher than that for overconsolidated clays with the same preconsolidation pressure. If the clay remains undrained for a longer period the pore pressure increases slowly and the long-term undrained shear strength becomes the same as for overconsolidated clays with the same preconsolidation pressure.

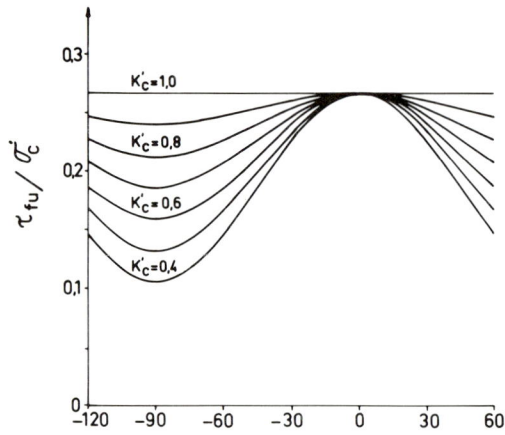

Fig 7. Undrained long-term shear strength in clay.

The given undrained shear strengths are only valid provided the material can take the maximum previous stresses at the same time as all frictional forces are mobilized without breaking down structurally. A number of clays, mainly high-plastic, have shown the ability to do this. Other clays undergo a structural breakdown when the stresses approach this stress combination. This does not seem to affect the anisotropy but the strength values become somewhat smaller.

4. DRAINED SHEAR STRENGTH IN FIBROUS MATERIALS

Fibrous soils, mud, muck and mucky peat, consist of plant fragments of different coarseness and various degrees of decay. Research concerning these materials is rather sparse as they have usually been avoided or replaced in construction practice. This is mainly due to the fact that these materials are very compressible and have not normally been subjected to any precompression. The materials contain a large number of short fibres of irregular shape and without any particular orientation. The tensile strength of the fibres is unknown but the friction between the fibres is high. The effective angle of friction fibre to fibre can roughly be estimated as $35°-50°$. In direct shear tests on these materials the shear strength can be expressed as $\tau = \sigma' \tan \phi'$ as for granular materials but the materials are very compressible and there is no empirical guidance as to what angle of friction can be mobilized for practical purposes.

At stress changes where the pressure in one direction is increased and the material drains fast enough there is normally no shear failure but only large compressions. Failure occurs if the stresses are changed so that tension occurs in any direction. When fibrous materials are consolidated for a vertical stress the increase in effective horizontal stress is very low so that the anisotropy in deformation characteristics and shear strength is considerable.

Mixed materials with mineral particles and organic fibres such as muddy clay and muddy silt form intermediate links between granular and fibrous materials. The organic content limit of 6% used to distinguish organic clay from mud at soil classification seems also to be useful as a limit for when the soil can be considered to be fibrous granular or fibrous only from strength considerations.

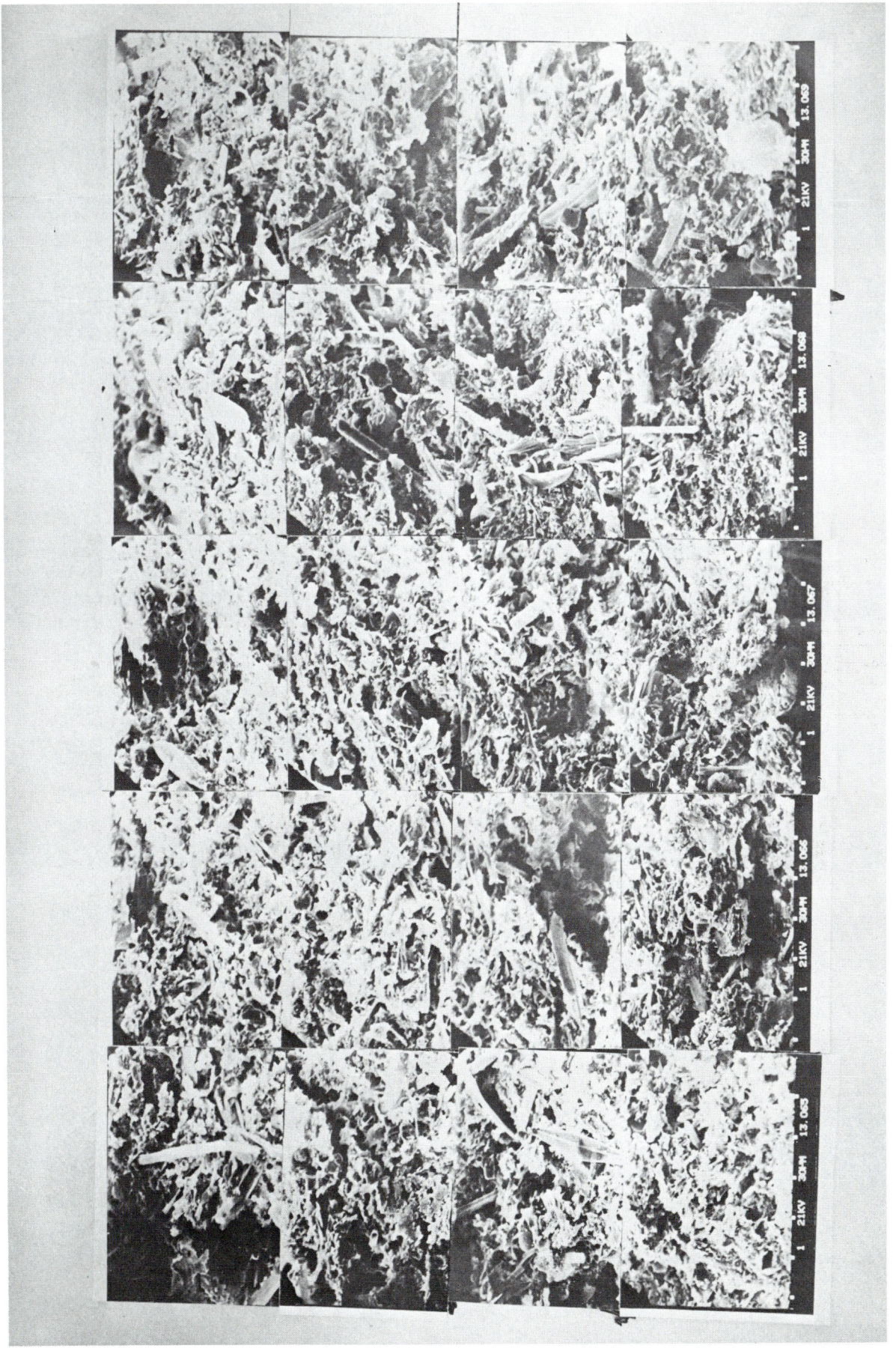

Fig 8. Micrograph of Välen organic clay, organic content ≈ 6%

5. UNDRAINED SHEAR STRENGTH IN FIBROUS MATERIALS

Fibrous materials have a low permeability so that the undrained shear strength is of great importance. The pore pressure changes seem to be similar to those in clay but no empirical guiding value to the stress level at which the material can be considered to be at a critical density has so far been established. If the material is loaded by increasing the stress in one direction shear failure does not occur until the pore pressure is so high that the effective stress becomes negative in another direction and tensile stresses occur. This can be seen in unconfined compression tests and standard undrained triaxial tests where no shear plane is established but the samples are split vertically. Due to the low effective horizontal stresses during consolidation for vertical loads the anisotropy is very large. The undrained short-term shear strength when the major effective stress is acting in a direction deviating by α^0 from the vertical direction can roughly be estimated as

$$\tau_\alpha = \sigma'_c(\cos^2\alpha + K'_c \sin^2\alpha)/2 \qquad (4)$$

6. STRENGTH PROPERTIES OF PEAT

Unhumified peat consists of fragments and pieces of branches, twigs and leaves intact enough to have a considerable tensile strength. The single parts are interwoven into each other but have an apparent horizontal orientation so that the tensile strength of the material is much greater horizontally than vertically. As the particles are interwoven into each other very high angles of friction are measured in simple shear tests. Effective angles of friction from $35°$ to $60°$ have been reported. This seems to be explained by the fact that the fibres do not break but have to be drawn out of each other. At a pressure increase in one direction no shear failure occurs but only large compressions unless the pressure increase is large and rapid enough to cause large tensile stresses due to high pore water pressures.

The permeability is normally so high that most constructions can be built as stage constructions without stability problems provided that the large compressions can be tolerated. In road construction care has to be taken so that the compressions under the road are uniform. Otherwise tension cracks may develop in the peat.

REFERENCES

(1) Bengtsson, P.E., Personal communication.

(2) Berre, T. and Bjerrum, L.(1973), Shear Strength of Normally Consolidated Clays. International Conference on Soil Mechanics and Foundation Engineering 8, Moscow 1973, Proceedings Vol 1.1 pp. 39-49.

(3) Bjerrum, L. (1973), Problems of Soil Mechanics and Construction on Soft Clays on Structurally Unstable Soils. General Report. International Conference on Soil Mechanics and Foundation Engineering 8, Moscow 1973, Proceedings Vol 3.

(4) Collins, K. and Mc Gown, A. (1974), The form and function of microfabric features in a variety of natural soils. Geotechnique 24 No 2, June 1974.

(5) Dansk Ingenjörsförenings norm for fundering. Teknisk Förlag. Köpenhamn 1977.

(6) Hansbo, S. (1975), Jordmateriallära. AWE/Gebers, Stockholm 1975.

(7) Helenelund, K.V. (1968), Compression, tension and beam tests on fibrous peat. Proceedings. 3 International peat congress. Quebec 1968.

(8) Janbu, N. (1970), Grunnlag i geoteknikk. Tapir Forlag, Trondheim 1970.

(9) Ladd, C.C., Foot, R., Ishihara, K., Schlosser, F. and Poulus, H.G. (1977), Stress-Deformation and Strength Characteristics. General Report. International Conference on Soil Mechanics and Foundation Engineering 9, Tokyo 1977. Proceedings Vol 2.

(10) Lade, P.V. and Duncan, J.M. (1973), Cubical Triaxial Tests on Cohesionless Soil. ASCE Proceedings Oct 1973.

(11) Lade, P.V. and Duncan, J.M. (1975), Elastoplastic Stress-Strain Theory for Cohesionless Soil. ASCE Proceedings Oct 1975.

(12) Lade, P.V. (1977), Elasto-plastic stress-strain theory for cohesionless soil with curved yield surfaces. Int 3 Solid Structures 1977, Vol 13.

(13) Landva, A.O. (1977), Contributions to Seventeenth Muskeg Research Conference. Saskatoon 1977.

(14) Larsson, R. (1977), Basic behaviour of Scandinavian Soft Clays. Swedish Geotechnical Institute, Report No 4, Linköping 1977.

(15) Pusch, R. (1970), Clay Microstructure. Document D8:1970, Nat. Swed. Build. Research.

(16) Pusch, R. (1973), Influence of organic matter on the geotechnical properties of clays. Nat. Swed. Build. Research, Document D11:1973.

(17) Rowe, P.V. (1962), The stress-dilatancy relation for static equilibrium of an assembly of particles in contact. Roy. Soc. Academy. Proceedings 296:500-527.

(18) Schofield, A. and Wroth, P. (1968), <u>Critical State Soil Mechanics</u>. Mc Graw Hill, London 1968.

STRESS-STRAIN RELATIONSHIPS FOR SILTY SOILS

L. Börgesson

Division of Soil Mechanics, University of Luleå, S-951 87 Luleå, Sweden

ABSTRACT

Triaxial tests show that there are three strain mechanisms at sub-failure stresses in silts: elastic deformations, plastic deformation with volume decrease and plastic deformation with volume increase. The stress history of a silt determines the extension of the elastic zone in the stress-space while the magnitude of the elastic and plastic deformations depend on the clay content and grain size distribution.

Since the failure envelope is curved, the Mohr-Coulomb parameters, the cohecion c' and angle of internal friction ϕ', depend on the stress level. In a silt the strength increase produced by an increased preconsolidation pressure is entirely due to dilatancy. Thus, c', which is obtained by extrapolating the failure envelope, only represents an apparent cohesion.

The importance of stress history, stress path and clay content is discussed in the paper and a model of the stress/strain/strength relationship is proposed for silty soils. This model resembles that of the Critical State Theory, developed in Cambridge. The main difference between the two concepts is the assumption of a plastic dilatant zone preceeding failure in the author's model.

SCOPE OF INVESTIGATION

The main object of this report is to illustrate the influence of stress history on the stress/strain/strength relationship of silty soils as observed by applying triaxial shear tests. The meaning and importance of cohesion and internal friction were of major interest in the study.

SOIL MATERIAL

The investigated soil in all the tests referred to in this report was a silt from Boden in northern Sweden. The main geotechnical properties are shown in Table 1. The samples used in the author's tests were taken from 3 m depth, that is well below the ground water level and the dry crust. Fig. 1 shows a micrograph of an undisturbed specimen taken by means of a scanning microscope. It can be seen that silt particles are dominant and that clay particles are attached to their surfaces. In a clayey silt the coarser particles are separated by clay particles to a certain extent. However, a considerable number of silt particles are in direct contact also at a fairly high clay content.

Table 1

Depth m	w_0 %	ρ g/cm³	% clay	% silt	% sand
3.0	36	1.89	9	86	5

Fig. 1 Micrograph of an undisturbed silt specimen. 1 cm equals 2μ.

TEST PROCEDURE

The triaxial tests were made by using undisturbed samples with a height of 10 cm and a diameter of 5 cm. Most of them were run as drained tests with a constant strain rate of 0.01%/min but some undrained tests were made as well. The axial stress and strain, the pore pressure, and the volume change were recorded.

The samples were consolidated under high σ_3' and σ_1' values and were then sheared at different σ_3' values. The consolidation eliminated the influence of previous stresses and made it possible to examine the influence of a known stress history on the stress/strain relationship.

RESULTS

Fig. 2 shows a typical result from a drained active triaxial test on an over-consolidated silt. The stress-strain curve begins with an almost elastic deformation with small strains up to a relatively high deviator stress. At the point marked EL the deformation increases and turns into a non-recoverable plastic strain. This strain increases with increasing deviator stress until failure is obtained, which is defined as the state where the maximum deviator stress is recorded. It is generally reached at an axial strain of more than 10%.

The curve representing the volume change shows that there is an initial volume decrease which turns into a volume increase at larger strain if the silt is over-consolidated. For a normally consolidated silt, on the other hand, the volume stays constant at failure. The

state where there is no change in volume is marked CV on the stress/
/strain curve. The increase of the shear stress after having passed
this point is caused by the work required to increase the volume.

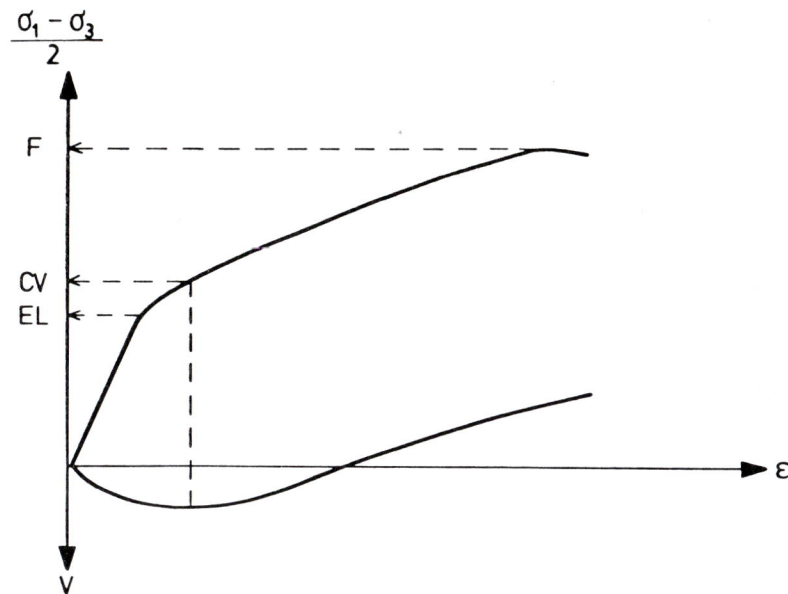

Fig. 2 Drained active triaxial test of an overconsolidated silt.

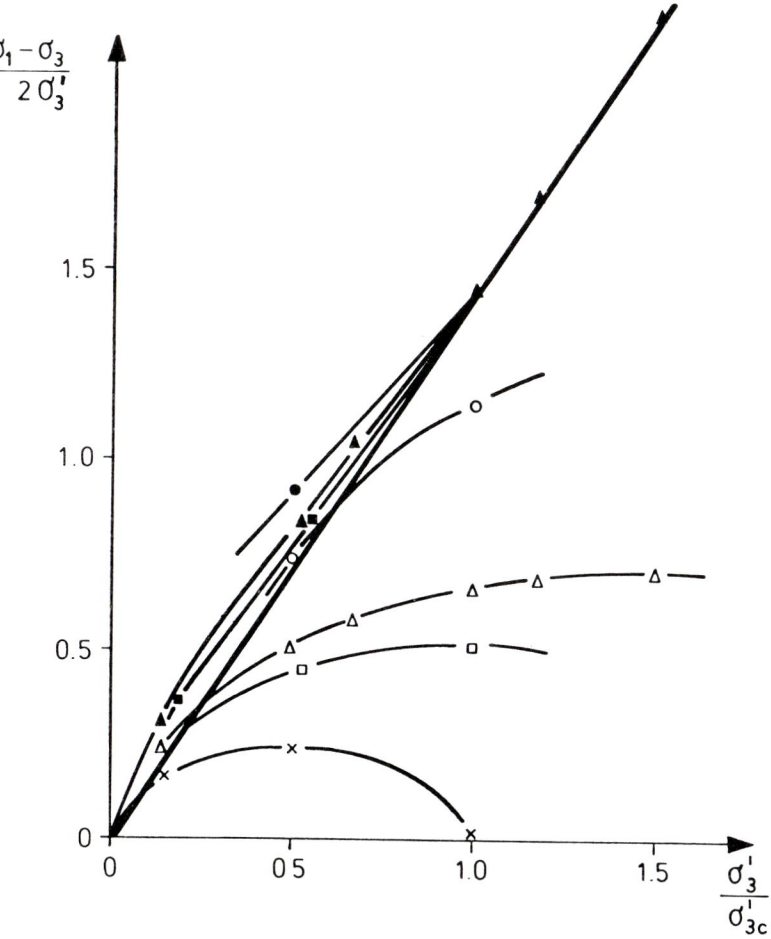

Fig. 3 Drained triaxial tests of silt preconsolidated under different
deviator stresses. × ◻ △ ○ correspond to the point EL in
Fig. 2. ■ ▲ ● correspont to the point F in Fig. 2. The point
CV in Fig. 2 is for all tests approximately situated on the
broad line.

Fig. 3 shows the results presented in a stress diagram. The silt samples have been preconsolidated under an isotropic pressure of 300 kPa and different deviator stresses. The three points (EL, CV and F) shown in Fig. 2 are recognized also in Fig. 3.

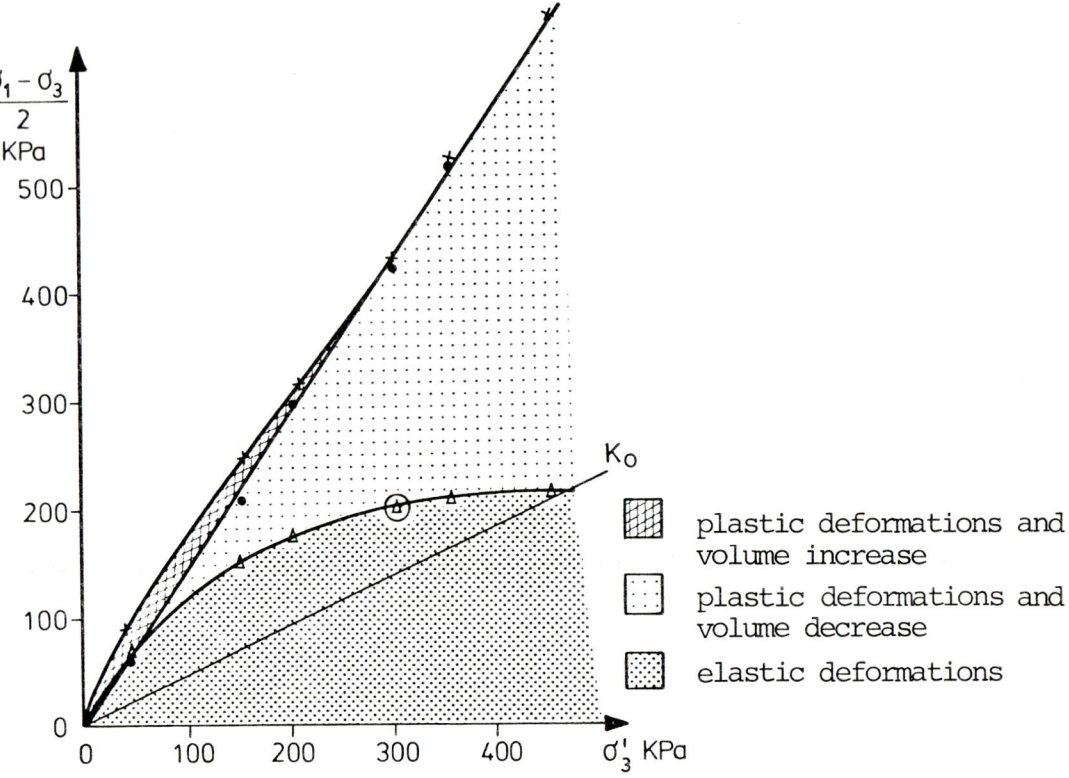

Fig. 4 The three zones created by overconsolidation of a silt.

Figs. 3 and 4 clearly show that the preconsolidation creates three zones with different stress/strain properties in the stress space. There is one zone with elastic deformations, one with plastic deformations and decreased volume, and one with plastic dilatant deformations. Fig. 4 shows the results from a test for which the consolidation condition can be written as $\frac{\sigma_1 - \sigma_3}{\sigma'_{3c}} = 1.3$. This figure also shows the stress path from a K_o-consolidation.

Tests made on normally consolidated silt show that the silt is plastic with a decreasing volume during the whole test until failure is reached. The failure envelope is a straight line through the origin. Fig. 3 shows that the preconsolidation creates an elastic zone which has the shape of an ellipse. It also increases the strength of the silt due to a plastic dilatant zone which lies above the straight line. Since dilatancy does not contribute to the strength at $\sigma_3' = 0$ this envelope also passes through the origin.

THE INFLUENCE OF STRESS HISTORY ON THE STRESS/STRAIN/STRENGTH RELATIONSHIP

An increased K_o preconsolidation pressure widens the elastic zone and the dilatant plastic zone. Since the shape of these zones does not seem to change, it is possible to normalize the stress space by dividing the stress values by the isotropic preconsolidation pressure (cf. Fig. 3).

If the silt is isotropically preconsolidated the elliptic elastic zone has its maximum at half the preconsolidation pressure (Fig. 3).

If the silt is unisotropically consolidated the elastic zone increases when the deviator stress increases but the elliptic shape is almost the same.

Fig. 3 shows that the deviator stress produced by unisotropical consolidaton influences the failure envelope as well.

THE IMPORTANCE OF THE STRESS PATH

If the silt is tested in an undrained condition, σ_3' will change during the test and the stress path will be as shown in Fig. 5. The investigated samples were normally consolidated under isotropic stress conditions and then sheared without drainage. It can be seen that the stress path approximately follows the border of the elliptic zone. At the top of the ellipse, near the K_o-line, the deformations become plastic and the stress path leaves the ellipse and approaches the straight failure envelope. Since the sample is not allowed to swell it cannot fail in the dilatant zone.

The tests presented here are active triaxial tests. If the sample is forced to follow a stress path with constant deviator stress and increasing σ_3', or if the test is run as a passive test, the transition from elastic to plastic deformations is rather diffuse.

THE MEANING OF THE MOHR-COULOMB PARAMETERS

The failure properties of a silt (and any soil) are often expressed by the Mohr-Coulomb parameters ϕ' and c'. This implies that the failure envelope is a straight line, which is rarely the truth, however. Thus, the parameters c' and ϕ' only describe the envelope within a limited stress interval while an extrapolation to zero effective stress yields a cohesion intercept which is only apparent. A true intercept only exists in cemented silts. Fig. 6 shows four possible ways of defining c' and ϕ'. Since strain is generally large at the maximum deviator stress a failure criterion which refers to a certain strain (e.g. 10%) is also reasonable.

STRESS/STRAIN/STRENGTH MODEL FOR SILT

The characteristic features of the stress/strain behaviour described in the preceeding text form the basis of a general silt model which is presented in Fig. 7. The main properties of this model are as follows.

An isotropic consolidation to point A results in an elliptic elastic zone from point A to origo. An anisotropic consolidation from point A to point B, C and D widens the ellipse, but its shape remains unchanged. This means that the extent of the elastic zone is independent of the point of the ellipse for which the sample has been preconsolidated.

The top points of all the ellipses for a certain silt are situated on a straight line which turns out to be the K_o-line for the investigated silt.

The failure envelope is drawn by using the maximum deviator stress as failure criterion. In a normally consolidated state the silt fails on the line ϕ'_{cv} where cv stands for constant volume. Between this line and the elastic ellipse the silt behaves as a plastic material which decreases in volume. In the over-consolidated state the strength increases due to the formation of the dilatant zone (Fig. 7).

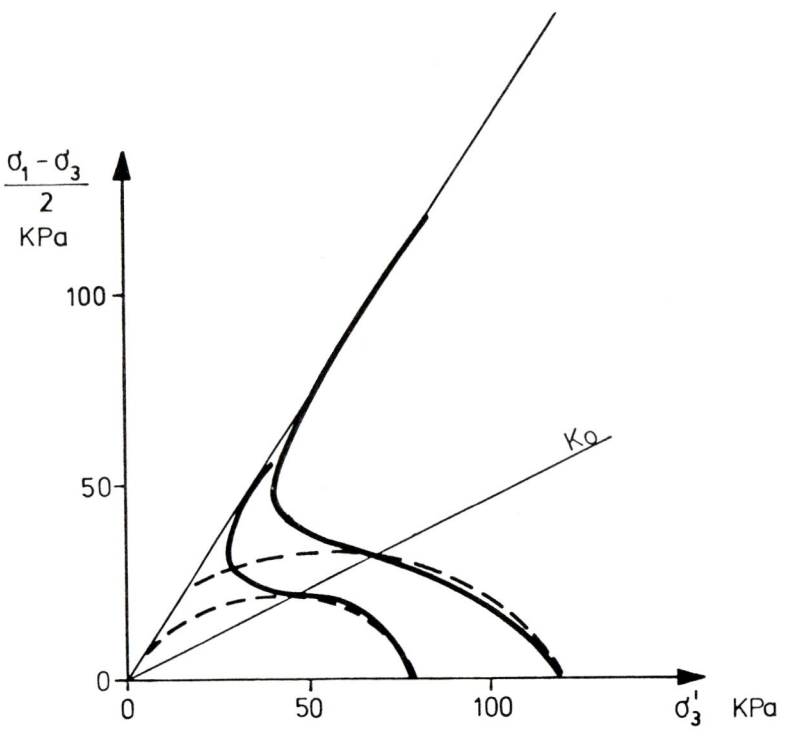

Fig. 5 Stress paths from undrained triaxial tests of silt.

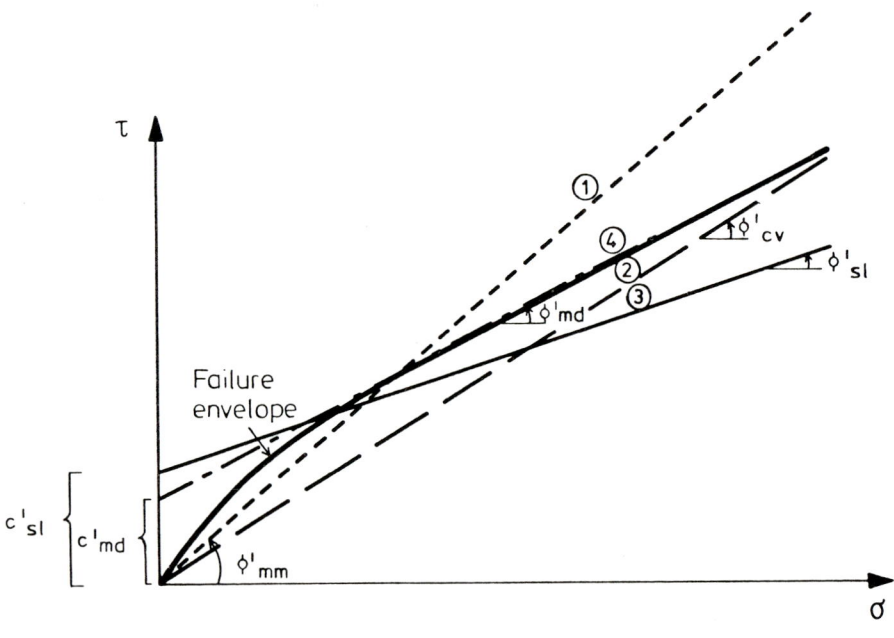

Fig. 6 Four possible ways to define c' and ϕ':

1. Maximum mobilized friction in one point. Gives ϕ'_{mm}.
2. Strength at critical void. Gives ϕ'_{cv}.
3. Strength defined as a strain limit. Gives ϕ'_{sl} and c'_{sl}.
4. Strength defined as maximum deviator stress. Gives ϕ'_{md} and c'_{md}.

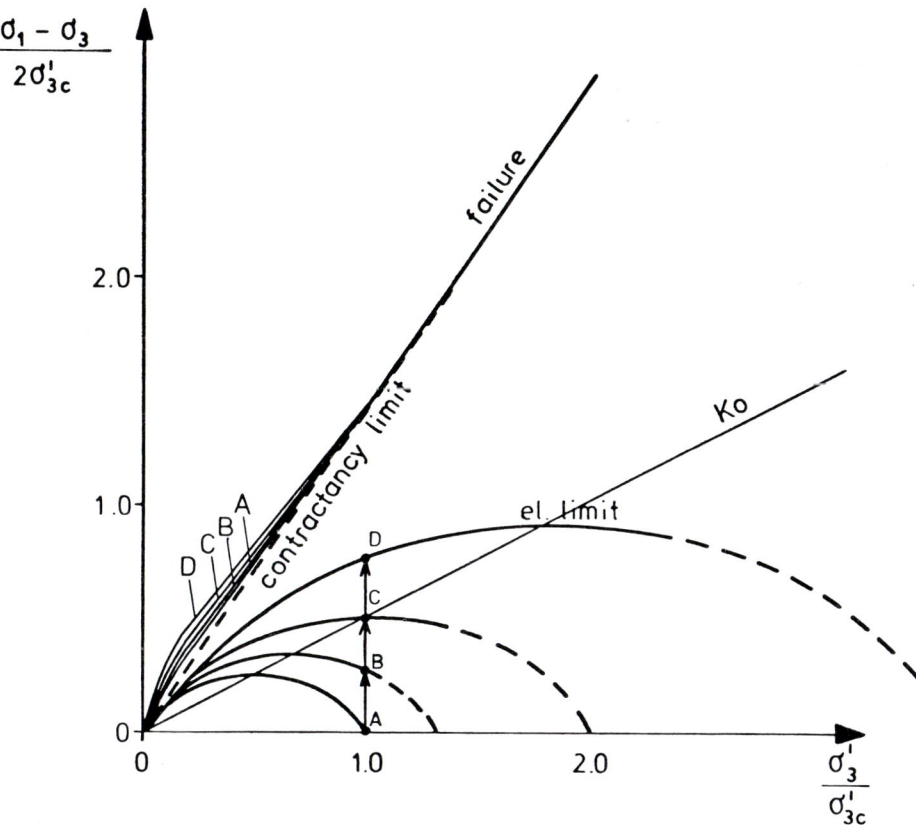

Fig. 7 Stress/strain/strength model for silt.

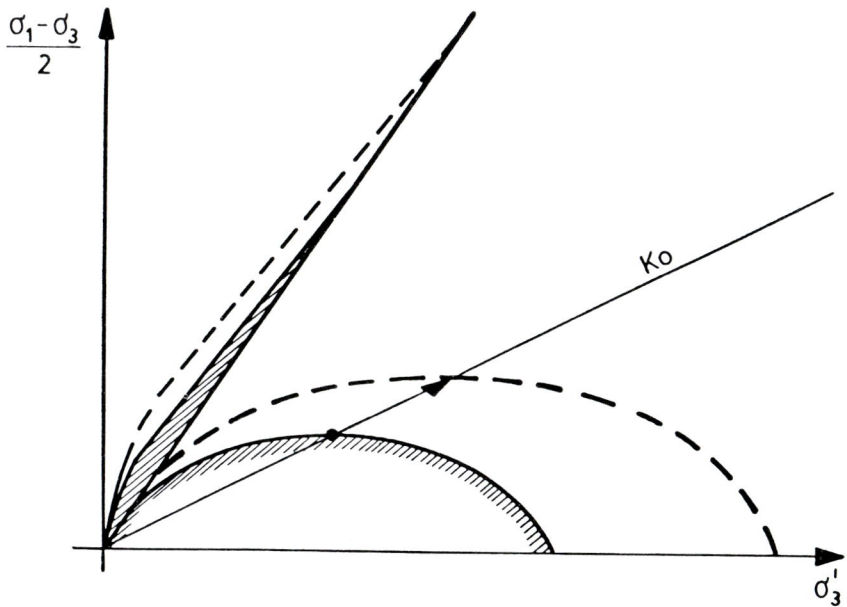

Fig. 8 The change in properties due to an increased K_o preconsolidation.

The magnitude of the elastic and plastic deformations and the value of ϕ'_{cv} are influenced by the physical properties of the silt.

The silt model is similar to the general model of the Critical State Theory, developed in Cambridge. The behaviour of the elastic zone during isotropic and anisotropic consolidation is found to be the same for silt as predicted by the Cambridge model. However, there are two very important discrepancies regarding the behaviour at failure. As can be seen in Fig. 9 the failure line for normally consolidated specimens goes through the top of the elastic zone for the Critical State model. This line is called the critical state line and corresponds to ϕ'_{cv} in Fig. 7. For a silt this line is situated outside the elastic zone and for the silt in the test series the K_o-line goes through the top of the elastic zone.

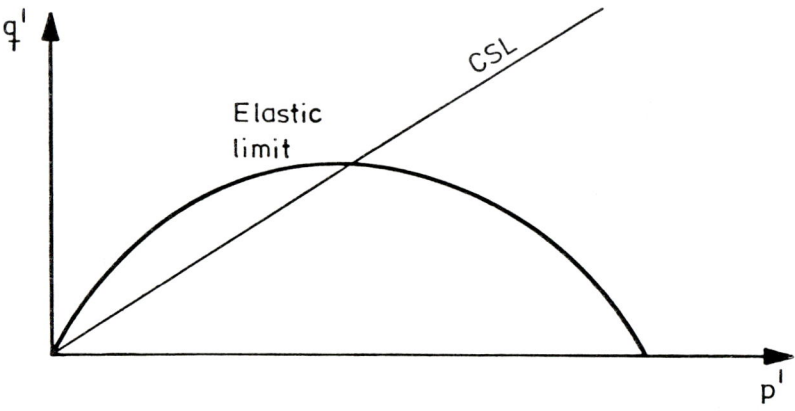

Fig. 9 The Critical State model. $q' = \sigma_1' - \sigma_3'$
$$p' = \frac{\sigma_1' + 2\sigma_3'}{3}$$

The Critical State Theory states that the zone outside the critical state line is elastic and that dilatancy does not begin until after failure. This is not true for silts. In the author's model the space above the ϕ'_{cv}-line implies plastic dilatancy before failure is reached. In this respect this model is in better agreement with the dilatancy theories [1], [2], [3].

REFERENCES

1   Rowe, P.W.,   Energy components during the triaxial cell and direct shear tests. Geotechnique Vol 14, 1964, No 3, p 247-261.

2   Bishop, A.W.,   Shear strength parameters for undisturbed and remoulded soil specimens. Stress-strain behaviour of soils, Proceedings of the Roscoe Memorial Symposium, Cambridge 1971.

3   Bjerrum, L.,   The effective shear strength parameters of sensitive clays. International Conference on Soil Mechanics and Foundation Engineering 5, Paris 1961, Proceedings Vol 1, p 23-28.

4   Schofield, A. and Wroth, P.,   Critical state soil mechanics. Mc Graw-Hill, London 1968.

CREEP MECHANISMS IN CLAY

R. Pusch

Division of Soil Mechanics, University of Luleå, S-951 87 Luleå, Sweden

ABSTRACT

The processes involved in creep are strongly dependent on the microstructural constitution of clays. Shearing produces a translation and rotation of rigid aggregates in connection with a break-down of a certain number of particle links which are transformed to slip units consisting of groups of parallel flaky particles. A physical clay model therefore implies that an element of clay contains a certain number of slip units in a given interval of the activation energy range. This suggests that a mathematical formulation of creep strain rate in clay can be based on stochastic models similar to those suggested for metals and alloys. The application is briefly discussed in the paper.

1. INTRODUCTION

Evaluation of inter-particle forces in clays requires that spatial and temporal fluctuations of the internal stresses are taken into consideration. Thus, the structural heterogeneity which is typical of natural clays invalidates the application of the simple Eyring theory [1, 2]. The use of stochastic models which allow for microscale heterogeneity may lead to a better understanding of the particle bond nature and eventually to a way of describing and even predicting creep strain.

This report describes the phenomenology of clay creep strain and an outline of some basic characteristics of an accordingly operating stochastic model of creep in clay.

The processes involved in creep are strongly dependent on the microstructural constitution. Confining ourselves to illitic clay we know from several studies that the main characteristic feature is aggregation, which is very strong in marine clays but less obvious in fresh water-deposited clays. Since the aggregation pattern is the very key to the understanding of creep mechanisms it will be discussed in some detail in the follwoing text.

2. CLAY MICROSTRUCTURE

The presently observed microstructural pattern of all sedimentary clays was formed already at their deposition. The governing factors were the concentration, mineral composition and grain size distribution of the suspension, the flow velocity, temperature and salinity of the water and the organic constituents in the water.

Aggregation is caused by several processes. One is the preferential attachment of small particles leading to big units. Their tendency to be connected to larger particles in the course of sedimentation implies an aggregate structure of the type shown in Fig. 1.

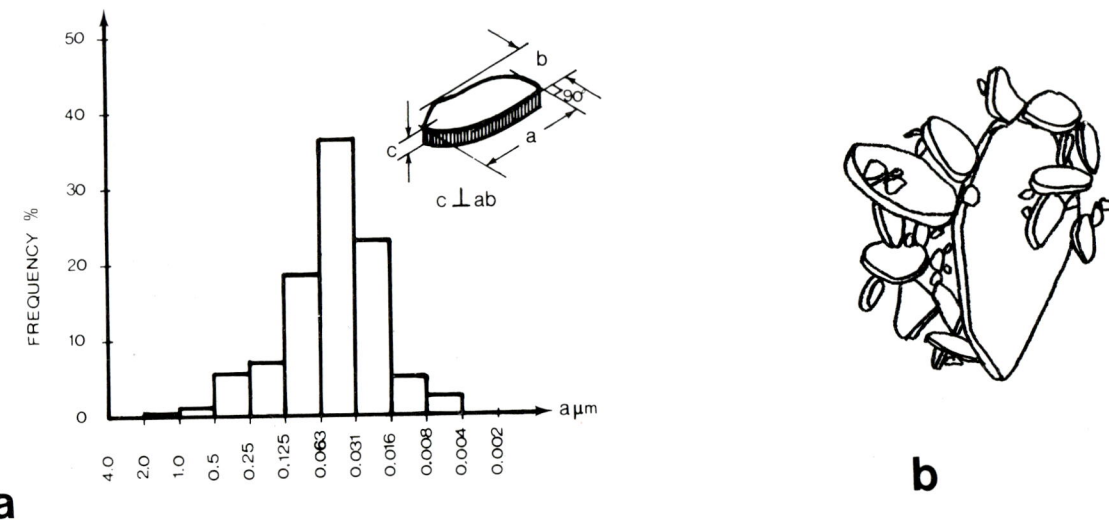

Fig. 1. a) Example of particle size distribution in the clay fraction.
b) Hypothetic picture of clay particle aggregate (cf. [1]).

In addition, a condition of minimum energy, stipulated by thermodynamics, implies a phase separation which leads to a system of particle aggregates separated by voids of various size and shape. Also, it is well established that electrolytes in the pore water govern the degree of particle aggregation. The current explanation, following classical colloid chemistry, is that the inter-particle distance is determined by the properties of electrical double-layers. Disturbance of the water lattice by certain cations which diffuse into the intercrystalline space and partially compensate the repulsion between adjacent particles may also contribute to their approach and, thus, produce aggregates.

The influence of electrolytes suggests that clays deposited in sea water should contain dense and large aggregates while fresh-water clays should consist of less dense aggregates and have a more uniform distribution of particles ("dispersion"). This difference is in fact very obvious as shown by a large number of microstructural investigations [4]. Fig. 2 illustrates the influence of salinity of the water in which deposition took place.

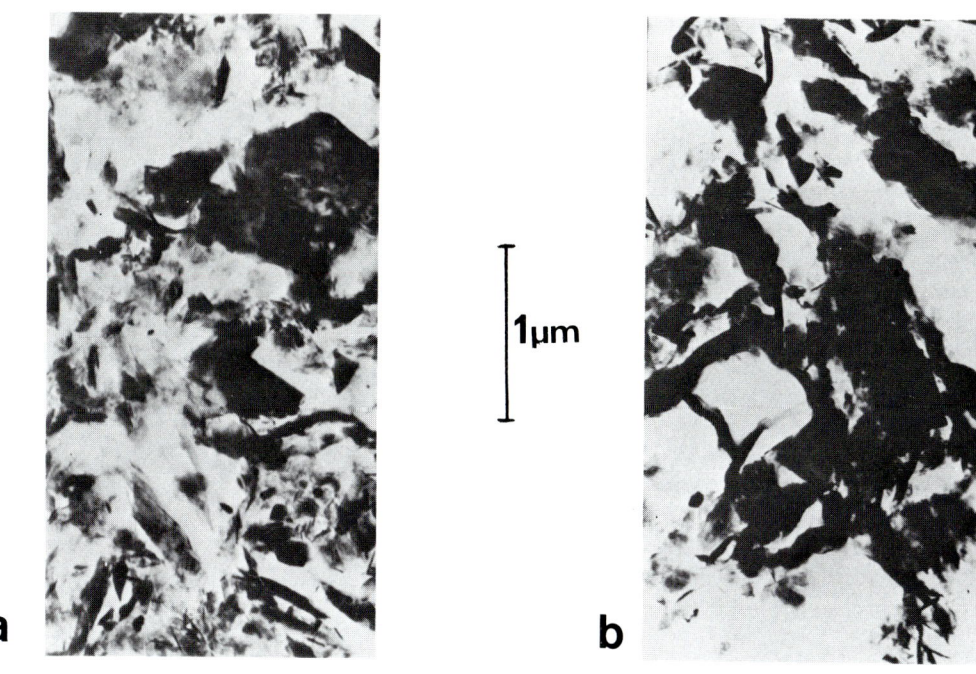

Fig. 2. Electron micrographs of ultra-thin sections of acrylate treated undisturbed soft clays with similar water content and shear strength. a) Fresh-water clay from Skå-Edeby. b) Marine clay from Lilla Edet.

Experimental work on microstructural influence on permeability [4] and on the amount of unfrozen water in freezing clay [5] confirms that soft natural illitic clays are aggregated and that the degree of aggregation strongly depends on the salinity of the water in which the sediment was formed. The average inter-particle distance in the aggregates is larger in fresh-water clays than in marine clays which means that the aggregates of the last-mentioned type of clay are much stiffer and stronger than those of fresh-water clays. This is of primary importance as concerns the bulk stress//strain behaviour of these clays.

Taking all these features into consideration it is concluded that the microstructural pattern of soft, undisturbed illitic clay is of the type shown in Fig. 3.

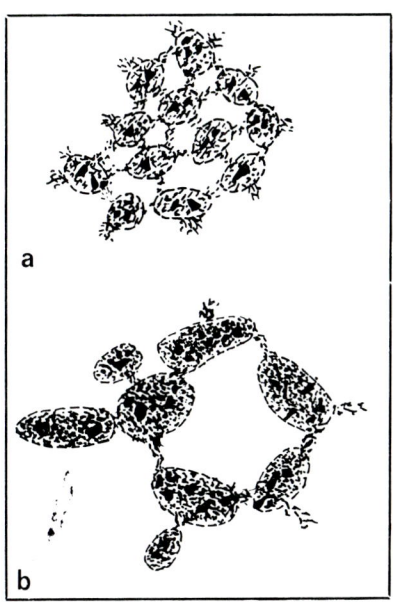

Fig. 3. Schematic clay particle arrangement. a) Clay deposited in fresh water, having small, relatively porous aggregates and small voids. b) Marine clay with large, dense aggregates separated by large voids.

3. SHEAR-INDUCED MICROSTRUCTURAL CHANGES

Fig. 3 implies that there is a considerable bond strength variation in aggregated clays. In principle the aggregates represent stronger structural elements while the particle links, which connect the aggregates, represent the weakest elements. When a sufficiently high deviator stress is applied, there will be an almost instant overcome of all energy barriers lower than a certain quantity by which plastic bulk strain is produced. Strong barriers, on the other hand, will give a negligible contribution to the bulk strain. The last-mentioned energy barriers dominate in dense aggregates while the low energy barriers are typical of particle links and porous aggregates. Accordingly, the detailed behaviour of a sheared clay specimen is characterized by a translation and rotation of rigid aggregates in connection with a break-down of a certain number of particle links. This break-down involves a reorientation leading to local groups of parallel particles (Fig. 4). This is a well-known effect - "domain formation" - which has been described by several investigators such as Emerson, Pusch and others [4, 6].

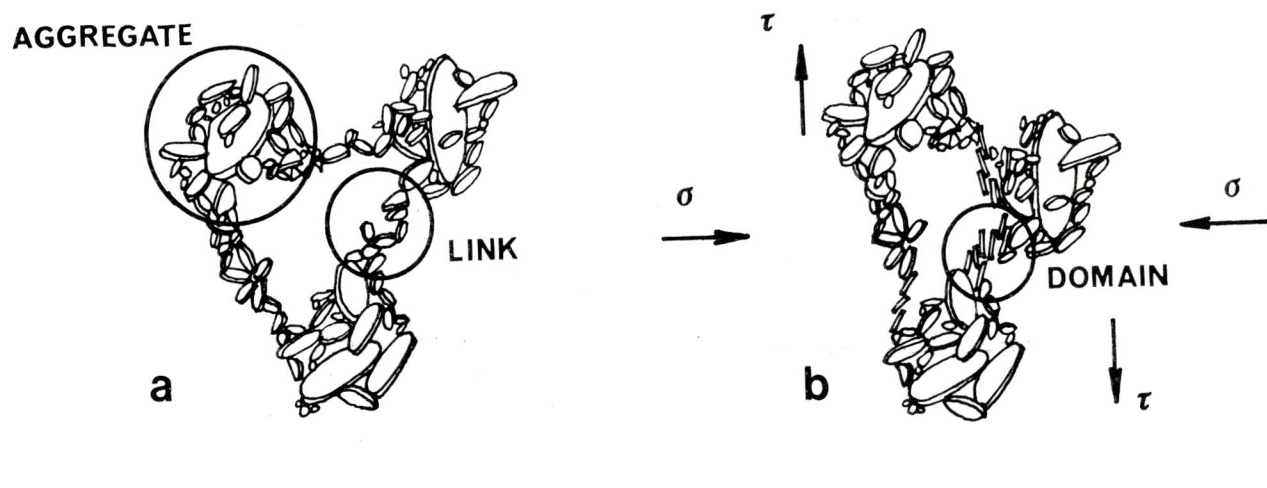

Fig. 4. Formation of domains in clay. a) Schematic detailed picture of natural microstructural network. b) Break-down of particle links resulting in domain formation. c) Electron micrograph of domains produced by simple shear in marine clay.

No doubt, the domains are slip units, the integrated movement of which gives the major contribution to the bulk shear strain. A number of interesting domain features are known from various electron microscope investigations:

- An increased deviator stress produces an increased number of domains.

- The frequency of domains is dependent on the bulk shear strain and on the aggregate size distribution. Thus, a fresh-water clay contains a much larger number of domains than a marine clay at the same shear strain, if their mechanical and physical properties are similar.

- All slowly consolidated soft, normally or slightly over-consolidated, natural illitic clays have only a very small content of domains despite the comprehensive internal shear processes which must have taken place in connection with the rather extreme compression in course of consolidation in nature.

- Domains formed when the deviator stress is increased are oriented in any direction. Not until bulk failure is approached the domains tend to join and merge into a macroscopic shear zone.

- Slip movement of domains leads to an interaction of larger aggregates.

The last point means that the shear resistance at large shear strain ("residual strength") is mainly caused by dilatancy and viscous effects. The first three points mean that both bulk stress and strain govern the density of domains, except for very low loading or strain rates which are known to produce only a small number of domains. In natural clays this is probably because the very low loading and consolidation rates may have been in phase with a delayed deformation of the aggregates and with a successive link reformation, meaning that no state of broken links and temporary rigidity of the aggregates ever occurred. This is contrary to illitic clay rapidly consolidated in the laboratory. Such clay shows numerous domains and this is also the case for clays rapidly consolidated in nature, such as the Mississippi Birdfoot Delta clays [7].

4. THE PHYSICAL CREEP MODEL

Confining ourselves to creep strain caused by an instant application of a specific deviator stress to an undrained clay specimen, the following basic properties of a physical model are suggested:

(1) The clay operates as a heterogeneous system of dense and strong structural elements (the aggregates) connected by links of various strength which are potential slip units.

(2) The formation of slip units and, thus, creep strain is insignificant at low deviator stresses (less than 1/3 of the deviator stress as determined in conventional shear tests).

(3) Stresses of high intensity (more than 2/3 of the deviator stress as determined in conventional shear tests) produce a critical number of slip units which leads to an accelerating creep rate followed by complete failure. The magnitude of such a critical stress and even its existence can certainly be debated, especially when very long periods of time are considered. The suggested limit is generally accepted, however (cf. [8]).

(4) Intermediate deviatoric stresses lead to the formation of local areas over which slip has taken place. The continuation of this slip process leads to local stress-relaxation, and hence to an _increase_ in the heights of the energy barriers for subsequent activated jumps. This is akin to work-hardening, e.g. as in metals.

Another effect arising from the redistribution of microstresses facilitated by slip will be the appearance of higher local stresses at points where previously they were relatively low. Hence some barriers to slip will decrease. The transient form of the creep suggests that the energy barrier increase outweighs this effect.

5. THE MATHEMATICAL MODEL

The behaviour of the physical model requires the following properties of the mathematical model:

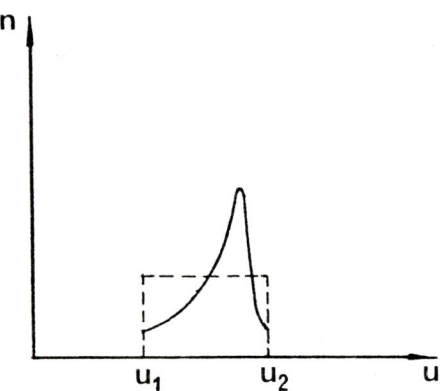

(1) There is a variation in link bond strength. This variation, which is equivalent to a variation in activation energy, is illustrated in Fig. 5. The magnitude of the energy spectrum interval and the shape of the distribution curve reflect the microstructural heterogeneity and the type of bonds (hydrogen bonds, primary valence bonds etc).

Fig. 5. Activation energy spectrum at a given time t after the onset of creep.

(2) Each element of clay contains a certain number of slip units in a given interval of the activation energy range.

(3) Displacement of a slip unit can be assumed to take place as the shifting of a patch of atoms or molecules as one unit along a geometrical slip plane.

(4) In course of the creep the low energy barriers are triggered early. New slip units come into action at the lower activation energy end of the spectrum in Fig. 5. This end represents a "generating barrier", while the high u-end is an absorbing barrier.

(5) While one can take both effects referred to in points 4 and 5 of Section 4 into account in the model, we shall, for the sake of simplicity and adequacy at this stage of development, disregard from "softening" mentioned in point 5, so that the model will allow only for jumps which, when they take place, bring the given slip unit up against a barrier by an amount δu higher than the previous one.

(6) A changed deviator stress affects the rate of shift of the energy spectrum to higher u-values.

Considering the case of a clay element subjected to a constant deviator stress ("intermediate") we assume the number of barriers of height u to be $n(u,t) \cdot \delta u$ where δu is the energy interval between successive jumps of a unit, and t the time. As a first approximation the initial distribution $n(u,0)$ at the application of the creep stress can be taken as constant (broken line in Fig. 5, "box shape"):

$$n(u,0) = \text{const.}, \qquad u_1 \leq u \leq u_2 \qquad (1)$$

The change of the activation energy in course of creep means that the number of slip units is determined by the outflux from any u-level into the adjacent higher energy interval and by a simultaneous influx into the interval from $u-\delta u$. To describe this process we can modify Feltham's theory [9] which was developed for metals and alloys but which should work for any material which is structurally changed in course of creep.

Following [9] but allowing for "uphill" rather than "downhill" jumps we obtain the expression for the rate of change of $n(u,t)$ with time:

$$\partial n(u,t)/\partial t = \nu\{-n(u+\delta u,t)\exp[-(u+\delta u)/kT]+n(u,t)\exp(-u/kT)\} \quad (2)$$

This differs from the corresponding expression in [9] only by the signs of the terms on the right-hand side. It can be written as:

$$\partial n/\partial t = -(\nu\delta u)\partial[n \exp(-u/kT)]/\partial u \quad (3)$$

where
δu = width of an energy spectrum interval
ν = vibrational frequency ($\sim 10^{11}$ per second)
k = Boltzmann's constant
T = absolute temperature

Let us now introduce Feltham's transition probability density N to describe the time-dependent energy shift:

$$N(u,t) = n(u,t)\exp(-u/kT) \quad (4)$$

By putting $r = \exp(u/kT)$ we find:

$$\partial N(r,t)/\partial t = -D\partial N(r,t)/\partial r \quad (5)$$

where
$D = \nu\delta u/kT$

The solution of Eq. (5) is of the type:

$$N = \emptyset(r-Dt) \quad (6)$$

This expression is of the "travelling thermal front" type with the shape of N remaining fixed but its magnitude changing with time. One simple N-spectrum suggested by Feltham is shown in Figs. 6 and 7:

$$N = N_o[(r-Dt)^o + C(r-Dt)^1], \quad C = \text{const.}, \quad r_1 \leq r \leq r_2 \quad N(u,t) > 0 \quad (7)$$

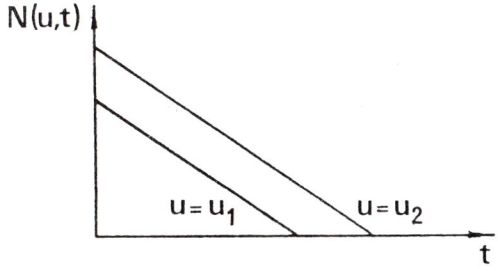
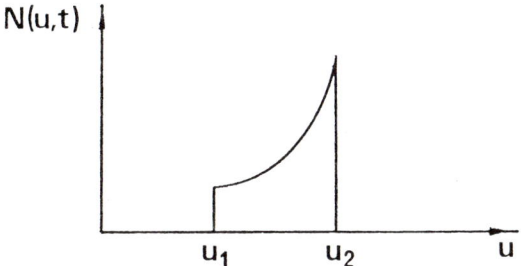

Fig. 6. Example of transition probability density versus time

Fig. 7. $N(u,t)$-distribution for $t = $ const

The criterion that low barriers are triggered early in the creep is allowed for by the model. Thus, for small r-values n will decrease rapidly for a given t, compared with n-values for larger u.

Again applying Feltham's theory we find that if each transition of a slip unit between consecutive barriers gives the same contribution to the bulk strain, then the (shear) strain rate is proportional to the integral of N(u,t) over the available u-spectrum:

$$\dot{\gamma} \propto \int_{u_1}^{u_2} N(u,t)du \qquad (8)$$

This yields with Eq. (7):

$$\dot{\gamma} \propto (1-t/t_o) \qquad (9)$$

with $t \leq t_o$ as boundary condition.

The appropriate constant of proportionality and the value of t_o depend on the deviator stress, temperature and structural details of the slip process.

The creep is then of the form:

$$\gamma = \alpha t - \beta t^2, \, (t \leq \frac{\alpha}{2\beta}), \qquad (10)$$

i.e. it starts off linearly and then dies out. This is in sufficient agreement with observations to form a basis for further work, which will comprise an extension of the theory to account also for activated jumps to lower barriers, drained conditions etc.

REFERENCES

[1] Pusch, R., Shear deformation of clay microstructure. Proc. VIIth Int. Congr, Rheology, Gothenburg, Sweden 1976 pp.270-271.

[2] Pusch, R., Rate process theory and clay microstructure. Constitutive equations of soils. Prepr. Spec. Sess. 9, IX Int. Conf. Soil Mech. a. Found. Engng. Tokyo 1977 pp. 223-227.

[3] Pusch, R., Clay particles. Handl. nr 40. Transactions. Byggforskningen 1962 pp. 16-40.

[4] Pusch, R., Clay microstructure. Document D8:1970. Nat. Swed. Build. Res. 1970.

[5] Pusch, R., Unfrozen water as a function of clay microstructure. Proc. Int. Symp. Ground Freezing, Ruhr-University Bochum, Germany 1978 pp. 103-107.

[6] Emerson, W.W., The structure of soil crumbs. J. Soil Sci., vol 10(2) 1959 pp. 235-244.

[7] Bennett, R.H., Bryant, W.R. and Keller, G.H., Clay fabric and geotechnical properties of selected submarine sediment cores from the Mississippi delta. U.S. Dep. of Commerce. Nat. Oceanic and Atm. Adm., Prof. Paper 9 1977.

8 Prevoust, J.-H., Undrained stress-strain-time behaviour of clays. Norges Geotekniske Institutt, Publ. Nr. 117, 1977 p.3.

9 Feltham, P., A simple stochastic model of low-temperature creep and stress-relaxation in solids. Proc. VIIth Int. Congr. Rheology, Gothenburg, Sweden 1976 pp. 166-167.

PLASTIC FLOW MECHANISMS AND RHEOLOGICAL PROPERTIES OF THE EARTH'S MANTLE

G. Ranalli

Department of Geology, Carleton University, Ottawa, Canada K1S 5B6

ABSTRACT

Creep of silicate polycrystals is governed by the mobility of lattice defects. The relative importance of bulk diffusivity, grain-boundary diffusivity, dislocation motion, grain-boundary sliding, and liquid diffusivity (in two-phase systems) is a function of temperature, pressure and grain size. Assuming an olivine (Fo_{90}) composition for the upper mantle, applied deformation maps are constructed that allow the prediction of creep laws, strain rates and effective viscosities as a function of depth for various geothermal gradients. Similar calculations for the lower mantle require an estimate of the variations of activation energy and activation volume with depth, and an evaluation of the effect of polymorphic phase transformations on rheological parameters. Using the best available estimates of these quantities, it is concluded that the whole mantle, with the possible exception of local low-viscosity zones within the asthenosphere, flows by power-law creep with effective viscosities that do not vary with depth by more than two orders of magnitude.

1. CREEP MECHANISMS IN MANTLE MATERIAL

The rheological properties of the mantle affect the Earth's response to surface loads and consequently could be inferred from the study of glacio-isostatic rebound; such studies, however, have led to conflicting results [1,2,3,4], which perhaps show that the rheology-induced differences in postglacial rebound are not sufficiently strong to be detected with the present data. Another approach is offered by the extrapolation of creep laws for silicates to mantle pressure, temperature, and strain-rate conditions [5,6,7,8]. Although strain rates in the laboratory are orders of magnitude larger than geological strain rates, the general nature of the governing equations, the opposite influence of temperature and strain rate on creep stress, and the similarity of flow textures in naturally and experimentally deformed materials [9], provide sufficient justification for this method.

Creep equations in solids have been reviewed elsewhere [7,8,10]. For stresses of the orders of magnitude likely to obtain in the mantle (from a few tenths to a few hundreds of bars; 1 bar = 10^5 Pa), the two main competing mechanisms are power-law creep (related to dislocation migration)

$$\dot{\varepsilon}_{ij} = \frac{3^{(n+1)/2}}{2} \frac{AD\mu b}{kT} \left(\frac{\tau}{\mu}\right)^{n-1} \frac{\sigma'_{ij}}{\mu} \qquad (1)$$

and newtonian creep (related to vacancy diffusion)

$$\dot{\varepsilon}_{ij} = 21 \frac{D^*\mu\Omega}{kTd^2} \left(\frac{\sigma'_{ij}}{\mu}\right) \qquad (2).$$

The diffusion coefficient in equation (1) is given by

$$D = D_o \exp\left(-\frac{E+pV}{RT}\right) \qquad (3)$$

and in equation (2) it is the sum of bulk and grain boundary diffusivity

$$D^* = D + \frac{\pi\delta}{d} D_{GB} \qquad (4).$$

In the above equations, $\dot{\varepsilon}_{ij}$ is the strain rate, σ'_{ij} the deviatoric stress, τ the square root of the second invariant of the deviatoric stress tensor, p and T pressure and temperature, μ the rigidity, and k and R Boltzmann and the gas constant, respectively.

Olivine (Fo_{90}) is the volumetrically predominant phase in the upper mantle; the values of its rheological parameters have been compiled from various sources [10] and are listed in Table 1.

Table 1: Rheological parameters for olivine.

Dorn parameter	$A = 0.7$
Dorn parameter	$n = 3.0$
Burgers vector	$b = 6.98 \cdot 10^{-10}$ m
Atomic volume (O^{--})	$\Omega = 1.15 \cdot 10^{-29}$ m^3
Grain size	$10^{-4} \leq d \leq 10^{-2}$ m
Grain boundary width	$\delta = 1.4 \cdot 10^{-9}$ m
Preexponential diffusivity	$D_o = 10^{-1}$ m^{2}s^{-1}
Activation energy (lattice)	$E = 5.4 \cdot 10^5$ J mole^{-1}
Activation energy (grain boundary)	$E_{GB} = 3.6 \cdot 10^5$ J mole^{-1}
Activation volume (O^{--} rate-controlling species)	$V = 11 \cdot 10^{-6}$ m^3 mole^{-1}

If a small fraction of liquid phase is present (e.g., incipient melting in the seismic low-velocity zone) an additional high-diffusivity path becomes available, and one further term should be added to the effective diffusivity in equation (4). However, liquid phase diffusivity depends on the viscosity of the melt which in turn may vary quite substantially for very narrow melt channels (electroviscosity): lack of knowledge of the geometry of partially molten zones in the mantle thus prevents an accurate estimation of the degree of change in rheological properties induced by the melting [7,8].

Another mechanism which enhances linear creep, grain-boundary sliding [11] or superplasticity, appears to require very high stresses and is therefore unlikely to be relevant to mantle creep except in narrow bands of stress concentration [8].

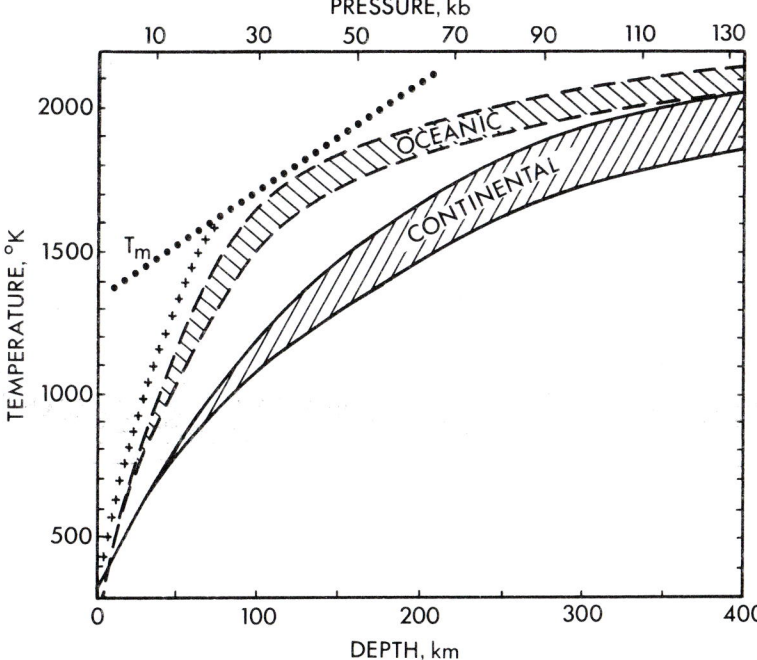

Fig. 1. Continental and oceanic geotherms; also shown are an orogenic geotherm (crosses) and upper mantle solidus (T_m).

2. RHEOLOGICAL PROPERTIES OF THE UPPER MANTLE

The strain rates (and thus the predominant creep mechanism) in dislocation and diffusion creep are a function of the rheological parameters of the material and of temperature and pressure. Fig. 1 shows the pressure-depth dependence and the range of continental and oceanic geotherms, according to most current estimates [12,13]; also shown are an "orogenic" geotherm (of the type for instance that may pertain to the Cordilleran thermal anomaly zone in Canada [14]) and the upper mantle solidus.

The variations of rigidity with depth are calculated using a first order p,T-dependence [7] down to a depth of $z = 400$ km; at larger depths a spherically symmetric Earth model [15] is adopted.

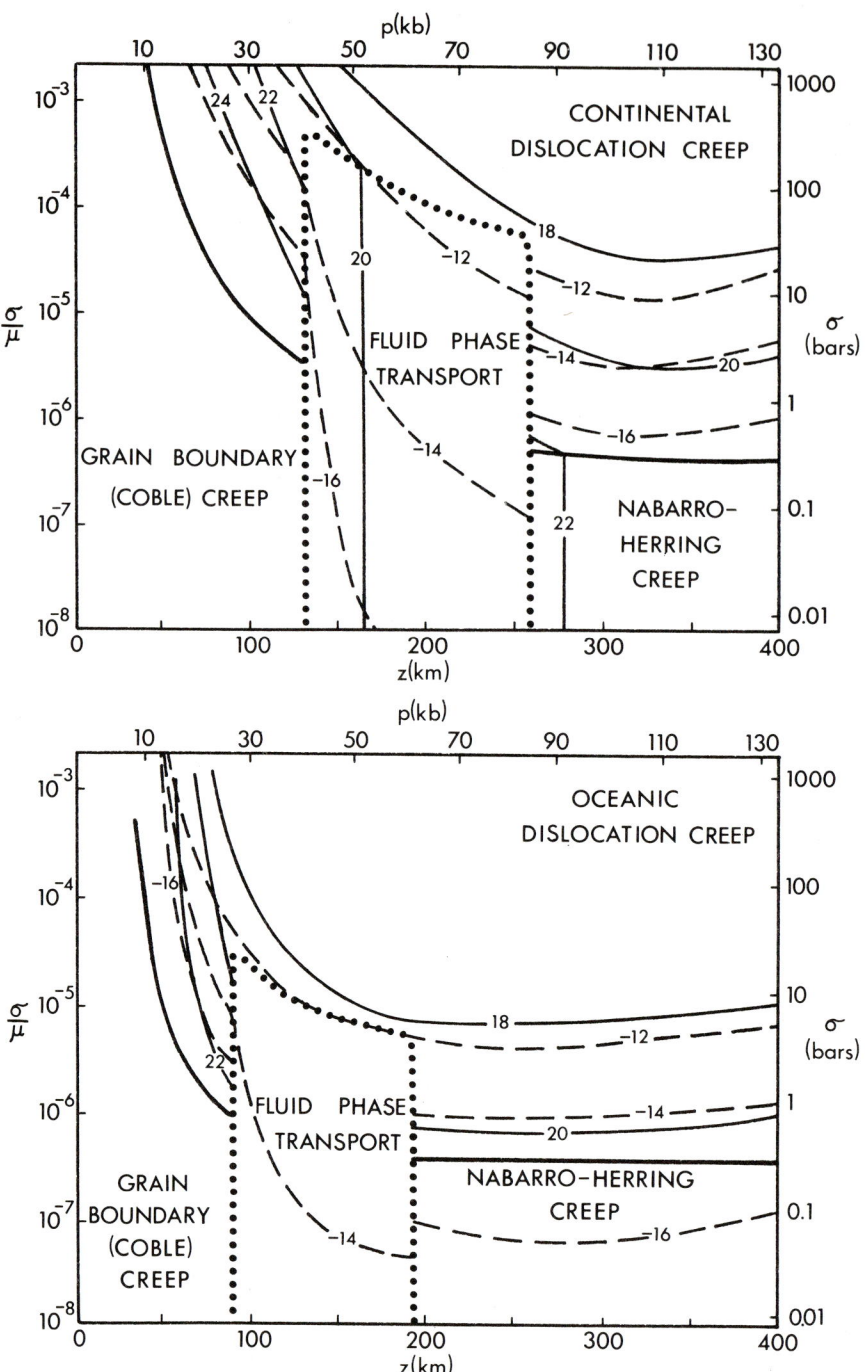

Fig. 2. Applied deformation maps for continental (top) and oceanic (bottom) upper mantle; see text for details.

Fig. 2 shows applied deformation maps for the upper mantle, constructed using equations (1-4) for continental and oceanic geotherms and d = 0.1 cm. Thick continuous lines separate the depth-stress fields where one creep mechanism is predominant; thin continuous lines represent loci of constant effective viscosity, and dashed lines loci of constant strain rate (both as powers of ten). These maps allow the determination of the rheology of the upper mantle as a function of depth and stress (shear stresses are assumed to be predominant), and have been introduced relatively recently in geological methodology [7]. The areas enclosed by a dotted line indicate the possible effects of 1% partial melt in the seismic low-velocity zone if complete wetting of grain boundaries occurs and electroviscous forces are negligible [8]. It is at present impossible to say if these effects are present; it is entirely possible that the seismic LVZ has no

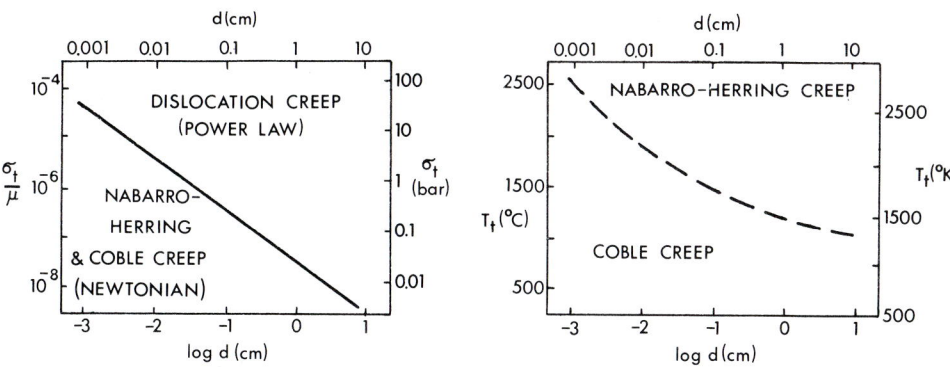

Fig. 3. Transition stress between power-law and diffusion creep (left), and transition temperature between Nabarro-Herring and Coble creep (right), as a function of grain size.

signature in steady-state rheology.

One important feature of the transition stress σ_t between dislocation and diffusion creep is that, since D^* tends to D with increasing T, σ_t tends to an asymptotic value that is a function of grain size only. In the linear field, the transition temperature between Nabarro-Herring creep (bulk diffusivity predominant) and Coble creep (grain boundary diffusivity predominant) is also a function of grain size. The d-dependence of σ_t and T_t is shown in Fig. 3. Although the mean grain size of the mantle is not known, for the most likely values σ_t is between 0.05 and 5 bar. For most stresses of geodynamic significance, therefore, the upper mantle should flow by power-law creep.

Another way of assessing the relative importance of diffusion and dislocation creep is by comparing viscosities in the two types of flow. For diffusion creep, the viscosity depends on grain size and can be derived from equation (2)

$$\eta = \frac{kTd^2}{42D^*\Omega} \tag{5}$$

For dislocation creep, the effective viscosity is strain-rate dependent and can be derived from equation (1)

$$\eta = \frac{2^{-2/3}}{3^{2/3}} \mu^{2/3} \left(\frac{kT}{ADb}\right)^{1/3} \dot{\epsilon}^{-2/3} \tag{6}$$

Equations (5) and (6) yield η as a function of depth. Fig. 4 shows $\eta(z)$ (1 Poise = 0.1 Pa s) for two different grain sizes in diffusion creep, and for three strain rates in power-law creep. If $d = 0.1$ cm is a representative grain size for the upper mantle, dislocation creep at geological strain rates is associated with the lower viscosity and is therefore predominant; the situation would be different for $d \lesssim 0.01$ cm, but grain sizes in the upper mantle are likely to be larger than this. Also, lateral viscosity differences of up to two orders of magnitude are present in the upper mantle, being a consequence of regional variations in geothermal gradients [8].

Fig. 5 shows η as a function of $\dot{\epsilon}$ at a depth of 145 km for continental and orogenic geotherms [16]. The shaded rectangle represents the values of these quantities usually thought to be associated with geodynamic processes such as glacio-isostatic rebound and convection currents. It is noteworthy that a transition from power-law creep (sloping part of $\eta(\dot{\epsilon})$) to newtonian (horizontal part) is a distinct possibility in regions of high geothermal

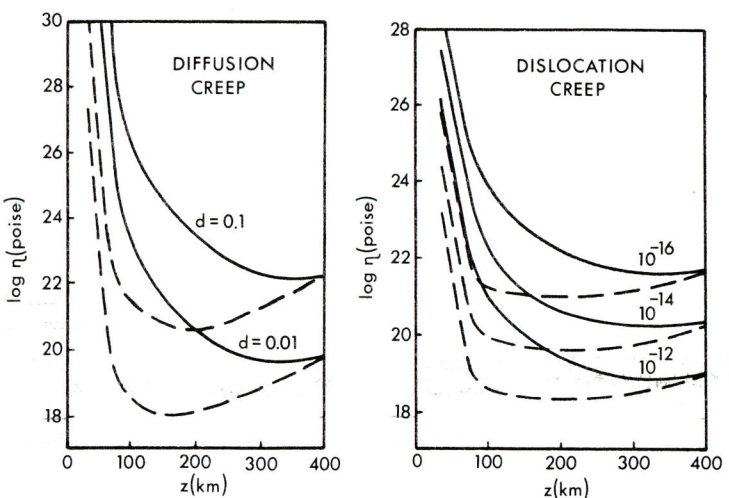

Fig. 4. Linear viscosity in diffusion creep (left) and effective viscosity in dislocation creep (right). Full lines: continental geotherms; dashed lines: oceanic geotherms.

gradient. Not only the effective viscosity, but also the creep law of the upper mantle may change in space and time during geodynamic processes.

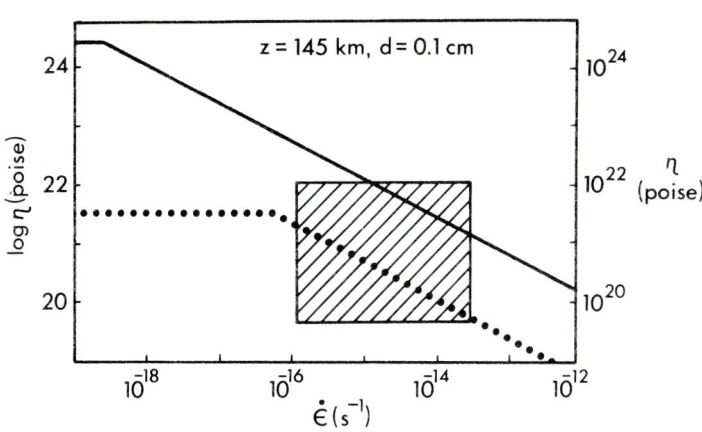

Fig. 5. Viscosity v. strain rate for continental (full line) and orogenic (dotted line) geotherms; see text for details.

3. RHEOLOGICAL PROPERTIES OF THE TRANSITION ZONE AND LOWER MANTLE

In the depth interval $400 \lesssim z \lesssim 900$ km, mantle material goes through a series of polymorphic phase transitions. The determination of the rheological properties in the transition zone requires the estimation of the activation energy of the phases involved. Also, the pressure dependence of the activation volume, neglected in the upper mantle, becomes important in the lower mantle due to the larger pressure range. These two factors greatly complicate the problem.

A first attempt to estimate η has been made simply by using the upper mantle values for olivine [10] : inasmuch as the dependence of E and V on depth is of opposite sign, there is some justification for this. Elastic continuum theory [17] has been used to show that V at the base of the mantle is reduced by a factor of two or more. Other indirect estimates [18] give an even larger decrease (to approximately $4 \cdot 10^{-6}$ m^3 mole^{-1}), and a total increase of E across the transition zone of about $0.6 \cdot 10^5$ J mole^{-1}.

In the following calculations, smoothed values for E and V [18] are used in conjunction with a spherically symmetric Earth model for p and μ [15]. The temperature is determined by assuming that T = 2000°K at z = 400 km, and taking a constant adiabatic temperature gradient in the mantle below. Fig. 6 shows the resulting newtonian viscosity for d = 0.1 cm and power-creep effective viscosity for $\dot{\varepsilon} = 10^{-14}$ s^{-1}. Upper mantle values are from the previous section; transition zone and lower mantle values are obtained from equations (5) and (6). The assumed geothermal gradients are shown on each curve in °K km^{-1}. For the chosen values of the parameters, dislocation creep is the predominant flow mechanism; the variation of viscosity with depth is not large, and indeed η may be approximately constant or even decreasing with depth if the adiabatic gradient is sufficiently large ($0.3 \lesssim dT/dz \lesssim 0.6$ is probably a reasonable guess). A mantle with nearly constant viscosity is also derived from analysis of glacio-isostatic rebound [2,3].

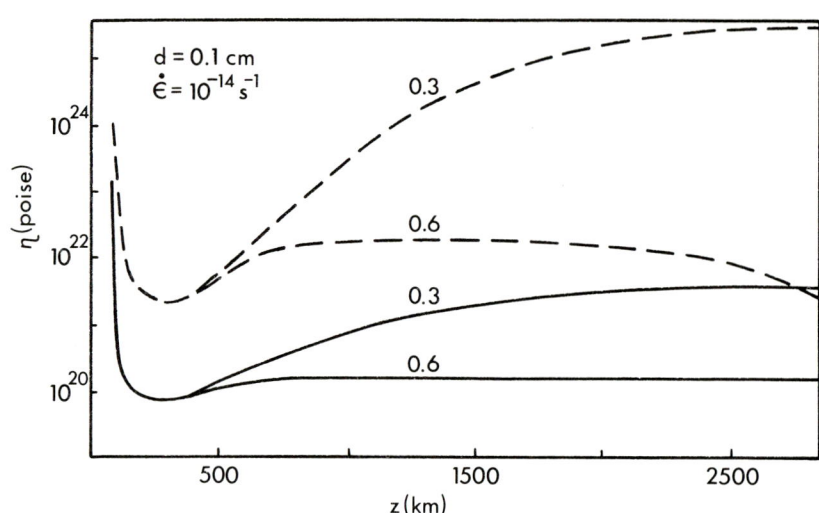

Fig. 6. Mantle viscosity taking into account variations of E and V with depth. Dashed line: newtonian creep; full line: power-law creep. See text for details.

4. SUMMARY

The two main results of the present work therefore are

(i) dislocation creep is the predominant mode of flow of the mantle for realistic values of the parameters and with the possible exception of zones of partial melting the rheology of which is at present difficult to assess; and,

(ii) the effective viscosity of the mantle below the low-viscosity zone (asthenosphere) does not increase with depth by more than one or two orders of magnitude and may well be constant ($\eta \simeq 10^{21}$ Poise) if the adiabatic gradient is sufficiently large.

Given the uncertainties in the quantities involved, it is the trends rather than the actual values of effective viscosity which are of importance. Even with all its limitations, however, the "direct" approach used here is a powerful tool in the study of mantle rheology.

This work was supported by NRC Canada under grant no. A7971. Thanks are due to W. Yzerdraat for help in the computations.

REFERENCES

[1] Walcott, R.I., Structure of the Earth from glacio-isostatic rebound. Ann. Rev. Earth Planet. Sci. 1 (1973), 15-37.

[2] Brennen, C., Isostatic recovery and the strain-rate dependent viscosity of the Earth's mantle. J. Geophys. Res. 79 (1974), 3993-4001.

[3] Peltier, W.R. and J.T. Andrews, Glacial-isostatic adjustment - I. The forward problem. Geophys. J.R. Astr. Soc. 46 (1976), 605-646.

[4] Crough, S.T., Isostatic rebound and power-law flow in the asthenosphere. Geophys. J.R. Astr. Soc. 50 (1977), 723-738.

[5] Weertman, J., The creep strength of the Earth's mantle. Rev. Geophys. Space Phys. 8 (1970), 145-168.

[6] Kirby, S.H. and C.B. Raleigh, Mechanisms of high-temperature, solid-state flow in minerals and ceramics and their bearing on the creep behaviour of the mantle. Tectonophysics 19 (1973), 165-194.

[7] Stocker, R.L. and M.F. Ashby, On the rheology of the upper mantle. Rev. Geophys. Space Phys. 11 (1973), 391-426.

[8] Ranalli, G., Regional models of the steady-state rheology of the upper mantle. In: N.-A. Mörner (Ed.), Earth Rheology, Isostasy and Eustasy, Wiley-Interscience, London (in press, 1978).

[9] Nicolas, A. and J.P. Poirier, Crystalline Plasticity and Solid State Flow in Metamorphic Rocks. Wiley-Interscience, London 1976, 444 pp.

[10] Ranalli, G., Steady-state creep in the mantle. Ann. Geofis. (in press, 1978).

[11] Twiss, R.J., Structural superplastic creep and linear viscosity in the Earth's mantle. Earth Planet. Sci. Letters 33 (1976), 86-100.

[12] Froidevaux C., G. Schubert and D.A. Yuen, Thermal and mechanical structure of the upper mantle: a comparison between continental and oceanic models. Tectonophysics 37 (1977), 233-246.

[13] Mercier, J.-C. and N.L. Carter, Pyroxene geotherms. J. Geophys. Res. 80 (1975), 3349-3362.

[14] Caner, B., Electrical conductivity structure in western Canada and petrological interpretation. J. Geomag. Geoelectr. 22 (1970), 113-129.

[15] Anderson, D.L. and R.S. Hart, An Earth model based on free oscillations and body waves. J. Geophys. Res. 81 (1976), 1461-1475.

[16] Ranalli, G., Regional rheology of the upper mantle in Canada as determined from solid-state flow laws of olivine. Can. Geophys. Union, 5th Ann. Meeting Abstracts (1978), 26.

[17] O'Connell, R.J., On the scale of mantle convection. Tectonophysics 38 (1977), 119-136.

[18] Sammis C.G., J.C. Smith, G. Schubert and D.A. Yuen, Viscosity-depth profile of the Earth's mantle: effects of polymorphic phase transitions. J. Geophys. Res. 82 (1977), 3747-3761.

THE SIMPLE FLUID WITH FADING MEMORY AS A RHEOLOGICAL MODEL FOR STEADY-STATE FLOW OF ROCKS

C. C. Ferguson

Department of Geology, Nottingham University, Nottingham NG7 2RD, U.K.

ABSTRACT

The phenomenological approach to determining constitutive equations has rarely shown itself capable of producing predictive descriptions of complex materials. The equations used in geology are essentially of the form

$$\underline{T} = -p\underline{I} + H(\underline{D})$$

where the stress $\underline{T}$ at a given $(\underline{x}, t)$ is entirely determined by the rate-of-deformation $\underline{D}$ at that instant in time. H may be linear or non-linear but deformations taking place at other times are assumed to be irrelevant. Rocks are manifestly more complicated than this and a more general axiomatic model - the simple fluid - is outlined in this paper in which the stress at $(\underline{x}, t)$ is determined by the entire history of $\underline{D}$. In particular we examine some restrictions imposed by deformation geometry, and by the 'fading memory' approximation for very slow flows. Some implications for rigid crystal rotations in porphyroblastic rocks are considered: the classical analysis here assumes a Newtonian rock matrix and we explore the effects of employing, instead, a simple fluid with fading memory model.

1. INTRODUCTION

Over the past decade geologists have increasingly appreciated the need to develop rheological equations that provide predictive bulk descriptions of rock behaviour during large permanent deformations, and which can be used for solving boundary-value problems. Previously a 'Newtonian' model was widely used although this choice was often dictated by the desire to retain mathematical tractability. Several different types of investigation can lead to useful rheological descriptions although geologists have largely favoured the phenomenological approach in which constitutive equations are derived from experimentally determinable bulk properties.

The most common non-Newtonian phenomenon observed by rheologists using viscometric tests (e.g. Couette flow, cone-in-plate flow) is shear dependent viscosity, usually following a so-called 'power law'. Thus in simple shear flow in a cartesian co-ordinate system x_i with velocity components

$$v_1 = \dot{\gamma} x_2, \qquad v_2 = v_3 = 0 \tag{1}$$

the power law is usually written

$$t_{12} = K \dot{\gamma}^n \tag{2}$$

in which $\dot{\gamma}$ is the shear rate, and t_{12} is the shear stress (t_{ij} are components of the stress tensor). When $n = 1$, (2) reduces to Newton's law with K equivalent to μ, the Newtonian viscosity; when $n < 1$, the fluid is referred to as pseudoplastic, and when $n > 1$ as dilatant (even though many such fluids do not dilate on shearing in the usual sense of the word). Viscometric tests on elasticoviscous fluids are characterised, in addition, by the presence of non-zero normal-stress differences. Thus in steady simple shear (1) the stress distribution for an elasticoviscous fluid can be written [1],

$$\left. \begin{array}{ll} t_{11} - t_{33} = \sigma_1(\dot{\gamma}), & t_{12} = \dot{\gamma} \, \eta(\dot{\gamma}) \\ t_{22} - t_{33} = \sigma_2(\dot{\gamma}), & t_{13} = t_{23} = 0 \end{array} \right\} \tag{3}$$

η is the viscosity function, σ_1 and σ_2 are the 'first' and 'second' normal-stress differences, and the three functions together are referred to as the viscometric functions. As we shall see in Section 3, fluids may be capable of a wide range of material responses that cannot be detected using viscometric experiments. The phenomenological approach is therefore rarely capable of producing adequate descriptions of complex materials.

Geologists now recognise a number of deformation mechanisms that are probably capable of producing natural steady-state (or pseudo steady-state) flow in rocks: these include mechanisms such as cataclasis and various types of diffusional creep [2; 3] that can sometimes be recognised by their distinctive microstructures [4 but cf. 5; 6] although little relevant experimental work is available. In contrast there is considerable phenomenological support - based on axially symmetric shortening experiments - for the contention that high temperature secondary steady-state creep of rocks can be represented by the well known power-law equation proposed by Weertman [7],

$$\dot{\varepsilon} = A \exp(-Q/RT) \sigma^N \tag{4}$$

in which the stress exponent N usually takes values between 2 and 8 [8]. Note that N corresponds to $1/n$ in (2) so that Weertman's equation is that of a pseudoplastic fluid.

T.E.M. studies of many experimentally deformed rocks show dislocation features such as subgrain walls, loops and dipoles that indicate climb as well as glide, thus supporting the idea that dislocation creep is the mechanism responsible for producing the observed power-law relationship. Also, a comparison of experimentally and naturally deformed rocks reveals a remarkable similarity in terms of operating crystallographic slip systems, nature and orientation of dislocations, and overall microstructural patterns [9, 10]. Consequently the power-law creep model has found wide support and has become the 'preferred' rheological model for rock deformation throughout much of the lithosphere.

Before discussing power-law creep of rocks further, we outline in the next section an alternative approach to rheology that is axiomatic in that material behaviour is deduced from constitutive assumptions that are regarded as definitions of certain classes of materials. The work of Truesdell and Noll [11], though of celebrated difficulty, is the most complete exposition of this approach; a brief but excellent introduction in the geological literature is given by Hobbs [12], an apparently neglected paper to which the writer is happy to acknowledge his indebtedness.

2. AXIOMATIC THEORY OF FLUIDS

If $\underline{X}$ denotes the position of a particle at time t_o and $\underline{x}(\underline{X},t)$ its position at time t, then the deformation gradient tensor $\underline{F}$ with components $\partial x_i/\partial X_A$ relates the initial span $d\underline{X}$ of a material element to its current span $d\underline{x}$ by

$$d\underline{x} = \underline{F}\, d\underline{X}.$$

The 'motion' is the function

$$\underline{x} = \underline{f}(\underline{X}, t)$$

which describes the trajectory of the particle initially at $\underline{X}$. $\underline{F}$ is thus identified with the gradient of the 'motion',

$$\underline{F} = \text{grad}\ \underline{f}$$

while the velocity $\underline{v}$ of the particle is

$$\underline{v} = d\underline{x}/dt.$$

If $\underline{L}$ (= grad $\underline{v}$) denotes the velocity gradient tensor, the rate-of-deformation tensor, $\underline{D}$, is defined as the symmetric part of $\underline{L}$,

$$\underline{D} = \tfrac{1}{2}(\underline{L} + \underline{L}^T).$$

The constitutive equations outlined in this section are expressed in terms of either $\underline{F}$ or $\underline{D}$. Also used are the Rivlin-Ericksen tensors $\underline{A}_1, \underline{A}_2, \ldots, \underline{A}_n$ in which $\underline{A}_1 = 2\underline{D}$, the succeeding tensors being obtained from the recurrence formula

$$\underline{A}_{n+1} = \underline{\dot{A}}_n + (\underline{L}^T \underline{A}_n + \underline{A}_n \underline{L}). \qquad (5)$$

Successive Rivlin-Ericksen tensors therefore involve higher time derivatives of $\underline{D}$ thus enabling successively more complicated time-dependent effects to be represented.

The most general fluid to be considered here is the simple fluid ('memory fluid'); indeed, the theory of simple fluids is so general that almost all the constitutive equations that have been proposed in the literature are special cases of it. Simple fluid theory supposes that the stress at a given point and time is determined by the entire history of $\underline{F}$ so that, if we assume no volume change (i.e. det $\underline{F} = 1$), the constitutive equation may be written

$$\underline{T} = -p\underline{I} + \mathcal{H}_{s=0}^{s=\infty} [\underline{F}] \qquad (6)$$

$\underline{T}$ is the total stress, p the pressure (which, as usual, is indeterminate for incompressible fluids), $s = t - t_o$, and $\mathcal{H}$ is a functional. Although the current stress in a memory fluid may depend on the entire history of deformation, we might expect this dependence to be less strong for ancient deformations compared with recent ones. This idea leads to the theory of fading memory that we return to later.

A Rivlin-Ericksen fluid is one in which the stress is a function of the first n Rivlin-Ericksen tensors (that is, it depends on the velocity gradient and its higher time derivatives),

$$\underline{T} = -p\underline{I} + H(\underline{A}_1, \underline{A}_2, \ldots, \underline{A}_n). \qquad (7)$$

H is an isotropic function, though not necessarily linear. It is not correct to assume that, with increasing n, the stress dependence on deformation history is, in the limit, completely represented (so that the $n = \infty$ Rivlin-Ericksen fluid is equivalent to equation 6). This is easily seen by considering an arbitrary motion that is suddenly brought to a halt. If (7) applies, all tensors become zero when the motion stops. Therefore, although many time-dependent effects can be represented by (7), the phenomenon of stress relaxation, for example, cannot.

The constitutive equation for a Reiner-Rivlin fluid assumes that the stress at a point at a given time is entirely determined by the rate-of-deformation at that instant of time,

$$\underline{T} = -p\underline{I} + H(\underline{D}). \qquad (8)$$

No restrictions of linearity are imposed on H but deformations taking place at other times are assumed to be irrelevant so that the time-dependent behaviour characteristic of elasticoviscous fluids cannot be represented. A Newtonian fluid is one in which the function H in (8) is linear. The slope of the straight line relating $\underline{T}$ and $\underline{D}$ is physically associated with the Newtonian viscosity, μ, so that the constitutive equation is usually written,

$$\underline{T} = -p\underline{I} + 2\mu\underline{D} . \tag{9}$$

The Weertman equation (4) and similar power-law equations used by geologists to represent dislocation creep are clearly special cases of the Reiner-Rivlin equation (8). And yet experimentally rocks exhibit phenomena, such as stress relaxation, which show that they are more complicated than Reiner-Rivlin fluids. We therefore make the assumption that the simple fluid is a more acceptable rheological model for high-temperature steady-state flow of rocks. This implies that the stress at a given time is influenced by the history of deformation, so that identical rocks arriving at the same finite strain state by different deformation paths might be expected to exhibit different stress states. Of course, the constitutive equation for the memory fluid is so general as to be useless for any practical application; it is inconceivable, for instance, that a set of experiments could be devised that would completely characterise the functional $\mathcal{H}$. We therefore proceed by examining the asymptotic approximations that result when special deformation geometries, or restricted deformation rates, are considered.

3. FLOW GEOMETRY AND RATE LIMITATIONS ON THE MEMORY FLUID

Flows for which the history of 'deformation' (history of stretch) does not depend on the instant of observation t, but only on the time lag $s = t - t_o$ are known as constant stretch history (c.s.h.) flows. That is, a flow has a c.s.h. if the velocity gradient is constant along the path of a material point [13]; thus, using D/dt for the material derivative,

$$D\underline{L}/dt = \partial\underline{L}/\partial t + \underline{v} \cdot \text{grad } \underline{L} = \underline{0}$$

Deformation by steady axially symmetric shortening involves a c.s.h. flow in which the velocity gradient $\underline{L}$ is symmetric and hence equal to $\underline{D}$. Viscometric flows are a subclass of c.s.h. flows in which material response can be completely characterised by the three viscometric functions (3); they include Poiseuille flow, Couette flow, channel flow and, of particular interest to the geologist, simple shear flow. Constant stretch history flows are important because, in such flows, the memory of a simple fluid (however elaborate it may be) is left very little to remember! The material response becomes much simplified in consequence and, indeed, they are almost the only flows for which rigorous analysis is possible. However, experiments using viscometric flows, for example, can determine no more about a simple fluid than the nature of the viscometric functions. They cannot yield information anything like sufficient to determine the response functional so that two materials that behave identically in viscometric flow might behave quite differently in some other situation. Thus in a viscometric flow such as inhomogeneous simple shear all simple fluids, no matter how complex their potential response, are indistinguishable from Rivlin-Ericksen fluids [14]. Also, in steady axially symmetric shortening experiments it is impossible to distinguish the class of simple fluids from the class of Reiner-Rivlin fluids [11, p.473]. In other words, the deformation experiments performed by geologists lead to Reiner-Rivlin constitutive equations even though the rocks may be much more complicated materials. The ubiquitous 'power-law' may be giving us more information about the experimental apparatus than about the rock! It is interesting to note that there is hardly any adequate theoretical foundation for the dislocation (power-law) creep equations [15,16]; they are almost entirely empirical and based on exactly those shortening experiments that are incapable of leading to constitutive equations more complicated than that of the Reiner-Rivlin fluid irrespective of how complicated the response functional of the rock may be. It follows that there is no justification whatsoever for using power-law equations for modelling the more general deformations involved in folding, boudinage, diapirism, and convective mantle flow, or for flow situations (such as the porphyroblast rotation model considered in Section 4) that involve flow past obstacles.

We now turn to another factor that limits the actual response that a simple fluid can exhibit for, although the stress depends on the history of deformation, it is expected that this dependence should be less strong for deformations in the distant past than for more recent ones. The theory of simple fluids with fading memory is based on this idea and, for such materials, one can approximate the equation of a simple fluid by an expansion known as the slow flow approximation [14]. Consider a sequence of deformation histories such that each history differs from a reference history in that the time scale is retarded by a factor α, ($0<\alpha<1$). That is, the deformation that occurs at time t is the same as that which occurred in the reference history at time αt. The stress can then be expressed as a power series in α and, if the series is truncated at the term in α^n, the resulting constitutive equation is said to be that of an nth-order (or nth-grade) fluid. It then turns out [14] that the zeroth-order fluid (of trivial interest only) is the ideal fluid in which the stress is always hydrostatic. The first-order fluid is the Newtonian fluid, and the second-order fluid is the Rivlin-Ericksen fluid with constitutive equation

$$\underline{T} = -p\underline{I} + \eta_0 \underline{A}_1 + \lambda_1 \underline{A}_1^2 + \lambda_2 \underline{A}_2 \tag{10}$$

in which η_0, λ_1 and λ_2 are material constants related to the viscometric functions. $\underline{A}_1$ and $\underline{A}_2$ are the Rivlin-Ericksen tensors defined in Section 2. Thus, if the flow is slow enough a simple fluid, no matter how complicated potentially, will behave as a Newtonian fluid while the first departure from Newtonian behaviour as the flow is speeded up is the second-order fluid (10). The viscometric functions for second-order fluids are [11, p.504]

$$\eta(\dot{\gamma}) = \eta_0$$

$$\sigma_1(\dot{\gamma}) = (\lambda_1 + 2\lambda_2)\dot{\gamma}^2$$
$$\sigma_2(\dot{\gamma}) = \lambda_1\dot{\gamma}^2$$

so that the viscosity function is constant as in Newtonian fluids. The normal-stress differences are not, however, zero.

It is important to appreciate that the incompressible Newtonian fluid has only one material parameter other than its density, namely the viscosity μ. A viscometric test serves, in principle, to determine μ and thereby to determine the constitutive equation <u>for all flows</u>. This special status of the Newtonian fluid extends to the second-order fluid (but no further) so that, from a single viscometric test the three constants η_o, λ_1 and λ_2 may be found and, again, these fully specify the constitutive equation <u>for all flows</u>. This suggests that a few experiments using a suitably modified Weissenberg Rheogoniometer (which instrument can measure normal-stress differences) may produce more valuable information about rock rheology than numerous axially symmetric shortening experiments of the type currently performed. Of course, we are particularly interested in how slow the flow must be before Newtonian, or second-order, fluids become good approximations for a simple fluid. Here we consider a material constant known as the natural time, Λ, defined as [17],

$$\Lambda = \lim_{\dot{\gamma} \to 0} \frac{\sigma_1(\dot{\gamma})}{\dot{\gamma}\tau(\dot{\gamma})}$$

where $\tau(\dot{\gamma})$ is the shear stress function. (Note that the natural time lapse, s_o, of Truesdell and Noll [11, p.437] is similar to Λ but is defined in terms of all three viscometric functions). Forming a dimensionless group - the Weissenberg number, We ($= \Lambda\dot{\gamma}$) - we have,

$$We = \lim_{\dot{\gamma} \to 0} \frac{\sigma_1(\dot{\gamma})}{\tau(\dot{\gamma})}$$

so that We can be regarded as a measure of the relative importance of normal-stress ('elastic') and shear-stress ('viscous') effects in slow viscometric flow. For those fluids (e.g. Newtonian fluids) showing no normal-stress effects (and for those only) we have We = 0. Roughly, then, we can say that the Newtonian approximation for simple fluids in slow viscometric flow is reasonable when We is very small (say 1/We >> 1). Unfortunately calculation of We requires the limiting value of $\sigma_1(\dot{\gamma})$ as $\dot{\gamma} \to 0$ while to measure this function accurately requires large shearings. Thus, although the second-order fluid seems intuitively reasonable as a model for steady-state natural flow of rocks, no information is available at the present time that is relevant to deciding which order approximation (if any) can be justified. In the next section we examine a kinematic model for rotations of rigid particles immersed in a fluid, for which we assume the second-order fluid as a suitable rheological model.

4. RIGID PARTICLE ROTATIONS IN A SECOND-ORDER FLUID

There is a considerable body of experimental evidence [18] supporting the idea that isolated, neutrally buoyant, rigid, axisymmetric particles in creeping Newtonian simple shear flow describe accurately closed periodic orbits in agreement with the theory of Jeffery [19]. The orbits traced by the ends of the particles are spherical elliptical orbits (fig.1) specified by the pair of angles (φ, θ) according to

$$\tan \varphi = r \tan [\gamma/(r + 1/r)] \tag{11}$$

$$\tan \theta = k(\cos^2\varphi + 1/r^2 \sin^2 \varphi)^{\frac{1}{2}} \tag{12}$$

where r is the axis ratio (for a prolate ellipsoid) or the 'equivalent axis ratio' for other suitable particles (any axisymmetric shape with a centre of symmetry). Particular orbits are specified by the orbit constant k which is determined by the initial orientation of the particle (i.e. $k = \tan \theta_o$ where θ_o is the value of θ when $\varphi = 0$). Several workers have used Jeffery's equations to predict orientation distributions of rigid particles such as elongate porphyroblasts, and this suggests an obvious approach to palaeostrain estimation in porphyroblastic rocks [20]. However, it is known both theoretically [21] and experimentally [22] that a non-Newtonian enclosing fluid induces a drift of the particle through the family of Jeffery orbits. In this section we assume that the enclosing rock matrix can be modelled as a memory fluid and we investigate porphyroblast rotation behaviour for suitable slow flows. Of course, if the flow is sufficiently slow, all memory fluids behave as Newtonian fluids and Jeffery's analysis applies. The first departure from Newtonian behaviour for faster flows is the second-order fluid. Rotations of rigid rods immersed in an elasticoviscous fluid undergoing Couette flow have been investigated experimentally [22] and, as we shall see shortly, this fluid behaves as a second-order fluid at low shear rates. Couette flow - produced in the annular region between two coaxial cylinders rotating with different angular velocities - is a viscometric flow very similar to simple shear except that it is not unidirectional; it is, however, invariant under translation along every streamline so that theoretical results such as Jeffery's equations continue to hold for Newtonian fluids [21]. The elasticoviscous solution used was prepared by dissolving 3% polyacrylamide (PAA) in water; the elastic behaviour of the solution, obvious in that it showed the Weissenberg climbing effect, was also evident in that small polystyrene spheres suspended in the solution in Couette flow showed translational recovery (in the direction opposite to flow) when the flow was suddenly stopped. Apparent viscosity, and the 'first' normal-stress difference, for this solution are plotted against shear rate in fig. 2. Slender rigid rods immersed in the PAA solution in Couette flow exhibit a marked orbital drift and the work of Leal [23] allows this experimental

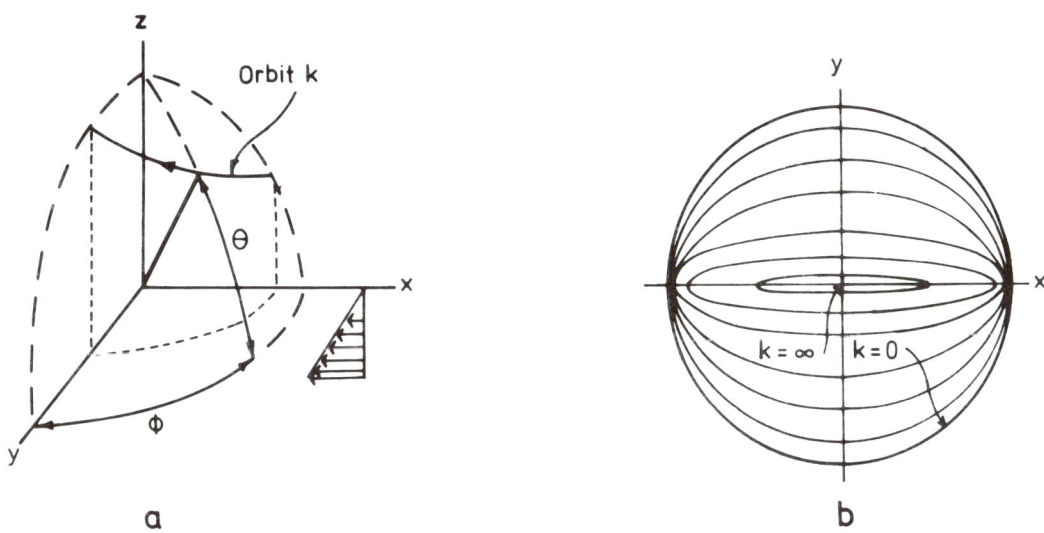

Fig.1.(a) Symmetry axis of prolate ellipsoid, one end of which is tracing the path of a spherical elliptical orbit (orbit constant k), whose projection onto the x y plane is an ellipse (fine broken line).
(b) Projections of Jeffery orbits onto x y plane for a rod with r = 16·1

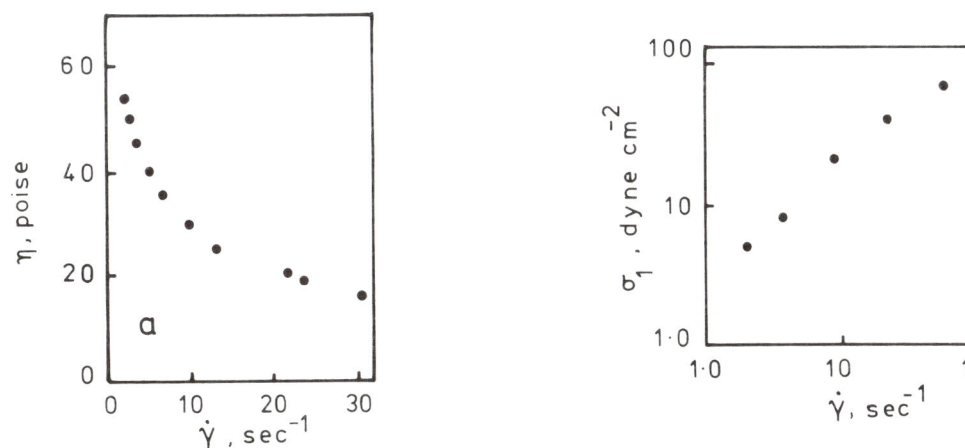

Fig.2.(a) Plot of apparent viscosity against rate of shear for 3% PAA solution.
(b) Log-log plot of 'first' normal-stress difference against rate of shear for same.
After Gauthier et al.[17]

drift to be checked against a theoretically derived particle path.

Leal's analysis is based on the slow flow approximation but is restricted (to retain tractability) to an incompressible second-order Rivlin-Ericksen fluid in which λ_1 and λ_2 in (10) are small so that the constitutive relationship differs only slightly from that of a Newtonian fluid. Leal's analysis is based on slender-body theory so that, strictly, the slender rods are treated as having infinite axis ratios. The rate of change of φ (see fig. 1a) with time is given by

$$\dot{\varphi} \approx \dot{\gamma} \cos^2\varphi + \beta\dot{\gamma} \sin^2\theta \sin\varphi \cos\varphi (\sin^2\varphi - \cos^2\varphi) \quad (13)$$

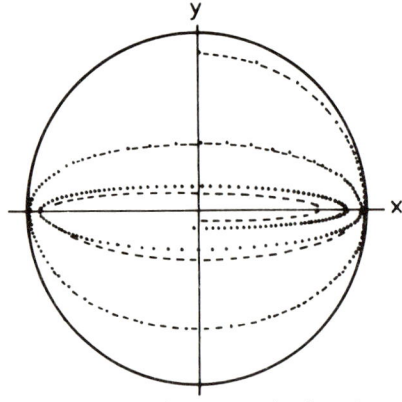

Fig. 3. Projection of one end of rod onto x y plane. 'Data points' define theoretical orbit from Leal's analysis with $r = 16.1$, $\dot{\gamma} = 0.53$ sec^{-1}, $\beta = 0.037$, $k_o = 0.46$; 'points' are spaced at 0.75 sec intervals. Broken line shows experimentally measured orbit based on Gauthier et al. [17, fig. 7]
After Leal [18, fig 7]

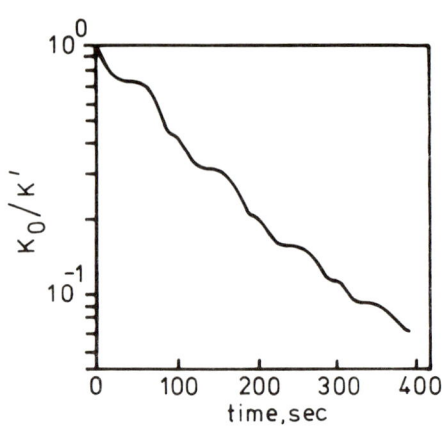

Fig. 4. Calculated orbit drift for analysis shown in figure 3. (see text)
After Leal [18, fig. 6].

where β is a constant for a given particle size and shape and for particular values of the viscometric functions. From (13) it is obvious that the particle never rotates through the shear plane ($\varphi = \pi/2(2n + 1)$) even in the Newtonian case ($\beta = 0$). In order to accommodate finite axis ratios a higher order correction term ($\dot{\varphi} \sim \dot{\gamma}/r^2$) is added so that the $\dot{\gamma} \cos^2\varphi$ term in (13) becomes $\dot{\gamma}(\cos^2\varphi + 1/r^2)$. This provides the small torque necessary to cause the particle to move through the shear plane.

In order to check his theoretical predictions against experiment Leal has inserted the values of axis ratio ($r = 16.1$) and shear rate ($\dot{\gamma} = 0.53$ sec^{-1}) used by Gauthier et al. [22], and also estimated the other values ($\beta \sim 0.037$, $k_o \sim 0.46$) from their experimental data. The qualitative agreement between theory and experiment (fig. 3) proves to be remarkable, especially in view of the limited experimental accuracy and the possible errors involved in estimating a value for β, and for k_o (the orbit constant at time $t = 0$). Note that the rate of orbital drift is greatest within about $\pm 10°$ of x, the end of the particle elsewhere almost following a constant Jeffery orbit. Also, plotting orbital drift [$\log (k_o/k')$] against time from his theoretical analysis, Leal finds a nearly linear relationship but with a small-scale wavy structure superimposed on it (fig. 4). Analysis of this curve shows that the wave pattern is due to the orbital drift (in the vicinity of x) being more rapid as the particle moves away from from x then towards x; note that the transient orbit constant (the Jeffery orbit in which the particle finds itself at any instant) is designated k' to distinguish it from a solution of (12). Let us consider this orbital drift in a geological context. On fig. 4 a large geological strain (say $\gamma = 10$) roughly corresponds to time $t = 19$ sec, by which time the particle would have drifted from $k_o \sim 0.46$ to $k' \sim 0.61$;

a further strain of 10γ would result in a further drift only to k' ~ 0.66. Thus, in spite of the apparently dramatic nature of the orbital drift, it would be barely detectable for most geological strains and would induce negligible error in calculating porphyroblast orientation distributions. Also Leal's analysis uses an axis ratio (r = 16.1) much greater than those (say 1<r<4) appropriate for most porphyroblasts. It is therefore anticipated that less orbital drift (and probably very much less) would occur for such particles although this cannot be quantified because the slender-body theory is not robust for small axis ratios (say, r<6).

5. SUMMARY AND DISCUSSION

The Weertman equation (4) and similar power-law equations describing dislocation creep are essentially phenomenological in nature. Their basis - the steady-state axially symmetric shortening experiment - allows only a very restricted material response, so that constitutive equations derived from such tests do not provide information relevant to flows without constant stretch history (or to other c.s.h. flows such as inhomogeneous simple shear). Indeed, it is difficult to think of boundary-value problems of interest to the geologist for which the power-law equations are relevant! Intuitively the simple fluid is a more realistic rheological model for natural steady-state flow of rocks and the fact that, experimentally, rocks exhibit stress relaxation lends support to this. Asymptotic approximations for simple fluids in very slow flows are the Newtonian fluid, and the second-order Rivlin-Ericksen fluid. Calculation of the Weissenberg number would help the geologist decide whether natural flow of rocks is sufficiently slow to justify using these approximations (at least for viscometric flows), but this requires knowledge of the viscometric functions. If the second-order fluid is appropriate, experimental determination of the viscometric functions would allow the constitutive equation to be specified for all flows.

The motions of rigid particles in creeping Newtonian simple shear flow can be analysed using Jeffery's theory [19]. The end of such a particle describes an accurately closed spherical elliptical orbit but, in a non-Newtonian fluid, suffers a progressive drift through the family of Jeffery orbits. Calculation of this drift is possible using Leal's analysis for a second-order fluid [23] but, again, this requires knowledge of the viscometric functions (in order to evaluate β in equation 13). However, a slender rod (r = 16.1) in a 3% PAA solution shows only slight orbital drift over the small strain intervals likely in porphyroblastic rocks. This solution exhibits very marked elastic effects and shear dependent viscosity, but behaves in slow flow ($\dot{\gamma}$ = 0.53 sec^{-1}) as a second-order fluid differing only slightly from a Newtonian fluid. It is anticipated that, if rock flow can be modelled by the second-order fluid, the small strains (and small axis ratios) relevant to porphyroblastic rocks would lead to negligibly small orbital drift so that Jeffery's analysis would continue to be appropriate.

Finally, it should be emphasised that <u>all</u> the constitutive equations mentioned in this paper are those of isotropic fluids. With only a few exceptions geologists have paid surprisingly little attention to the rheology of anisotropic materials even though isotropic theories cannot be expected to provide even rough approximations to the behaviour of strongly anisotropic rocks. Continuum theories for highly anisotropic materials have been developed in the past decade and will form the subject of a later paper. Unfortunately theories appropriate to weakly or moderately anisotropic rocks are less well developed although geologists have made some progress in this field [24; 25].

ACKNOWLEDGEMENTS

I thank Professor B. E. Hobbs and Dr. T. G. Rogers for critically reading the manuscript.

REFERENCES

[1] Walters, K., <u>Basic concepts and formulae for the rheogoniometer</u>. Sangamo Controls, Bognor Regis, 1968, 34p.

[2] Ashby, M. F. & Verrall, R. A., Micromechanisms of flow and fracture, and their relevance to the rheology of the upper mantle. <u>Phil.Trans.R.Soc.Lond.</u> A 288 (1978) 59-95.

[3] Rutter, E. H., The Kinetics of rock deformation by pressure solution. <u>Phil.Trans.R.Soc.Lond</u>. A 283 (1976) 203-219.

[4] Higgins, M. W., Cataclastic rocks. <u>Geol.Surv.Prof.Pap</u>. 687 (1971) 1-97.

[5] Bell, T. H. & Etheridge, M. A., Microstructure of mylonites and their descriptive terminology. <u>Lithos</u> 6 (1973) 337-48.

[6] Elliott, D., Diffusion flow laws in metamorphic rocks. <u>Geol.Soc.Am.Bull</u>. 84 (1973) 2645-64.

[7] Weertman, J., Dislocation climb theory of steady-state creep. <u>Trans. ASME</u>, 61, (1968) 681-694.

[8] Heard, H. C., Comparison of the flow properties of rocks at crustal conditions. <u>Phil.Trans.R.Soc.Lond</u>. A 283 (1976) 173-186.

[9] Carter, N. L., High-temperature flow of rocks. <u>Revs. Geophys.Space Phys</u>. 13 (1975) 344-349.

[10] Nicolas, A. & Poirier, J. P., <u>Crystalline plasticity and solid state flow in metamorphic rocks</u>. Wiley, New York, 1976, 444p.

[11] Truesdell, C. & Noll, W., Non-linear field theories of mechanics. Handbuch der Physik (ed. S. Flügge) vol. III/3, Springer-Verlag, Berlin 1965, 602p.

[12] Hobbs, B. E., Deformation of non-Newtonian materials in simple shear. AGU Geophys. monograph. no. 16 (1972) 243-258.

[13] Huilgol, R. R., A class of motions with constant stretch history. Quart.Appl.Math. 29 (1971) 1-15.

[14] Coleman, B. D. & Noll, W., An approximation theorem for functionals, with applications in continuum mechanics. Arch.Rat.Mech.Anal. 6 (1960) 355-370.

[15] Ashby, M. F., A first report on deformation mechanism maps. Acta Metall 20 (1972) 887-897.

[16] Kelly, A., Introduction (to a discussion on creep of engineering materials and of the earth). Phil.Trans.R.Soc.Lond. A 288 (1978) 3-8.

[17] Astarita, G. & Marrucci, G., Principles of non-Newtonian fluid mechanics. McGraw-Hill, London 1974, 289p.

[18] Goldsmith, H. L. & Mason, S. G., The microrheology of dispersions. In F. R. Eirich (ed.) Rheology, Theory and Applications, vol. IV. Academic Press, New York 1967 pp. 85-250.

[19] Jeffery, G. B., The motion of ellipsoidal particles immersed in a viscous fluid. Proc.R.Soc.Lond. A 102 (1922) 161-179.

[20] Harvey, P. K. & Ferguson, C. C., A computer simulation approach to textural interpretation in crystalline rocks. In D. F. Merriam (ed.) Recent advances in geomathematics. Pergamon, Oxford 1978, pp. 201-232.

[21] Bretherton, F. P., The motion of rigid particles in a shear flow at low Reynolds number. J.Fluid Mech. 14 (1962) 284-304.

[22] Gauthier, F., Goldsmith, H. L. & Mason, S. G., Particle motions in non-Newtonian media I: Couette flow. Rheol. Acta 10 (1971) 344-364.

[23] Leal, L. G., The slow motion of slender rod-like particles in a second-order fluid. J.Fluid Mech. 69 (1975) 305-337.

[24] Biot, M. A., Mechanics of incremental deformation. Wiley, New York 1965, 504p.

[25] Cobbold, P., Mechanical effects of anisotropy during large finite deformations. Bull.Soc.geol.France 18 (1976) 1497-1510.

CRAZES AND THE FRACTURE OF GLASSY THERMOPLASTICS

G. W. Weidmann* and W. Döll**

*Faculty of Technology, The Open University, Milton Keynes MK7 6AA, U.K.
**Fraunhofer-Gesellschaft, Institut für Festkörpermechanik,
7800 Freiburg i. Br., FRG

ABSTRACT

A characteristic of the fracture process in glassy thermoplastics is the formation of a zone of crazed material at the crack tip. These craze zones are described and the results of measurements of their size and shape as functions of molecular weight, temperature and loading are discussed. By applying a fracture mechanics model, the Dugdale model, to these results the stress-strain behaviour of the crazed material can be evaluated and, hence, the strain energy of craze deformation up to the point of fracture.

1. INTRODUCTION

In many glassy thermoplastics the process of fracture is preceded by the formation of a characteristic zone of inelastically deformed material at the crack tip. This zone, which is known as a craze zone, is similar in appearance to a crack, with well-defined, specularly-reflecting boundaries, but it is, nonetheless, still load-bearing. Just ahead of the crack tip, where the stresses are particularly concentrated, molecular chains of the polymer are drawn out of their amorphous arrangement in the bulk material into bundles under the action of the tensile stress component acting normal to the crack plane. These bundles are interspersed by voids, leading to a material with rubberlike properties. Crazing thus differs significantly from conventional plasticity in other materials which is essentially a volume-conserving, shear phenomenon.

Owing to their structure, crazes have a lower density and, hence, a lower refractive index than the bulk polymer. Thus, it is possible to measure their sizes and shapes using optical interference, provided that the separations involved are comparable in order of magnitude to the wavelength of light, that the refractive index of the craze is significantly different from that of the bulk material and that the boundary between the two regions is sharp. This was first demonstrated for poly(methyl methacrylate) or PMMA by Kambour(1) and Bessonov and Kuvshinskii (2). Their work has subsequently been extended by Ward and his co-workers (3-5), Vavakin et al (6) and ourselves (7-10) to other materials, to the effects of stress intensity factor, molecular weight and temperature, and to the applicability of a fracture mechanics model, the Dugdale model (11), to the craze at the tip of a crack.

In this paper we would like to consider the way in which such studies of crazing at the tips of cracks can help establish a link between microstructural aspects of the polymeric material and continuum mechanics. In particular we would like to discuss the way in which a study of the micromechanics of craze deformation can contribute an insight to the energetics of the fracture process.

2. THE CRACK TIP CRAZE

A highly schematic representation of a crack tip in a glassy thermoplastic is shown in Fig. 1. The main features to be noted are that there is an approximately wedge-shaped craze zone (indicated by the vertical lines of "molecular bundles") ahead of the crack tip; that the crack opening approximates to a narrow parabola; and that there is a layer of crazed material left on both fracture surfaces (this layer gives rise to the characteristic interference colours on freshly broken fracture surfaces of PMMA when viewed in white light). If such a crack tip region is observed in a reflection microscope using an optical arrangement such as that depicted in Fig. 2 two sets of interference fringes (Fig. 3) are seen. One of these sets arises from the interference between light reflected from the upper and lower craze boundaries. The other is due to a similar process between the reflections from the two crack faces. The length of the craze zone in Fig. 3 is about 40 μm.

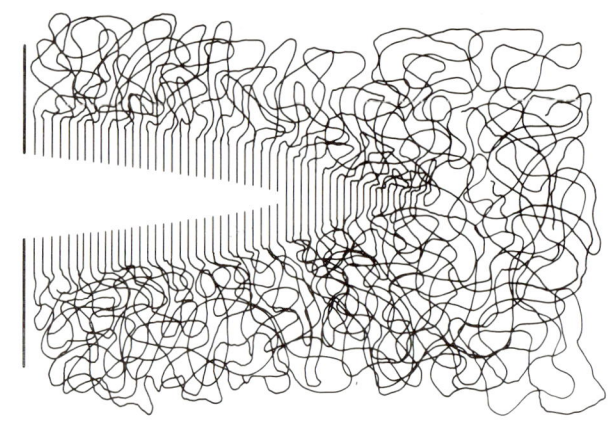

Fig. 1 Schematic representation of crack tip region in glassy thermoplastic.

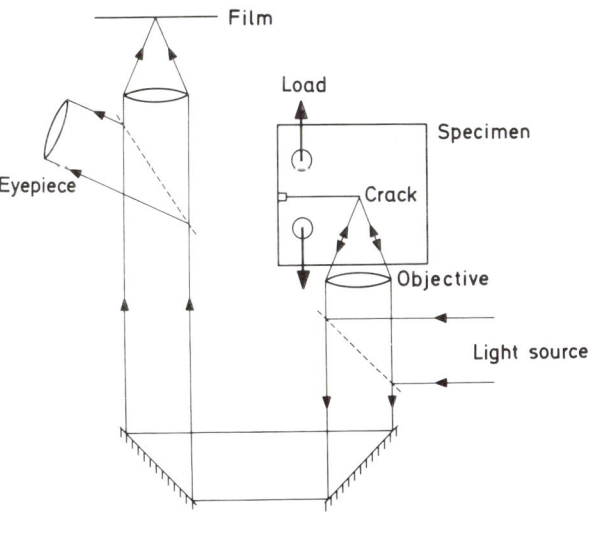

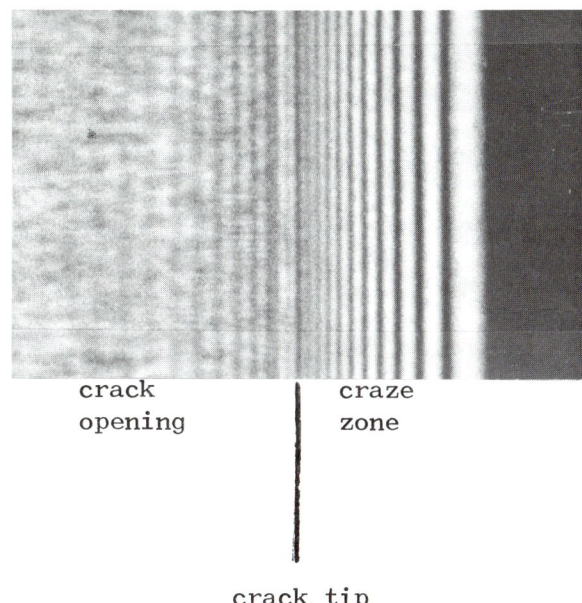

Fig. 2 Sketch of optical arrangement

Fig. 3 Interference fringe pattern in PMMA

Details of the experimental procedure and the method of evaluating the interference fringe patterns are given in Ref. 7. The shapes and sizes of the craze zone and crack opening in PMMA at the onset of slow crack propagation are shown in Fig. 4 for two different molecular weights and at two different temperatures. It is important to note that the vertical scale is a factor of ten larger than the horizontal scale. The actual shapes of the craze zones are shown at the bottom of Fig. 4. It can be seen that the craze zone in the lower molecular weight material is scarcely affected by temperature whilst that in the high molecular weight material increases in size by about 50% between $20^°$ and $70^°C$. Also the craze zones are much smaller in the PMMA of lower molecular weight.

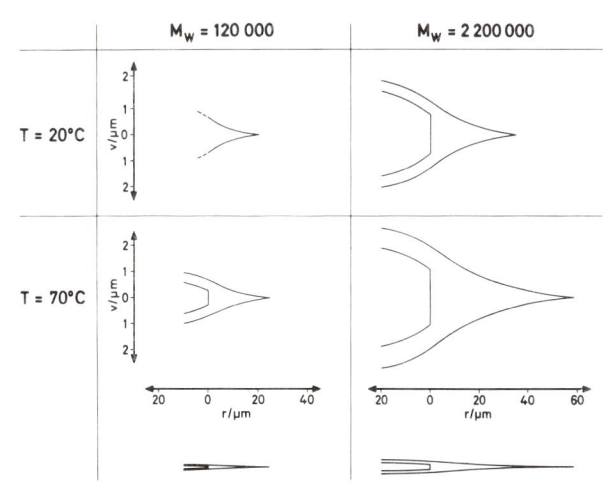

Fig. 4 Craze zones and crack openings in PMMA

A useful parameter to characterise the craze zone is its extension normal to the crack plane at the crack tip, which we call $2v_c$. At fracture $2v_c$ represents the maximum length to which the bundle of molecules can be stretched before separation occurs. The way in which $2v_c$ at the onset of slow crack propagation varies with molecular weight in PMMA at two temperatures is shown in Fig. 5. At $20^\circ C$ $2v_c$ first rises steeply with increasing molecular weight and then becomes almost constant at molecular weights above about 500000. If the value of $2v_c$ at a particular molecular weight is expressed as a ratio to the maximum stretched length of the polymer molecule at that nominal molecular weight, then the value of this ratio is constant at about 4 in the steeply rising part of the $2v_c$ versus M_w curve, and then decreases steadily to about 0.1 at $M_w = 8000000$. This suggests a transition from a chain sliding mechanism of fracture at low molecular weights to a chain breaking mechanism at high molecular weights.

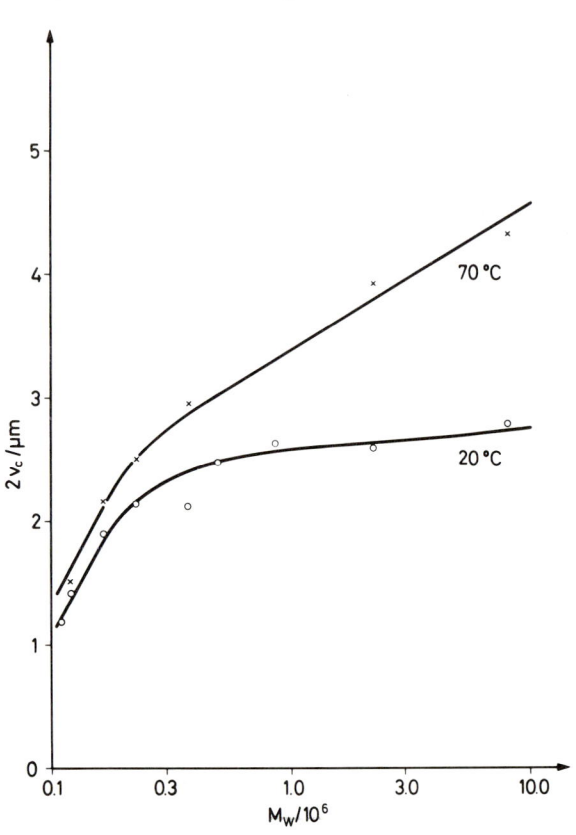

Fig. 5 Variation of maximum craze extension $2v_c$ with molecular weight M_w.

The results at $70^\circ C$ support the idea of such a transition since there is very little change of $2v_c$ with temperature at the lower molecular weights. This implies that the size of the craze zone is limited by the molecular chain length in this region, whilst at higher molecular weights not only is $2v_c$ larger than at $20^\circ C$ but also its rate of change with molecular weight is much higher.

Since the materials investigated in this study all had fairly broad molecular weight distributions and it is not yet known what contribution the molecules, particularly those at the high end of the distribution, make to the fracture process, it is not possible to draw any specific conclusions as to how many molecular lengths span the craze zone at the crack tip just prior to fracture from these results.

3. CRAZE DEFORMATION AND THE DUGDALE MODEL

Dugdale (11) developed a model for the plastic zone at the tip of a crack based on applying cohesive stresses across the tip region of an elastic crack so as to cancel out the elastic stress singularity at the crack tip. Rice (12) and Goodier & Field (13) derived solutions for the displacements 2v of the elastic-plastic boundary of the Dugdale model in terms of the yield stress σ_F and Young's modulus E of the material and the length s of the plastic zone. Using such solutions it has been shown (3, 6, 9) that the Dugdale model provides a very good description of the size and shape of the crack tip craze in glassy thermoplastics. Further, the values of macroscopically measured yield stress and Young's modulus agree very well with those determined by fitting the Dugdale model to the craze zone. An example of this is shown in Fig. 6 for PMMA and poly(vinyl chloride) or PVC. Here the yield stresses σ_{DF} obtained by fitting the Dugdale model to craze zones at crack tips in the two materials at the onset of slow crack propagation and over a range of temperatures are compared with those obtained from tensile tests on the same materials.

By applying the Dugdale model to crazes evaluated at a series of loads between zero and that required for the onset of slow crack propagation it is possible to determine the stress-strain behaviour of the crazed material at the tip of a crack (10).

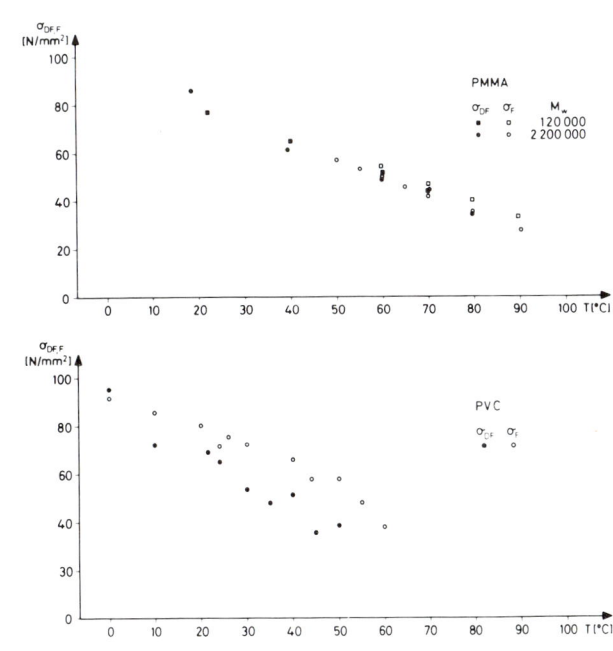

Fig. 6 Comparison of Dugdale model yield stress σ_{DF} and tensile yield stress σ_F at different temperatures

Highly linear stress-strain curves are obtained for PMMA and PVC with maximum strains at fracture of up to 500%. Young's modulus of the crazed material is of the order of 1% of that of the bulk polymer.

By integrating under the stress-strain curve the strain energy U_O in the craze at fracture can be calculated. This is shown in Figure 7 for PMMA of two different molecular weights and over a range of temperatures.

Also plotted is the temperature variation of the critical strain energy release rate G_{IO} for the onset of slow crack propagation. These results show that, for the lower molecular weight PMMA, U_O is about 30% of G_{IO}, whilst for the higher molecular weight the fraction is about 40%. Earlier work by one of us (14) has shown that approximately 60% of the energy of fracture of PMMA is converted into heat. This percentage agrees very well with the results of calorimetric measurements of the heat generated by the cold drawing of PMMA (15). It is, therefore, plausible to postulate that, of the energy required to propagate a crack in a glassy thermoplastic (i.e. G_{IO}) the larger part is dissipated in the plastic work of craze formation and the remainder is dissipated in the work of craze deformation up to fracture. This postulate

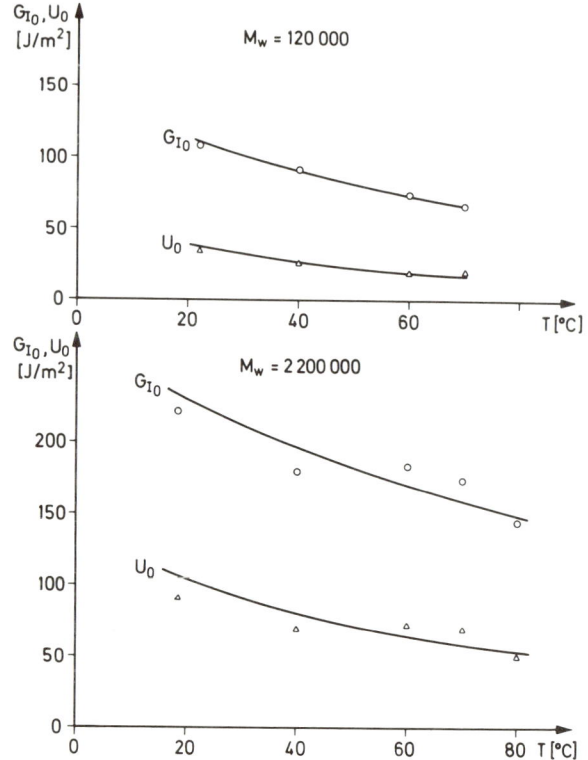

Fig. 7 Craze strain energy U_O and critical strain energy release rate G_{IO} for PMMA as functions of temperature

does, however, require that the strain energy of craze deformation is not recoverable as elastic strain energy of the bulk material after the craze has broken.

ACKNOWLEDGEMENTS

The experimental work on which this paper is based was carried out at the Institut für Festkörpermechanik, Freiburg i. Br. and was financially supported by the Deutsche Forschungsgemeinschaft.

REFERENCES

(1) Kambour, R.P., J. Polym. Sci. A-2 4 (1966) 349

(2) Bessonov, M.I. and Kuvshinskii, E.V.,
Sov. Phys. - Solid State 3 (1961) 950

(3) Brown, H.R. and Ward, I.M. Polymer 14 (1973) 469

(4) Morgan, G.P. and Ward, I.M. ibid 18 (1977) 87

(5) Fraser, R.A.W. and Ward, I.M. ibid 19 (1978) 220

(6) Vavakin, S.A., Kozyrev, Yu. I. and Salganik, R.L.,
Izvest. akad. Nauk SSSR, Mekh. Tverd, Tela 11 (1976) 111

(7) Weidmann, G.W. and Döll, W.
Colloid & Polym. Sci. 254 (1976) 205

(8) idem Third Int. Conf. on Deformation, Yield and Fracture
of Polymers, Cambridge (1976)

(9) idem Non-Crystalline Solids, Ed. G.H. Frischat,
Trans Tech Publications, 1977, p.606

(10) Döll, W. and Weidmann, G.W. Colloid & Polym. Sci. (to be published)

(11) Dugdale, D.S. J. Mech. Phys. Solids 8 (1960) 100

(12) Rice, J.R. Fracture - An Advanced Treatise,
Ed. H. Liebowitz, Academic Press, 1968, Vol. 2, p. 264

(13) Goodier, J.N. and Field, F.A. Fracture of Solids
Ed. D.C. Drucker and J.J. Gilman, Gordon and Breach, 1963, p. 103

(14) Döll, W. Colloid & Polym. Sci. 256 (1978)

(15) Müller, F.H. Struktur und physikalisches Verhalten der Kunststoffe,
Ed. R. Nitsche and K.A. Wolf, Springer Verlag, 1962, Vol. 1, p. 428.

A THEORY OF MASS-TRANSFER-BUCKLING DEFORMATION IN FINITE AMPLITUDE SINUSOIDAL MULTILAYERS

D. W. Durney

School of Earth Sciences, Macquarie University, N.S.W. 2113, Australia

ABSTRACT

The effects of preferred diffusion paths on mass-transfer deformation in microfolded anisotropic media are analysed for a model of high diffusivity layer boundaries in a folded multilayer system. The divergence of diffusive flux is shown to be coupled with interlayer slip and layer buckling, giving an approximate bulk mass-transfer-buckling strain rate ($\dot{e}_{STB}$) whose importance relative to simultaneous flattening deformation ($\dot{e}_{FL}$) is an inverse function of (i) the ratio of diffusional stiffness (S) to mechanical viscosity (η) and of (ii) the wavelength cubed, and depends on (iii) normalized amplitude (a_o). Effects similar to those predicted are found in certain crenulated rocks.

1. INTRODUCTION

Laminated or strongly anisotropic rock types commonly undergo a style of geological deformation known as <u>crenulation</u> [1]. This consists of short wavelength (usually less than a few cm), sub-sinusoidal, harmonic folding [1] and is typically accompanied by a segregation of minerals into the hinge zones and inflexion zones [1,2].

Microscopic observations [2,3,4,5] suggest that a mass-transfer process (pressure-solution, or solution-transfer) accompanies folding in such a way as to assist in the fold development. The fold initiation has been explained in terms of the Biot-Ramberg theory of buckling of anisotropic and multilayered media [4,5]. Several hypotheses have been suggested for the mechanism of mass-transfer but a detailed theoretical account has not as yet been published. A previous theoretical attempt by Fletcher [6] omitted to take into account continuity of the transport and deposition process [7]. Another treatment has been communicated in abstract by Casey [8]. Raj and Ashby [9] have analysed sliding across sinusoidal corrugations and their result is somewhat analogous to the hinge displacement expression given here ($\underline{\dot{n}}_o$) except that it lacks a wavelength dependence.

It has been suggested [10, case 3 p.235; 5, p.277-282] that two general factors may significantly influence the direction of bulk mass-transfer in closed polycrystal systems: the intensive thermodynamic factor of <u>mean stress variations</u> (as along folded layers [11]) and the kinetic factor of <u>preferred diffusion paths</u> (as along folded anisotropic fabrics

[5]). Here, it is proposed to deal only with a special case of the latter, for which an analytical mass-transfer solution has been found.*

The model is a semi-infinite multilayer system composed of layers of uniform isogonal thickness (h) and viscosity (η) separated by interlayer surfaces of relatively high grain boundary diffusivity (D_w). The layers are assumed to be initially folded, the folds being finite in amplitude, sinusoidal, symmetric and harmonic, and remaining harmonic during deformation. For simplicity the boundary stresses are assumed to be applied at right angles and parallel to the axial planes of the folds. However, this is not a general restriction on the model.

It will then be shown that a continuous transfer of matter takes place along the interlayer surfaces, whose divergence induces removal of the solid from the walls of the surface around the inflexions and deposition around the hinges of the folds. In response to these changes the layers undergo a series of relative displacements. But to satisfy continuity this can only occur when a buckling and slip of the layers accompanies the mass-transfer displacements, rather analogous to the coupling of diffusion with grain boundary sliding in compact granular materials [12]. Likewise, simple buckling and slip alone are not capable of producing continuous deformation in folds of the assumed geometry subjected to the given boundary stresses. Although additional continuous deformations (homogeneous strain and simple shear parallel to the axial planes) are not necessary for continuity of the mass-transfer displacements, it is nevertheless shown that a flattening deformation is also generally coupled with the mass-transfer-buckling deformation above.

An example of interlayer mass-transfer in a natural crenulated multilayer is shown in Fig. 1. The effects illustrated appear to agree reasonably well with the detailed predictions of eq. (8) and to represent a situation where mass-transfer-buckling deformation predominates over flattening deformation.

The stress within the multilayer system may be considered to be composed of three *a priori* independent fields:

(a) A field which acts only within the layers and has no component of shear or normal stress across the layer boundaries - Fig. 2a. This corresponds to the case of a free, curved, semi-infinite plate having the shape assumed when in a state of stress due to a force F_x per unit depth distributed over its ends.

(b) A homogeneous plane stress, with principal components σ_x and σ_y, acting through and across the layers in the profile plane (Fig. 2b).

(c) A homogeneous uniaxial stress σ_z which acts down the layers normal to σ_x and σ_y.

2. SOLUTION-TRANSFER ALONG A SINGLE SINUSOIDAL INTERLAYER SURFACE

Surface Geometry

Consider a single cylindrical cosinusoidal surface

$$y = a \cos(x/L_o) \qquad (1)$$

of amplitude a and wavelength $2\pi L_o$ referred to cartesian axes xyz

---

* The author's Fig. 3 and an exponential version of his eq. (8) were initially communicated by D. Gray at the 25th Internat. Geol. Congr. in Sydney, 1976.

(Fig. 3a). $\underline{x}$ is the absolute distance and x/L_o the trigonometrical distance (in radians) in the direction of the wave train.

The slope $\underline{dy/dx}$ at any point $(\underline{x,y})$ on this surface is

$$dy/dx = -a_o \sin(x/L_o) = \tan \theta \qquad (2)$$

where $\underline{\theta}$ is the angle of slope at $(\underline{x,y})$ and $\underline{a}_o = a/L_o$. $-\underline{a}_o$ gives the maximum slope at the inflexions ($x = \pi L_o/2$ etc.) and represents a measure of the amplitude/wavelength relationship of the fold.

Surface Stresses

Of the three separate stress fields considered in Section 1 above, only field "b" gives rise to components of stress on the interlayer surface. These components are determined solely by field "b" and by the local inclination θ of the surface. Also the normal component on this surface $\underline{\sigma}$ is independent of all components acting along $\underline{z}$.

For convenience let $\sigma_x > \sigma_y (\lessgtr \sigma_z)$ be principal components (taking compressive stresses to be positive). We define a subsidiary mean stress for the profile plane $\bar{\sigma} = (\sigma_x + \sigma_y)/2$ and a deviatoric stress $\underline{\sigma'} = \sigma - \bar{\sigma}$. The deviatoric normal stress on the fold surface is then

$$\sigma' = (\sigma'_y - \sigma'_x) \cos^2 \theta + \sigma'_x \qquad (3a)$$

which is a periodic function of position $\underline{x}$ through relation (2)

$$\sigma' = (\sigma'_y - \sigma'_x)/(1+a_o^2 \sin^2(x/L_o)) + \sigma'_x \qquad (3b)$$

Interlayer Solute Concentration

Owing to the likely presence of water and other mobile components along grain boundaries in rocks [13] the mass-transfer will be expressed here in terms of 'solute' components, that is, a "solution-transfer" process.

Using the first order thermodynamic equilibrium condition for a pure mineral at constant temperature $(\underline{T})$ and constant and uniform grain boundary 'fluid' pressure $(\underline{p})$ [10 eqn.10] we relate the equilibrium activity $\underline{a}$ of its mobilised 'solute' phase at a point in the grain boundary surface to the normal stress acting on the boundary of the solid at that point:

$$RT \log_e (a/a_o) = \nu(\sigma - \sigma_o) \qquad (4a)$$

(R = gas constant, ν = crystal molar volume, subscripts $\underline{o}$ refer to a reference state and unsubscripted variables to actual states). Provided the variation of activity coefficient $\gamma = a/X$ is small for the range of a, we may replace a/a_o in (4a) by the mole fraction ratio X/X_o; and at low X_o (typical of aqueous solubilities of minerals at low to moderate $\underline{T}$ and $\underline{p}$ [14]) this approaches the concentration ratio $\underline{c/c_o}$. Attention will be restricted to low increments of stress $(\sigma-\sigma_o)$, less than about 10 or 20 MPa, so that (4a) will usually be approximated to within 10% by

$$RT (c-c_o)/c_o = \nu(\sigma - \sigma_o) \qquad (4b)$$

When the reference stress σ_o is chosen equal to the subsidiary mean $\bar{\sigma}$, we have $\sigma-\sigma_o = \sigma'$ and hence by (3b) a function for interlayer concentration $\underline{c}$ in terms of position $\underline{x}$ similar in form to σ' (Fig. 3b):

$$c = c_o + \frac{c_o \nu}{RT}[(\sigma'_y-\sigma'_x)/(1 + a_o^2 \sin^2(x/L_o)) + \sigma'_x] \qquad (5)$$

Solution-Transfer Fluxes

According to the assumption stated at the beginning of the paper, the interlayer diffusive flux will effectively control long range transport of material in the system. Now the variation of c with x implies a concentration gradient along the interlayer surface dc/dL ($dL = dx \sec \theta$). Assuming virtual equilibrium between the 'solute' and the solid phase and applying Fick's First Law*, the interlayer flux J_w (Fig. 3a) is

$$J_w = \frac{-D_w c_o \nu}{RT} \cdot \frac{a_o^2 \sin(2x/L_o)(\sigma_x' - \sigma_y')}{L_o (1 + a_o^2 \sin^2(x/L_o))^{5/2}} \qquad (6)$$

(D_w is the interlayer diffusivity times interlayer thickness, dimension $L^3 T^{-1}$). However, there is no component of flux along z because σ and hence c remains constant with z. The fluxes are therefore all directed in the xy plane. At the hinges ($x = 0$, etc.) and at the inflexions ($x = \pi L_o/2$, etc.) J_w is zero while at intermediate positions it is always directed towards the nearest hinge. (If $\sigma_y > \sigma_x$, however, the sign of J_w and all consequent terms will be reversed.)

The zero surface flux at hinges and inflexions implies zero exchange of matter across these points. Consequently, it is deduced that all transfer of matter is confined within quarter wavelength units of symmetry of the system. These are therefore closed subsystems within the total system.

Additionally, J_w is a function of x: it accelerates or decelerates along the surface. Assuming the reservoir capacity of the grain boundary solution to be negligible this divergence of the flux must be compensated by a flux of material J_s (dimension $ML^{-1}T^{-1}$) across the walls of the surface [10 eqn.11]

$$J_s = dJ_w/dL = \cos \theta \, dJ_w/dx \ . \qquad (7)$$

When J_s is positive (when J_w accelerates) the walls dissolve, and when it is negative they grow by deposits of material precipitated from the interlayer diffusive current. A rate of displacement $\dot{n} = -J_s/\rho$ normal to the surface therefore accompanies the transverse flux, and because this occurs at a solid-solid contact it causes a relative ($\dot{n}$) displacement of layers. (ρ is the density of the solid and $\dot{n}$ is positive for outward or growth displacements.) Expanding $\dot{n}$ we have

$$\dot{n} = \frac{D_w c_o \nu}{RT \, \rho} \cdot \frac{a_o^2 (\sigma_x' - \sigma_y')}{L_o^2 (1 + a_o^2 \sin^2(x/L_o))^3} \cdot \left[-\frac{5}{2} \frac{a_o^2 \sin^2(2x/L_o)}{1 + a_o^2 \sin^2(x/L_o)} + 2\cos(2x/L_o) \right]. \qquad (8)$$

This function is represented graphically in Fig. 3c. Maxima of $\dot{n}$ coincide with hinge points and minima occur in pairs located symmetrically about the inflexion points.

Loci of points of zero displacement $\dot{n} = 0$ form a series of z-lines on the fold surface at positions intermediate between the hinges and the inflexions found by equating the square bracket term in (8) to zero. Their positions are independent of the stresses and material constants but do depend on a_o. They separate hinge regions undergoing dissolution (pressure-solution) from limb regions undergoing growth (pressure-growth) which are completely linked to each other by the interlayer flux J_w across the $\dot{n} = 0$ lines.

---

* Here dc/dL is proportional to chemical potential gradient.

3. DISPLACEMENT AND STRAIN RATES

Continuity Conditions

It is clear by inspection of (8) and Fig. 3c that the multilayer as a whole cannot remain continuously connected when only normal displacements $\dot{n}$ take place across the layer boundaries. Three general conditions must be satisfied in order for the system to remain perfectly harmonic and connected during deformation:

(a) Equivalent points in the layers, initially aligned parallel to the axial plane, must remain parallel to the axial plane.

(b) The longitudinal strain between equivalent points initially aligned in the y-direction, must be uniform throughout the system.

(c) The longitudinal strain in the z-direction, $\dot{e}_z$, must be uniform throughout the system.

A fourth condition is imposed by the special symmetry of the present model:

(d) Rotation and shearing of the layers must be zero at the hinges. This also implies that the equivalent points in (a) remain parallel to y, and in (b), $\dot{e}_y$ must be uniform.

Additionally it will be assumed that the layers are flexible and shearable but incompressible and (to begin with) inextensible along their central surfaces.

Basic Relations for a Layer Element

Consider a narrow element of layer, h by δL ($\delta L \to 0$) whose bottom left corner is located at x and is one of a series of similar elements lying one above the other at x (Fig. 4a). We have, for example, the following identities:

$$n \equiv h \cos \theta \quad (9a), \qquad m \equiv h \sin \theta \quad (9b)$$
$$\delta x \equiv \delta L \cos \theta \quad (9c), \qquad \delta h \equiv \delta L \sin \theta \quad (9d)$$

Two permissible and a priori independent motions in the series, conforming to condition a, are rotations of the elements $d\theta$ about z and extensions dh parallel to y (Fig. 4b). Each of these two motions makes a partial contribution to the normal and shear displacements dn and dm across the layers when the other is fixed. Differentiating (9a) and (9b) w.r.t. (with respect to) time, we have

$$\dot{n} = \cos \theta \cdot \dot{h} - h \sin \theta \cdot \dot{\theta} \qquad (10a)$$
$$\dot{m} = \sin \theta \cdot \dot{h} + h \cos \theta \cdot \dot{\theta} \qquad (10b)$$

By condition d, $\dot{m}$ and $\dot{\theta}$ are both zero at the hinges, hence the y-displacement at these points is just $\dot{h}_o = \dot{n}_o$ (the subscripts "o" denote positions $x = 0$, etc.). Also, by condition b, $\dot{h} = \dot{h}_o$ so that

$$\dot{h} = \dot{n}_o, \text{ whereupon} \qquad (11)$$
$$\dot{\theta} = \dot{n}_o \cot \theta / h - \dot{n} \csc \theta / h, \text{ and} \qquad (12)$$
$$\dot{m} = \dot{n}_o \csc \theta - \dot{n} \cot \theta \qquad (13)$$

All of the motions $\dot{\theta}$, $\dot{h}$, $\dot{n}$ and $\dot{m}$ are therefore interrelated and determined by $\dot{n}$ and $\dot{n}_o$ (8).

Deformation of the Equivalent Element

We now consider the average deformation [15] representing the motions of an element. For convenience we take an equivalent element with corners a, b, c, d located at the centres of the two adjoining layers (Fig. 4a) and this undergoes an <u>equivalent deformation</u> to a',b',c',d' (Fig. 4c).

The <u>longitudinal strain rate along y</u> is simply:

$$\dot{e}_y = \dot{n}_o/h \qquad (14)$$

Differentiating (9c) w.r.t. time with δL constant (previous page), then substituting for δL and $\dot{\theta}$, the <u>longitudinal strain rate along x</u> is,

$$\underline{\dot{e}_x} = \delta\dot{x}/\delta x = \dot{n} \sec\theta/h - \dot{e}_y .$$

Repeating the procedure with (9d) and subtracting the contribution to $\delta\dot{h}$ due to $\dot{e}_y \delta h$, the <u>shear component along y</u> is $\dot{e}_{yx} = \dot{e}_y \cot 2\theta - \dot{n} \csc\theta/h$. By conditions a and d the shear component along x is $\dot{e}_{xy} = 0$. The <u>shear strain rate</u> is therefore $\dot{\gamma}_{xy} = \dot{e}_{yx}$, and it may be verified that $\dot{e}_x$, $\dot{e}_y$ and $\dot{\gamma}_{xy}$ simultaneously satisfy the compatability condition

$$\partial^2 \dot{e}_x/\partial y^2 + \partial^2 \dot{e}_y/\partial x^2 = \partial^2 \dot{\gamma}_{xy}/\partial x\,\partial y .$$

It is interesting to note (Fig. 5) that, whereas $\underline{\dot{e}_y}$ remains constant, $\underline{\dot{e}_x}$ and $\dot{e}_{yx}$ vary continuously with periods equal, respectively, to L_o and $2\pi L$. At the hinges $\dot{e}_x = \dot{e}_{yx} = 0$ leaving only a simple extension $\underline{\dot{e}_y}$ parallel to the axial plane. At junctions between the regions of growth and dissolution ($\dot{n} = 0$) a constant area strain arises solely due to slip and internal shear of the layers ($\dot{m}$). The maximum differential strain rate occurs just beyond this point, rather than at the inflexions of the folds (Fig. 5).

The <u>total x-strain rate</u> for a complete quarter wavelength unit is

$$(\dot{e}_x)_{tot} = \frac{2}{\pi L_o} \int_0^{\frac{1}{2}\pi L_o} \dot{e}_x dx = \left|\frac{-2 J_w}{\pi L_o h\rho}\right|_0^{\frac{1}{2}\pi L_o} - \dot{e}_y = -\dot{e}_y \qquad (15)$$

The <u>volumetric strain rate</u> is given simply by $\dot{\Delta} = \dot{e}_x + \dot{e}_y = \dot{n}\sec\theta/h$ (Fig. 5) and provides a measure of the rate of accmulation of the mobile components at every position x in the system (negative around the inflexions and positive around the hinges).

4. LAYER BUCKLING

Due to the variation of $\dot{\theta}$ with $\underline{x}$, $d\dot{\theta}/dx$ (the bending rate), there arises a stretching of the outer arcs of the layers $\dot{e}_{out}$ and a contraction of the inner arcs $\dot{e}_{inn}$ whose difference is given by

$$\dot{e}_{inn} - \dot{e}_{out} = h \cos^2\theta \; d\dot{\theta}/dx \qquad (16)$$

Close to the hinges of the folds it has the value

$$(\dot{e}_{inn} - \dot{e}_{out})_o = -(\dot{n}_o/L_o)(31\,a_o/2 + 4/a_o) \qquad (17)$$

(assuming $(1 + a_o^2 \sin^2(x/L_o))^{\frac{1}{2}} \simeq 1 + \frac{1}{2} a_o^2 \sin^2(x/L_o)$). This implies the existence of a buckling moment at the hinges due to the force $\underline{F_x}$ associated with <u>stress field "a"</u> (Introduction):

$$F_x = -(\dot{e}_{inn} - \dot{e}_{out})_o h^2 \eta/3a \qquad (18)$$

(F_x per layer and per unit depth in $\underline{z}$; η is the viscosity of the layers; plane strain has been assumed). The equivalent boundary pressure for field "a" is $\underline{P_x} = F_x/h$.

Solving (17) and (18) for differential stress ($\sigma_x - \sigma_y$ and P_x), summing those stresses, and substituting (17), (14) and (15), we obtain the approximate expression for coupled <u>solution-transfer-buckling creep</u> ($\dot{e}_{STB}$)

$$\dot{e}_{STB} = (\dot{e}_x)_{tot} = \frac{-a_h^2 (P_x + \sigma_x - \sigma_y)/\eta}{(5.17 a_o^2 + 1.33 + L_o^3 S_w a_h/2\nu a_o \eta)} \tag{19}$$

($P_x + \sigma_x - \sigma_y$ is the total differential stress applied to the system and a_h the amplitude/thickness ratio of the layers. S_w is the "<u>diffusional stiffness</u>" of interlayer transport; a parameter defined as $S_w = \rho RT/D_w c_o$ having the dimensions of viscosity and being independent of the geometry and dimensions of the transport system, provided the interlayer thickness $\underline{w}$ is constant).

But $P_x + \sigma_x - \sigma_y$ would also induce a <u>flattening strain</u>, which in the hinges would be

$$\dot{e}_{FL} = -(P_x + \sigma_x - \sigma_y)/4\eta \tag{20}$$

(plane strain has been assumed and $\dot{e}_{FL}$ is measured along $\underline{x}$). We are therefore obliged to relax the inextensible condition (Section 3) whence the total <u>combined strain rate</u> in $\underline{x}$ becomes

$$\Sigma \dot{e} = \dot{e}_{STB} + \dot{e}_{FL} \tag{21}$$

5. DISCUSSION

The possible effects of $\dot{\underline{e}}_z$ have not been considered here.

The chief problem which remains is to determine the values of $\underline{S_w}$ and $\underline{\eta}$ and their effect on the relative and absolute values of $\dot{\underline{e}}_{STB}$ and $\dot{\underline{e}}_{FL}$. At present nothing definite can be stated about the value of S_w in rocks as estimates of grain boundary diffusivity vary by some orders of magnitude [16,17]. Nevertheless it may be interesting to consider what effects the terms in (19) and (20) would have on the <u>relative significance of</u> $\dot{\underline{e}}_{STB}$ and $\dot{\underline{e}}_{FL}$.

(i) For flat layers $\dot{\underline{e}}_{STB}$ is zero because $\underline{a}_h = 0$. At moderate values of $\underline{a}_o$, $a_h > a_o$ causing $\dot{\underline{e}}_{STB}$ to become significant. Then at high $\underline{a}_o$, $a_h < a_o$ causing $\dot{\underline{e}}_{STB}$ to diminish in significance.

(ii) When $L_o^3 S_w/\nu\eta$ is small (<1; i.e. at relatively low diffusional stiffnesses and especially at short wavelengths), and for moderate values of a_o (0.25-2.5), $\dot{\underline{e}}_{STB}$ dominates over $\dot{\underline{e}}_{FL}$ but only up to a fixed ratio determined by $\underline{a}_h$ and $\underline{a}_o$. When $L_o^3 S_w/\nu\eta \gg 1$, $\dot{\underline{e}}_{STB}$ tends to vanish.

(iii) The relative values of $\dot{\underline{e}}_{STB}$ and $\dot{\underline{e}}_{FL}$ should be little influenced by the differential stress if the layers behave in a Newtonian manner. Non-Newtonian layer behaviour, i.e. dislocation creep, should cause $\dot{\underline{e}}_{FL}$ to increase in relative importance with increase of total differential stress due to the effect on the ratio S_w/η in (19).

Finally, what are the <u>limitations</u> of this model?

(iv) Strictly the equations give only the rates of deformation and mass-transfer for a short interval of time during which the fold geometry remains nearly constant. During progressive folding, non-sinusoidal shape changes are introduced as well as the progressive changes of $\underline{a}$, $\underline{h}$ and $\underline{L}_o$.

(v) The analysis assumes that buckling accommodates a mass-transfer which in turn is predetermined by a homogeneous partial stress field. An alternative assumption would be that the mass-transfer accommodates a predetermined buckling deformation. These two approaches appear to correspond to the Reuss and Voight approximations respectively of elastic aggregate theory [18] in which case the true behaviour should lie between the two different estimates. The present approach, however, has the advantage of tractibility and provides a theoretical example of pure path dependence uncomplicated by mean stress variations.

(vi) The model deals strictly with high diffusivity surfaces. In actuality they are high diffusivity layers (c.f. clay seams in Fig. 1). A reasonable correspondence between theory and observation would therefore be expected only when the high diffusivity layers are thin compared with low diffusivity, soluble layers.

I am grateful to J.R. Bishop for checking most of the mathematical derivations and suggesting improvements.

REFERENCES

[1]   Rickard, M.J., Geol. Mag. 98 (1961) 324-332.

[2]   Nicholson, R., Nature 209 (1966) 68-69.

[3]   Williams, P.F., Am. Journ. Sci. 272 (1972) 1-47.

[4]   Cosgrove, J.W., Journ. Geol. Soc. Lond. 132 (1976) 155-178.

[5]   Gray, D.R., unpubl. Ph.D. Macquarie Univ. 1976 325 p.

[6]   Fletcher, R.C., Geology 5 (1977) 185-187.

[7]   Durney, D.W., Geology 6 (1978) 68-69.

[8]   Casey, M., in K.R. McClay, ed., Conf. on Pressure Solution and Coble creep in rocks. Geol. Soc. Lond. Tect. Stud. Group, Abstract 1976.

[9]   Raj, R. and M.F. Ashby, Metall. Trans. 2(1971) 1113-1127.

[10]   Durney, D.W., Phil. Trans. R. Soc. Lond. A 283 (1976) 229-240.

[11]   Stephansson, O., Tectonophys. 22 (1974) 233-251.

[12]   Ashby, M.F. and R.A. Verrall, Acta. Metall. 21 (1973) 149-163.

[13]   Vernon, R.H., Metamorphic Processes. Murby, Lond. 1976 247 p.

[14]   Holland, H., In H.L. Barnes, ed., Geochemistry of hydrothermal ore deposits. Holt, Rinehart and Winston, N.Y. 1967 pp. 382-436.

[15]   Ramsay, J.G., Phil. Trans. R. Soc. Lond. A 283 (1976) 3-25.

[16]   Fisher, G.W. and D. Elliott, In A.W. Hofmann et al., eds., Geochemical transport and kinetics. Carnegie Inst. Publ. 634 Washington 1974 pp. 231-241.

[17]   Rutter, E.H., Phil. Trans. R. Soc. Lond. A, 283 (1976) 203-219.

[18]   Hearmon, R.F.S., An introduction to applied anisotropic elasticity. Ox. Un. Press, Oxford 1961 136 p.

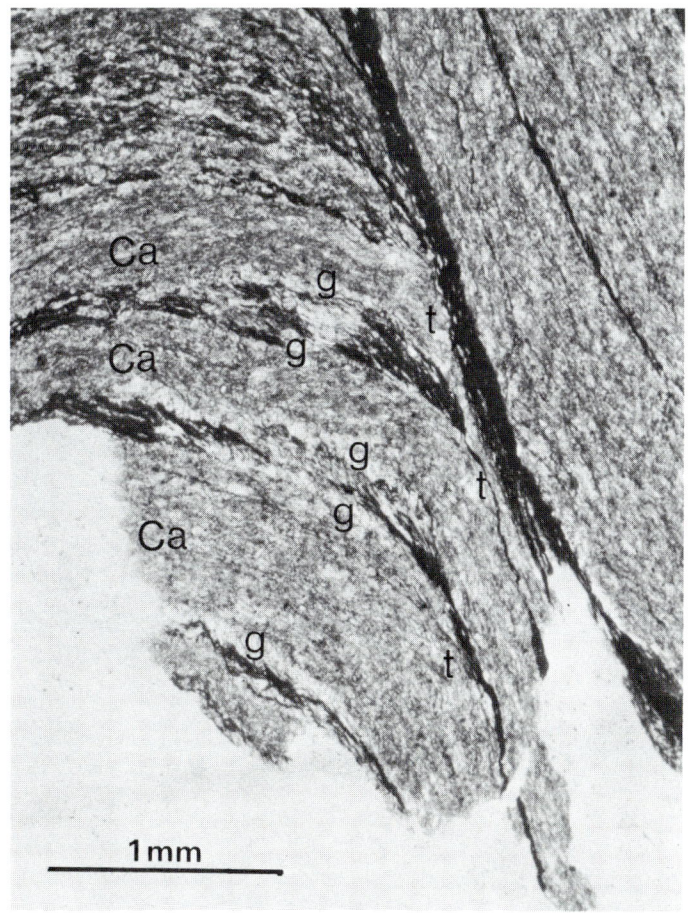

Figure 1 A crenulated calcite (Ca) - clay seam (black) multilayer from Six Armaille, Switzerland (spec. 55f). Preferential transport of $CaCO_3$ along layer boundaries shown by low angle truncation of calcite layers (t) and calcite overgrowths (g).

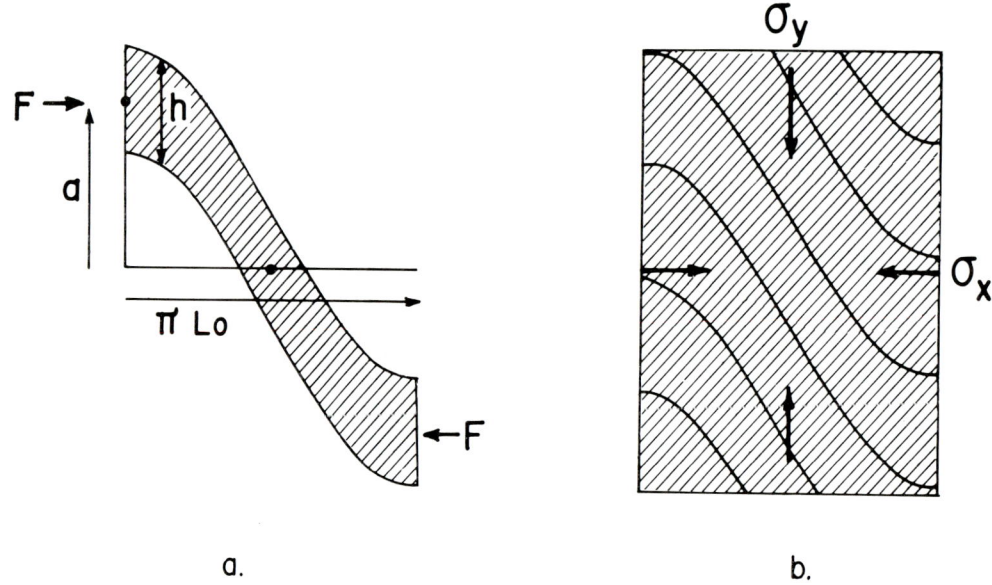

Figure 2 a: Stress field "a" in a single layer.
b: Homogeneous stress field "b" in the multilayer system.

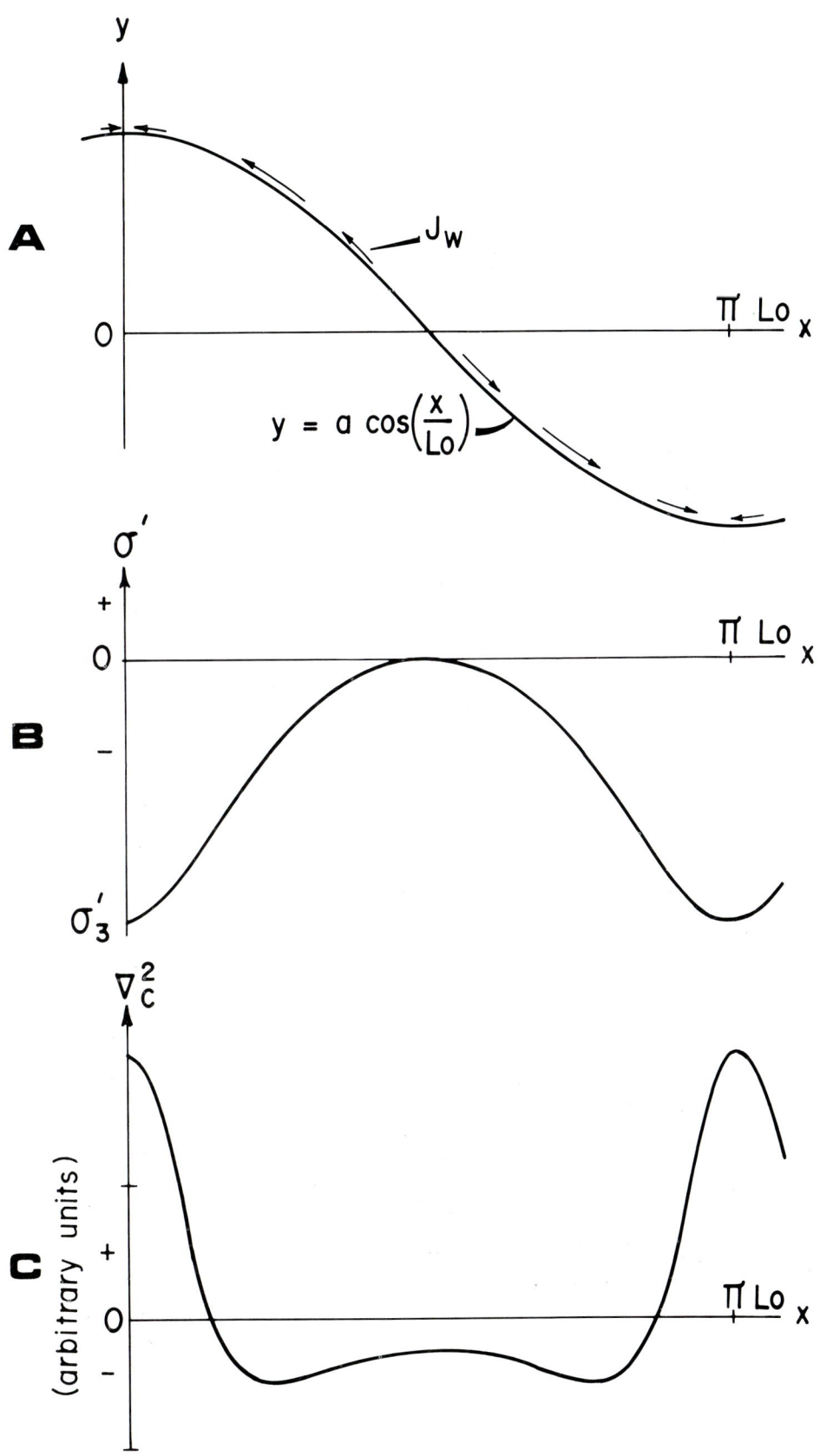

Figure 3 a: Interlayer surface geometry and surface flux (for $\bar{a}_o$ = 1). b: Deviatoric normal stress on interlayer surface. c: Second differential of σ' and c along interlayer surface.

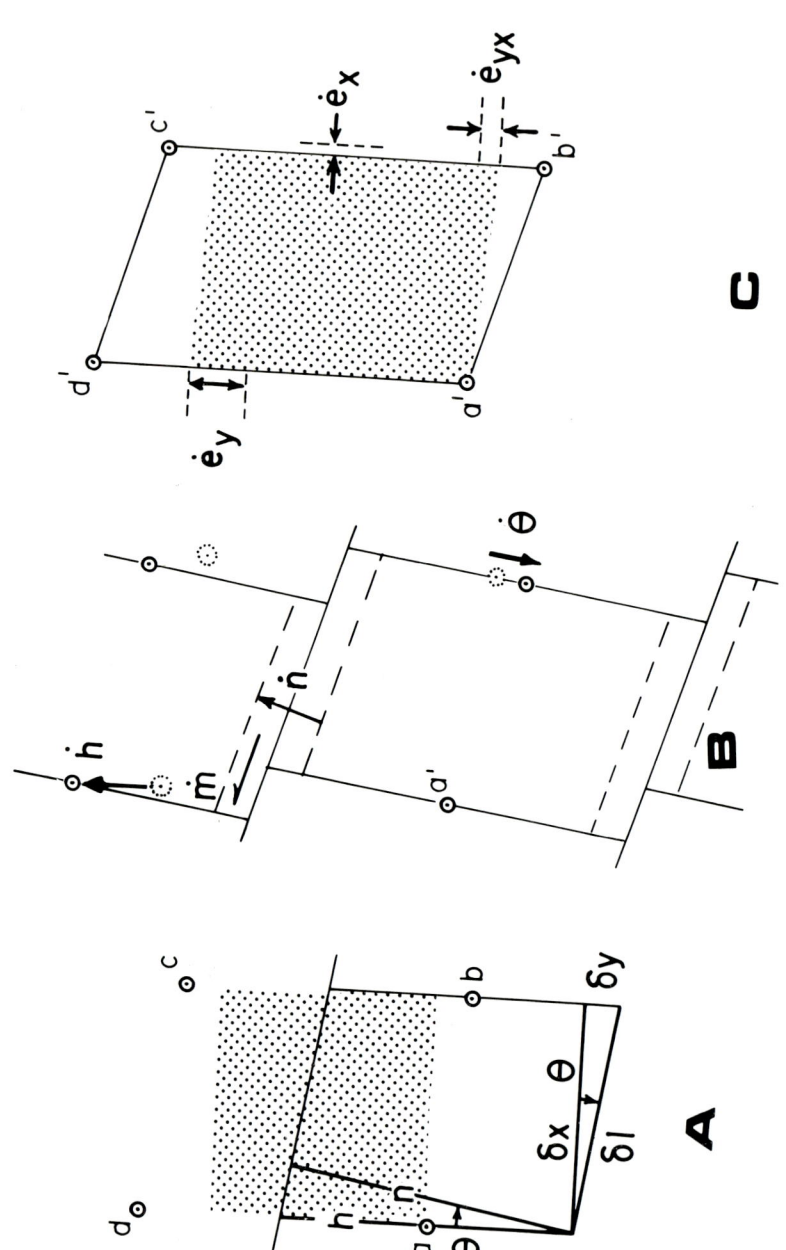

Figure 4 a: Initial material element (h x δL) and its equivalent element (a b c d). Shaded area is an equivalent unit square. b: Material displacements (ṁ only pure slip for clarity). c: Strains in the deformed equivalent element (a'b'c'd').

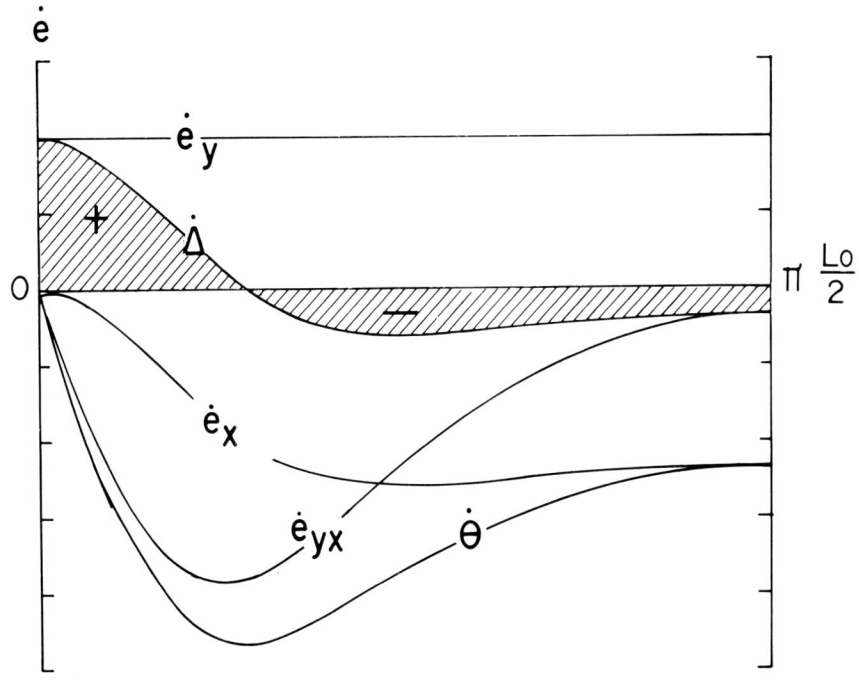

Figure 5 Local variations of equivalent strain rates ($\dot{e}_y, \dot{e}_x, \dot{e}_{yx}, \dot{\Delta}$) and layer rotation rate ($\dot{\theta}$) when $a_o^y = \dot{n}_o^{yx} = 1$, $a_h = 2$.

A THERMODYNAMIC MODEL OF CONSOLIDATION IN COHESIVE SOILS

L. Keinonen

Department of Civil Engineering, Tampere University of Technology, SF 33101 Tampere 10, Finland

ABSTRACT

A simple thermodynamic pattern of compression of cohesive soils is presented and tested. Soil is supposed to be composed of mineral particles and of bound pore water. The free energy of the bound water is the decisive factor determining the compression of cohesive soils. A simple relation between the compression stress and the thickness of the bound water films surrounding the soil particles is given by means of Gibbs' free energy of pore water. On the basis of this relation the compression of soil can be calculated. The method is valid within the ranges of customary pressures in the geotechnics.

1. INTRODUCTION

When considering the deformation of cohesive soils little attention has been paid to the effects of pore water. In clay, however, there is often water 60...75 % of the volume and only 25...40 % mineral particles. In addition to that, water is very firmly bound to the soil particles. The exceptional state of water appears, i.a., in the fact that a great part of water freezes not earlier than in the temperature of -1...-5 $^{\circ}$C.

2. THEORY OF CONSOLIDATION

After the cemented bonds which join the grains together have been broken the effect of the pore water upon the deformation begins to be decisive. This effect depends in the first place on two factors. Firstly, on the fact how strongly the water films are bound to the soil particles, and secondly, how close the particle pass by each other, i.e., how deep they penetrate into each other's water films when moving. The last mentioned circumstance depends on the water content of soil and on the stress under which the deformation takes place.

The following explanation of the consolidation phenomenon is based on the energy state of the pore water. By consolidation there is meant the compression of soil due to the flowing-off of the pore water from the soil.

In the description of energy state there is used Gibbs' free energy g

$$g = u + pv - Ts, \qquad (1)$$

in which u is the internal energy of water
    p is pressure
    v is volume
    T is temperature
    s is entropy

It is not necessary to know the absolute magnitude of the free energy of the pore water, but it is sufficient to know its relative value compared with the free energy of free water. The relative free energy of the pore water per unit volume can be determined by tensiometer, centrifuge and vapor pressure methods [1]. The result is usually presented in form of a pF curve, Figure 1.

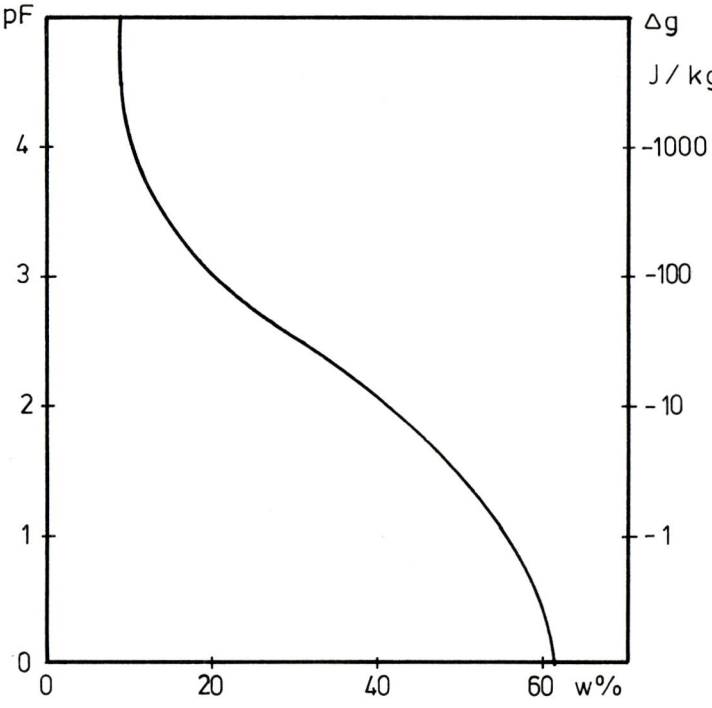

Figure 1. The relation between the water content of a soil and the relative free energy of the pore water.

There one may see that the smaller is the water content of soil, that is to say, the thinner are the water films on the faces of soil particles, the smaller is the free energy of pore water.

The effect of the pressure increase Δp on the free energy of pore water can be derived from the equation (1):

$$\Delta g \sim v \Delta p. \qquad (2)$$

The equation holds good quite exactly because the consolidation takes place in a constant temperature and the compression of water is small under usual conditions.

The magnitude of the free energy of water in different poits in soil determines the movements of pore water. Water is always flowing into the direction of the smaller free energy. When there is prevailing equilibrium, e.g. when the consolidation is finished, the free energy in soil and outside of it must be equal. This information can be used in the determination of the magnitude of consolidation.

If the total surface area of the soil grains to which the water films are bound is known there can be calculated by means of the pF curve the free energy of the water films at different thicknesses. When to this there is added the increase of the free energy due to the loading of soil calculated by means of the formula (2), there can be seen at which thickness of the water film the free energy of pore water becomes equal to the free energy of free water, i.e. when tho consolidation is finished.

3. TEST RESULTS

By presuming that the soil has a certain structure (e.g. Figure 2), there can be calculated on the grounds of the thickness of the water film the magnitude of consolidation.

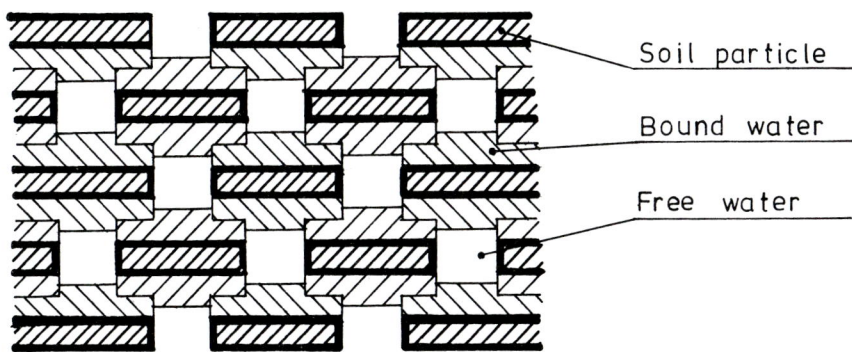

Figure 2. The soil structure used in the calculation of the compression.

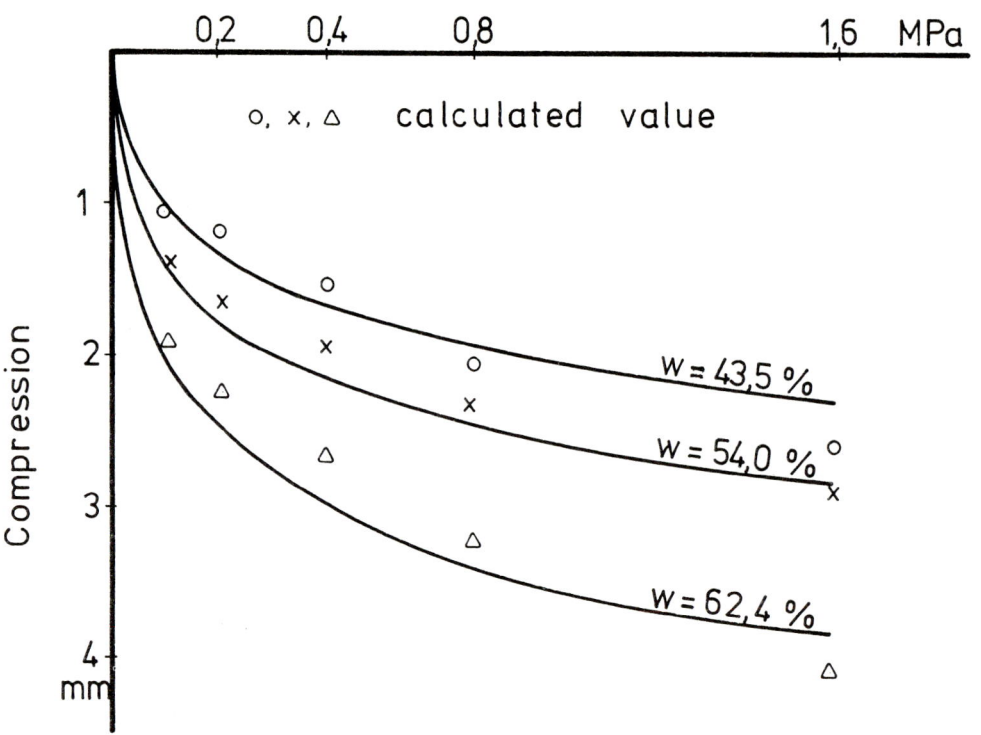

Figure 3. Ödometer curves and corresponding calculated compression.

In Figure 3 there are presented a few values of the onedimensional compression derived by calculation from the pF curves and by ödometer tests in disturbed cohesive soils.

By means of the exceptional energy state of pore water the ice lensing in soil has become possible to treat mathematically [2, 3]. Obviously through it the shear strength of the cohesive soil could be enlightened in a new way, too.

REFERENCES

1   Schofield, R. E., The pF of the water in soil. Trans. 3rd int. congr. soil sci, Vol. 2 pp. 37-48. 1935.

2   Keinonen, L., The physical basis of the growth of ice lenses in soils. Symp. on frost action on roads, Paris 1973. OECD. Vol. I, pp. 311-316.

3   Proc. int. symp. on frost action in soils. Univ. of Luleå, Sweden 1977. Vol. 1-2.

Session III

DISCUSSION

D. Shockey: Prof Hult defines a damage parameter, ω, which is a function only of the net load-bearing area A_n, and which therefore takes no account of microdamage morphology. This means that a penny-shaped crack will have the same effect as a spherical void of identical radius, even though the stress amplification ability of the crack is much greater than that of a sphere. Have you found that defect geometry has little effect?

G. Weidmann: In this connexion, what is the nature of the defects in CDM? Is there any experimental evidence that rupture occurs at the same value of s* irrespective of load history? Does the theory differentiate between an assembly of defects of total area A and a single defect of area A? (i.e. how does the theory interface with fracture mechanics?).

J. Hult: In the single parameter damage theory due to Kachanov no account is taken of the geometrical form or distribution of cavities. It is simply assumed that voids and cavities are forming with a certain distribution of shapes, typical of the material, which cause the rate of strain and damage to accelerate. In contrast to FM, which deals with the stability of macrodefects (finite cracks), CDM deals with the influence of microdefects.

A finite crack may be shown to be less stable due to the presence of continuous damage. Such interaction between macro- and micro-defects has been studied by Janson, cf work quoted in the paper.

Clear experimental evidence of a constant rupture stress s* would require an unambigious method of a post mortem measurement of damage in a ruptured specimen. No reliable such method has as yet been developed.

J. Kratochvíl: CDM and singular fracture approach can be probably coupled by the localization of deformation effect. Localization of deformation (or damage) to a band may occur under completely homogeneous conditions as an instability (bigureation) due to certain types of non-linearity of a material constituative equation (see J. Riec, XIV Int. Congr. Theor. and Appl. Mech., Delft 1976, Proc. vol. I, 207-220).

J. Hult: So far CDM has mainly been developed for uniform or nonuniform, but stable damage fields. This suggestion seems to offer an interesting idea of describing the creation of a macrodefect through the localization of microdefects along a certain band or plane. The exploitation of this idea should be well worth trying.

J. Kratochvíl: The nature of the fracture phenomenon suggests that effect of induced anisotropy occur. It would be therefore desirable to formulate the problem in a multiaxial anisotropy approach.

J. Hult: Microscopic examination of ruptured specimens has shown marked anisotropy in the shape and orientation of microcracks. Various attempts at formulating a multiaxial damage theory have taken this anisotropy into account. Important work in this area has been made at the University of Leicester (Hayhurst, Leckie) and O.N.E.R.A., France (Chaboche).

K. Axelsson: Dr Kratochvíls paper deals with the very important question of making a proper and relevant choice of the constitutive model for solving a certain boundary value problem in plasticity. I have two short questions:

1) For just how complex constitutive models have you been able to carry through the identification test and the solution of the corresponding inverse problem?

2) Have you been dealing with the rate-dependent problem as well?

J. Kratochvíl: 1) The presented note is a conceptual analysis, i.e. a program rather than a final answer. Now only the simplest model is adjusted and used. Experimental data are presently available for specification of two other described models of higher stress effects and delay effect (e.g. data from prof Ohashi's laboratory). But the concrete solution of inverse problem in these two cases has not yet been done. If it is possible to fully specify even more flexible models and overcome difficulties of identification testing and numerial solution remains an open question.

2) For modelling of elastic-visco-plastic properties the similar approach can be used. The principal difference is that additional flexibility represented at least by our new parameter has to be introduced. The new parameter covers the viscous character of the response.

G. Swan: My question concerns Dr H. Andersson's paper. We have measured K_{IC} and G_{IC} simultaneously on circumferentially notched rock cylinders, $a/R \simeq 0.2-0.4$. The results obtained, when used in conjunction with the elastic relationship between G and K give discrepancies of the order of 2. Does this reflect in some way upon your analysis? Would you speculate?

H. Andersson: In an elastic case the correspondence between K and G should be good. Discrepancies may occur due to the ways in which K and G are measured, or due to the fact that energy is dissipated in plates other than the crack tip, e.g. by the creation of many small cracks or over the rock cylinder.

G. Weidmann: I have a comment on the paper of Johannesson and Sjöblom. The distribution of strain energy between the different phases in models of composites depends on the assumptions made. Thus if the phases are equally stressed (series model) the strain energy per unit volume will be the same in each phase; if they are equally strained (parallel model), the strain energies will be different. Could you comment on this in the context of your model?

P. Sjöblom: Since we are interested in the effect of contiguity the model is not exactly a parallel nor a series model. If the contiguity is zero the model turns into a series model and as the contiguity increases the model tends towards a parallel one.

K. Easterling: You attempt in your paper to interpret fracture behaviour of cemented carbides in terms of the stress distribution across the phases and contiguity of the hard phase. However, isn't the fracture of the hard phase also dependent upon microstructural effects such as stress concentrations at phase boundaries due to slip in the soft phase?

P. Sjöblom: Because Youngs modulus of the hard phase is much higher than that of the soft phase and because the flow stress of the soft phase can be assumed to be lower than the fracture stress of the hard phase, these stress concentrations should not have a great influence.

O. Stephansson: Could Dr Berglund give a brief description of the distribution of the mean stress around his stiff inclusion?

K. Berglund: If the mean stress is to be interpreted as the sum of the principal micro-stresses divided by three i.e. the hydrostatic pressure of the stress, we get a high pressure at the poles which decreases with increasing θ to approximately zero at the equatorial zone.

A.W.B. Siddans: Do comparable equations exist for the displacement field in and around ellipsoidal stiff particles?

K. Berglund: Yes, they do. They are in principle easy to obtain, but the calculations of the stress and strain requires rather much work.

K. Berglund: I have a comment on Dr Bodare's paper. If the wave length is small compared with the internal structure you can not make a direct comparison between E_{dyn} and E_{stat} since E_{dyn} then not can be interpreted as a macromodulus.

A. Bodare: One of the intentions of this paper is to show how the determination of E_d sometimes is made. So if we make a dynamic measurement like this it is perhaps wrong to call it a dynamic elasticity modulus.

K. Berglund: Which wave-length compared to the internal structure are you thinking of?

A. Bodare: I'm not thinking of the microstructure inhomogeneities. If this simple calculation is applicable to reality it must be to inhomogeneities with a characteristic length of the order of meters. In particular I'm thinking of big two-dimensional discontinuities which completely block the waves.

B. Broberg: First, I'd like to say how much I enjoyed your presentation. I assume that when drawing direct quantitative conclusions from your model one has to consider carefully whether or not the material is stratified. In a three-dimensional arrangement of different components with considerable differences in E and ρ, waves should tend to travel across the "bridges" of the stiffer component. (?).

A. Bodare: Yes, I'm thinking of big two-dimensional discontinuities, as clay-filled cracks.

G. Ranalli: To have $E_d/E_s \sim 10$, one needs $E_1/E_2 = 1000$; is this not a bit too high for rock, except in the very upper crust?

Thus after all a frequency-dependent E may be a better hypothesis for rocks below the topmost crust.

A. Bodare: Yes, I don't know in what degree this phenomenon contributes to the difference between the static and dynamic elasticity modulii. I'm referring to the problems of engineering geologists and their problems are situated in the upper crust.

O. Stephansson: Bodare has found that a small amount of a soft material in the bulk material leads to a high value of the E_d/E_s ratio. This result is of great importance for understanding the static and dynamic behaviour of a rock mass, where blocks of rocks are surrounded by discontinuities or joints. Are there any field measurements supporting your results?

A. Bodare: I'm sorry to say no. But more work on this idea is going on.

J. Kratochvíl: A question on the paper of Kubát et al. There are now two theoretical models for one physical phenomenon. Can you suggest an experiment which shows the correct approach?

M. Rigdahl: We are working with generation of noise by external deformation which we hope will be fruitful, but so far we have not succeeded.

D. Shockey: I have a question for Dr G. Andersson. If the quartz bar does not split, does the gauge give an accurate measure of the pressure pulse? If the measurement is not reliable, can your analysis be used to obtain an accurate result from the guage signal?

G. Andersson: In so far as we know, the answer is yes. The gauge appears to be reliable until the crystal is broken. Once the crystal is broken, the guage may be discarded. I do not know the answer to the second question since no attempts have been made to use the analysis other than to predict the conditions for failure.

I.J. Smalley: I have a question for Prof Rosenqvist. The investigations by Beland (Quebec) on the material involved in the Nicolet landslide in Quebec indicated that the slide took place in a "clay" soil but that it contained no clay minerals. Can the high sensitivity of such a soil be explained by your theory or must an alternative explanation be sought?

I. Rosenqvist: No; but I doubt that the Nicolet clay does not contain a considerable amount of hydromicas, chlorite etc. If "clay mineral" is used solely for smectite, mixed layer minerals, etc the answer is: Yes!

I.J. Smalley: The closing paragraph in the paper of Prof Rosenqvist indicates that by chemical treatment (i.e. addition of KCl) the strength of a quick-clay could be increased by four times. This is an impressive engineering application of chemical knowledge and its success suggests that electro-chemical and colloid-chemical factors may, as Rosenqvist proposed, be critical in controlling quick-clay properties.

If any alternative model of quick-clay nature were to be applicable it would have to be able to allow for chemical stabilization of the material; and it appears that the inactive particle short-range-bond model can be constituted to allow for chemical stabilization.

Rosenqvist's picture of a quick-clay is of an essentially clay-mineral material composed of non-expanding clay minerals. My picture of a quick-clay is of an essentially primary mineral material composed of clay-sized (~2-5 μm) particles of quartz and feldspar with minor clay minerals and carbonates. When Beland examined the material of the Nicolet landslide in Quebec he found no clay mineral material and it is difficult to see how a clay mineral based model can account for the behaviour of a quick-clay which contains no clay minerals. Our investigations in Leeds of material from the 1971 St Jean Vianney landslide have shown ~9% of an illitic mineral - most of the particles were feldspar and quartz.

How then could such a material be strengthened by salt additions? Two factors should be considered in conjunction with how the added ions could react; (1) the feldspar particles of clay dimensions can act as relatively inactive clay mineral particles and respond to changes in the geometrical environment and (2) the small amount of actual clay mineral material present will respond to the added ions. The inter-particle interactions will be intensified by the cation additions and the bonds between the feebly negatively changes soil particles will be enhanced.

It must be emphasized that the basic bond type in the Smalley quick-clay model is the short range bond. The system does not have long range particle interaction and this explains the very low observed plasticities of quick-clays (Canadian quick-clays tend to have single figure PI's)
(ref: Beland, J. 1956. Proc. Geol. Assoc. Canada 8, 143-156).

I.J. Smalley: This question concerns Prof Hansbo's paper. Over the years there has been much discussion about whether clay mineral particles in soils are actually in contact or not. Can we now say with some certainty that the clay mineral particles are not in direct contact?

S. Hansbo: If a contact exists this contact probably exists only via a thin water film. But if we call this a direct contact I would say that there certainly exists a large number of direct contacts.

R. Pusch: Just one remark. Complete stress relaxation or (as you put it) a state where the stresses in the bridges are nil, requires that the aggregates are in direct contact, thus forming a system of mutually supporting structural members. Your reasoning can easily be made perfectly valid by assuming this slightly changed model of aggregate positions, still assuming the bridges to link the aggregates together.

S. Hansbo: Of course, a complete stress relaxation in the bridges is possible only if there exists a "direct contact" between the aggregates but when I speak of stress relaxation I am referring merely to shear stresses in the bridges. So I agree with what you say.

I. Rosenqvist: How can Dr Larsson explain the apparant consolidation having never been subjected to excess effective stresses?

R. Larsson: There are a number of reasons such as secondary compression, cyclic loading by variation of ground water level and chemical changes in the aggregates making them more stress resistant.

I.J. Smalley: Two of your statements about dilatancy require a brief comment. In an ideal situation maximum positive dilatancy would not depend on particle size; small spheres pack in the same way as large spheres. Significant positive dilatancy does not occur in clay because the contacts are "soft" - not because the particles are small. Cemented Canadian quickclays are believed to exhibit positive dilatancy (and they have "hard" cemented contacts between the clay-size particles).

R. Larsson: It is quite true that particle size should not have influence if the material was ideal and inelastic. Soils are not ideal materials and elastic deformations occur and counteract the dilatancy. These elastic deformations become larger the smaller the particles are. This might be what you mean with soft contacts in clay.

As I do not know what the nature of "cementation" is I can not comment on your second question. It is stated in the article that it deals with the most common Swedish soils which are certainly not cemented. Experience has shown that all findings about soils in a certain region can not be exported and directly applied elsewhere.

P. Smart & M.A.O. Bannaga: We have a comment on R. Larsson's paper, p. 275, Fig 5.

We have measured tan ϕ' for an almost normally consolidated laminated silty clay from Grangemouth, Scotland. The observations were fitted approximately by

$$\tan \phi' = 0.543 + 0.199 \sin^2(i-60°)$$

where i is the inclination of the samples. Note that there was a minimum value of tan ϕ' at 60° inclination, when the failure plane coincided with the direction of the layers. An alternative failure criterion will be reported in due course.

I.J. Smalley: Concerning Dr Börgesson's paper, I observe with great interest that some of the silt particles in your Fig 1 appear as flat plates. During your SEM examinations of the silty soild did you observe a significant number of plate silt particles? If you did can you account for their formation and suggest how they might affect the properties of the soil?

L. Börgesson: I have not done any systematical microscope investigation on silt. However a silt particle is defined by the measured grain size and it is obvious that the clay minerals also exist to some small amount in the silt fraction. The clay minerals influence on the properties is very strong. If the silt consists of flat and irregular particles which are not clay minerals I think the effect will be that the silt has a strong tendency to dilate.

I.J. Smalley: I think that a new definition of silt is required. The size range of 2 to 60 μm contains far too large a range of particle types. Silt should really be thought of as 20-60 μm - this material should certainly be distinguished from one with a range from 2 to 6 μm.

L. Börgesson: I don't think the particle size has a big influence on the properties compared to the influence of the clay mineral particles.

I. Rosenqvist: With respect to Prof Pusch's paper, bond energy in clay aggregates consist of at least two groups i.e. 100-200 kJ/mole and 15-30 kJ/mole. Is this not important in analysing the cohesional part of the strength?

R. Pusch: Certainly. This fact could and should be introduced in the choice of a proper energy spectrum in my model. It is important, as you say, not to believe that there is just one specific type of bonds in clays.

I.J. Smalley: Should we perhaps have concentrated less on illitic clays and silts (which are essentially Northern phenomena) and given more attention to other widespread soil materials - in particular laterite. An added benefit of the study of laterite is that it could be considered a material intermediate in type between soil and rock.

O. Stephansson: I should like to ask Dr Ranelli, what can we gain about plastic flow mechanisms and the rheology of the mantle from seismology today.

G. Ranalli: The zone in which seismology is most relevant is the asthenosphere: the decrease in seismic wave velocity can be interpreted in terms of a few % partial melting (although that is not the only possible explanation). Whether partial melting changes the creep properties of the material depends on the degree of grain boundary melting and the possible presence of electroviscous effect in the melt.

I. Rosenqvist: Is it possible that the postglacial rebound is not rheological but based on phase transformations in the upper part of the deeper mantel?

G. Ranalli: Yes, it is possible, and it has in fact been proposed. But if such processes are taking place, they do not make the estimate of mantle rheology from postglacial rebound any simpler.

J. Knott: Could the presence of micro-fissures in rocks give a significant volume fracture of a "phase" with zero modulus, thus giving a large difference between E_1 and E_z?

G. Ranalli: In practice, such fissures will be closed.

With reference to Dr Ferguson's paper, I should like to say that the axiomatic approach can be critizied on the grounds that it is not based on physical models and does not result in tractable equations. More particularly, the "simple fluid" model predicts newtonian flow for sufficiently slow deformation. However, rock specimens which come from the upper mantle show textures indicative of dislocations creep i.e. power law creep.

J.F. Knott: I referred in my paper to Ashby's deformation maps, where various types of thermally-activated flow are related to physical models of deformation to give appropriate power laws. I must confess that I become rather unhappy when rate equations are used, in the absence of physical models for the mechanisms of deformation.

C. Ferguson: The power law equation is based on particular constant-stretch-history deformation. The only possible justification for supposing that it is relevant to natural rock flowage is a crudely similar set of microstructure, say {M}. Now even if we have dislocation creep ⇒ {M}, it does not follow that {M} is a sufficient condition for dislocations. (There are, after all deformation mechanisms that leave us distinctive microstructure). Nor can we say that dislocation creep ⇔ power law relationship for all deformation geometries.

As for tractability, we can already see more progress towards solving geologically interesting problems using second-order Richin-Ericksen constitutive equations (see Hobbs, 1972 and Leal, 1975 references in my paper). Furthermore, numerical solutions present less severe difficulties, and I believe several people are currently working on Finite-Element models for flow of simple fluids.

I. Rosenqvist: I have a comment in connexion with Prof Keinonen's paper. The thermodynamic potential μ, i.e. the partial free energy of water is the same for all water in a given clay/water system. It is differing by changing the mineral water ratio.

L. Keinonen: The partial free energy of the pore water is equal at every point in the soil when the water is in equilibrium. The free energy is composed in different ways at different points of the pores. Near the surface of the soil particle the inner energy, the specific volume of water, the pressure and the entropy are different from those in the middle of the pore. However, the free energy has the same value in every point in the pore water.

In the compressed soil there are water films of different thickness and the free energy seems to be smaller in the thinner films than in the thicker ones. But the compressive pressure is acting there and it will increase the free energy so that this will be equal to the free energy of the thicker water films.

During the consolidation the change in partial free energy of the mineral phase is so small that it can be ignored and the solution of the consolidation can be based on the partial free energy of the pore water.

E. Suoninen: (Chairman)

There have been quite a few papers presented in this session giving a mathematical model for the mechanical behaviour of materials. Such models necessarily require parameters which describe the material considered. I would like to emphasize the need for intensified efforts to motivate these parameters by reasoning based on considering the physical processes lying behind the phenomena studied. It is well known that such motivation is in most cases very difficult and often considered impossible. Unfortunately, this does not change the fact that it is often absolutely necessary to make the model meaningful. I would like to mention the papers presented by Drs Bodare and Ranalli as positive attempts to meet the above requirements.

J. Hult:

If you would allow me, Mr Chairman, to challenge your own suggestion, I would say that what we call a physical interpretation of a parameter is more or less a matter of habit and of taste. Take Poisson's ratio for example. This parameter has been around for such a long time now that we tend to rank it as physically acceptable, even though it is a purely phenomeno-logical quantity. Is it not time now that we stop drawing that peculiar border line between "physical" and "non-physical" quantities in materials science? The only thing which really counts must be whether a certain quantity is found to be useful.

B. Broberg:

My friend Jan Hult likes to forward provocative questions. Like John Knott, I feel provoked: it does cost a lot of money to find material properties - those often needed in practice are not listed in the manufacturers' brochures. If we could arrive at some kind of a physical understanding we could also, perhaps, eventually, be able to make at least reasonable estimates. This applies, of course, to several fields - engineering is one example, studies of the Earth's interior another (you haven't been there, cannot even approach the high pressures in laboratory (static) test, etc).

RAYMOND H. FOG[...]
DATE

BOOKS A[...]
RECALL

MA[Y]

folio
TA
417.6
M43
1979

FEB 2 0 1980